HOW IT WORKS, ILLUSTRATED:

everyday devices and mechanisms

A POPULAR SCIENCE BOOK

HOW IT WORKS, ILLUSTRATED:

everyday devices and mechanisms

by Rudolf F. Graf and George J. Whalen

and the editors of *Popular Science*

Illustrated by Farmlett, Barsanti, & Wood, Inc.

SOUVENIR PRESS

First published in America by
Popular Science/Outdoor Life Book Division,
Times Mirror Magazines, Inc

First British Edition published 1975 by
Souvenir Press Ltd, 95 Mortimer Street, London W1N 8HP
Reprinted 1976
Reprinted 1979

ISBN 0 285 62173 4

Printed in Great Britain by
Fletcher & Son Ltd, Norwich

Contents

Preface

Several years ago, when this book was conceived, we set out to explore the inner workings and operation principles of a broad spectrum of everyday devices and mechanisms—to find out "what makes them tick." Our assignment, then, was to nourish and gratify the curiosity of the reader by describing the operation of each product, using illustrations lavishly.

We disassembled scores of products and laid bare to searching examination the ingenuity of the world's leading manufacturers. The result is an illustrated guide to man's inventive genius—an expose of the inner world of familiar products set forth in easy to understand pictures and words.

The reader will soon discover that the fundamental operating principles at work under the "skins" of a broad spectrum of devices are merely the mechanical execution of simple mechanisms arranged and orchestrated with imagination. This is *not* a "fix-it" book but rather a clear-cut description of the way things work and *why*.

We are indebted to Hubert P. Luckett, Editor-in-Chief of *Popular Science,* for permission to use material from the magazine's original "how-it-works" series and to Erick H. Arctander its principal contributor.

To William B. Sill, our publisher, and to Neil Soderstrom, our editor, go heartfelt thanks.

We also wish to acknowledge by name some of the scores of individuals whom we consulted during the preparation of the book. Their particular expertise, so freely shared, ensured the accuracy of every entry. Our thanks go out to Max Alth, Jean Anwyll, R. E. Babros, Bernard Banks, Arthur E. Belyea, Lee Ann Blystone, C. C. Chiang, Tony Chiu, Donald A. Dery, Charles P. Farley, Marian Finney, R. C. Good, John S. Hutchinson, William Lems, Leonard C. Lindgren, Robert Lyman, Art Osolin, George Marenzana, Vern G. Pedersen, Christian Petersen, James B. Rudden, Martin Satloff, Joel H. Schwartz, Hy L. Siegel, Joseph Star, James E. Stout, Blake Stretton, H. Gary Underhill, Everett C. Wilcox, and Len Wurzel.

Special words of appreciation go to Jeff Fitschen, who directed the art, and to the fine artists of Farmlett, Barsanti, & Wood, Inc., who produced the majority of illustrations. Thanks go also to Dan Todd, Ray Pioch, James Wright, Frank Schwarz, Carl DeGroot, and Robert Kupchin, whose combined talents created some of the illustrations that grace this book. Credit for art mechanicals goes to Frank DeMarco, Rocky Cipriano, Don Rowe, Ruth Markert, and Dave Houser. Deserving praise for their advice and technical skill are the typographic staff at Ruttle, Shaw & Wetherill, Inc., including Frank Righter, Jack Blaese, and Bill Hummer. The index was compiled by Susan Popkin.

Finally, we are grateful for the assistance provided by the following manufacturers:

Acme Air Appliance Co., Inc. (Div. of Ideal Corp.); Aeolian Corp.; AMF Voit, Inc.; AT&T; Automatic Orange Juicer Corp.; Berol Corp. (Berol Products Div.); Black and Decker Mfg. Co.; Bulova Watch Co.; Colt Industries (Fairbanks Morse Pump Div.); Continental Scale Corp.; Control Data Corp.; Dacor Corp.; Delta Products, Inc.; Dictran International Corp.; Disston, Inc.; Dymo Products Co.; Eaton Corp. (Climate Control Div.); EBCO Mfg. Co.; Edwards Power Door Co.; Emerson Electric Co.; Empire Exposure Meter Service; Endura Appliance Corp.; GAF Corp. (Consumer Photo Div.), (Office Systems Div.); General Electric Co.; Gilbarco Inc.; Gestetner Corp.; The Gillette Co. (Papermate Div.), (Toiletries Div.).

Hanson Scale Co.; Heath Co.; Honeywell Inc.; International Business Machines Corp.; Koh-I-Noor Rapidograph, Inc.; Kollsman Instrument Corp.; Leigh Products, Inc.; Leonard Valve Co.; The Marbelite Co., Inc.; Medeco Security Locks, Inc.; Melnor Industries; McGraw-Edison Co.; Montgomery Ward; Napco Security Systems; N-Con Systems Co., Inc.; North American Philips Corp.; The Parker Pen Co.; Petersen Mfg. Co., Inc.; Polaroid Corp.; Radio Shack; Rain Bird Sprinkler Mfg. Co.; Rival Mfg. Co.; Rockwell Mfg. Co.

Sanyo Electric, Inc.; Schick Electric, Inc.; Scoville Mfg. Co. (Hamilton Beach Div.), (Nutone Div.), (Oakville Div.), (Sewing Notions Div.); Sears Roebuck & Co.; Skil Corp.; Stihl American, Inc.; Taylor Instrument Co.; Tekmar Corp.; Tokheim Corp.; The Toro Co.; Turner Corp. (Div. of Olin Corp.); U.S. Divers Co.; U.S.M. Corp.; Victor Comptometer Corp.; Waste King Universal; West Bend Co.; Weston Instruments, Inc.; Whirlpool Corp.

New Rochelle, N.Y.
January 1974

Rudolf F. Graf and George J. Whalen

HOW IT WORKS, ILLUSTRATED:

everyday devices and mechanisms

Electric Grass Trimmer

Trimming the grass around flower beds, fences and walks with a hand trimmer has long been the bane of gardeners. With either of two types of electric grass trimmers shown, the dreaded chore is made easy. One type of machine has a set of fork-like shears protruding from the front of the trimmer. The other employs a rotary blade located beneath the motor housing. Both machines run on battery power. The batteries are located in the handle and can be recharged by plugging the trimmers into a wall socket overnight. About 500 recharges, each furnishing one-hour's use, are possible.

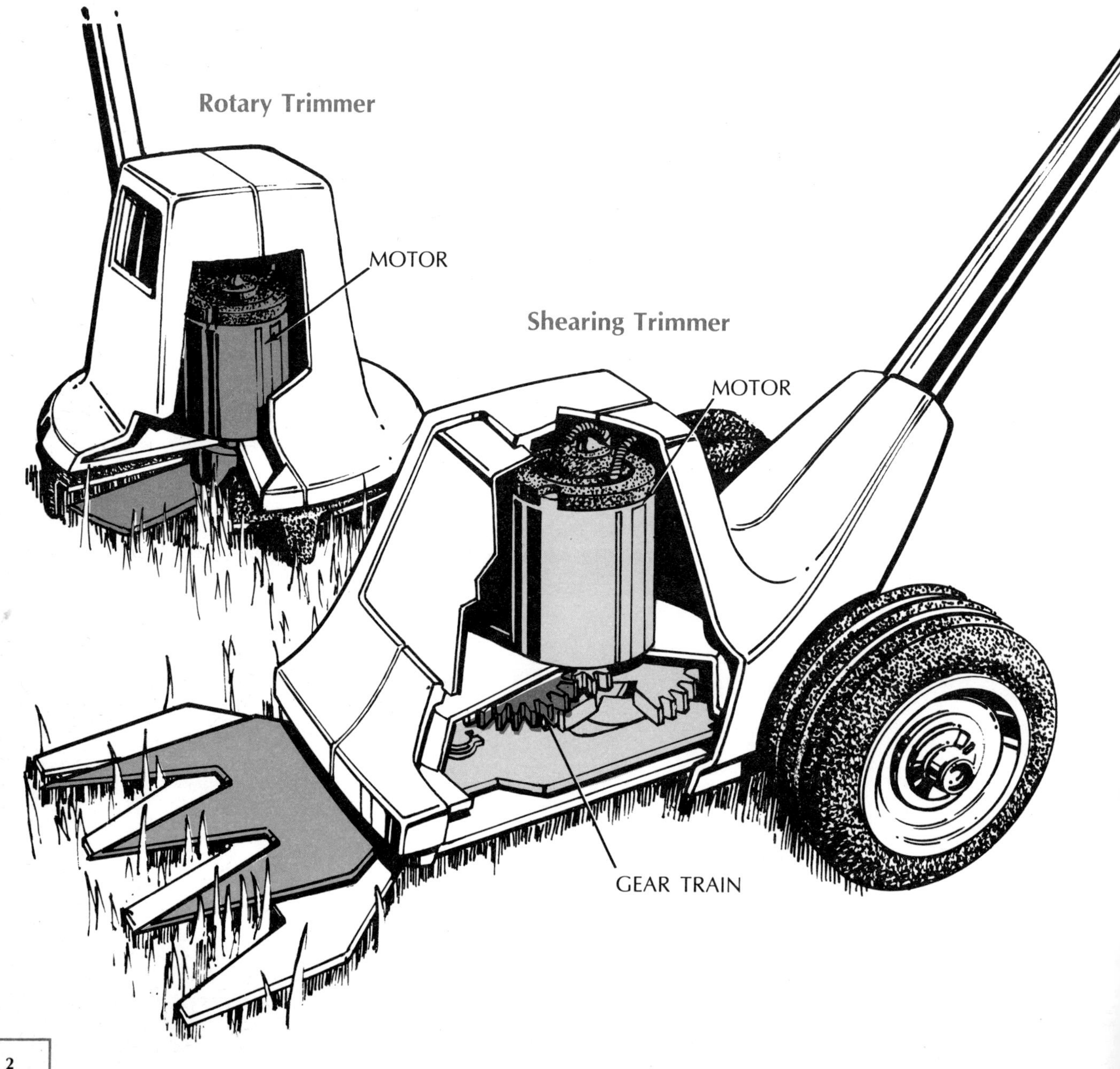

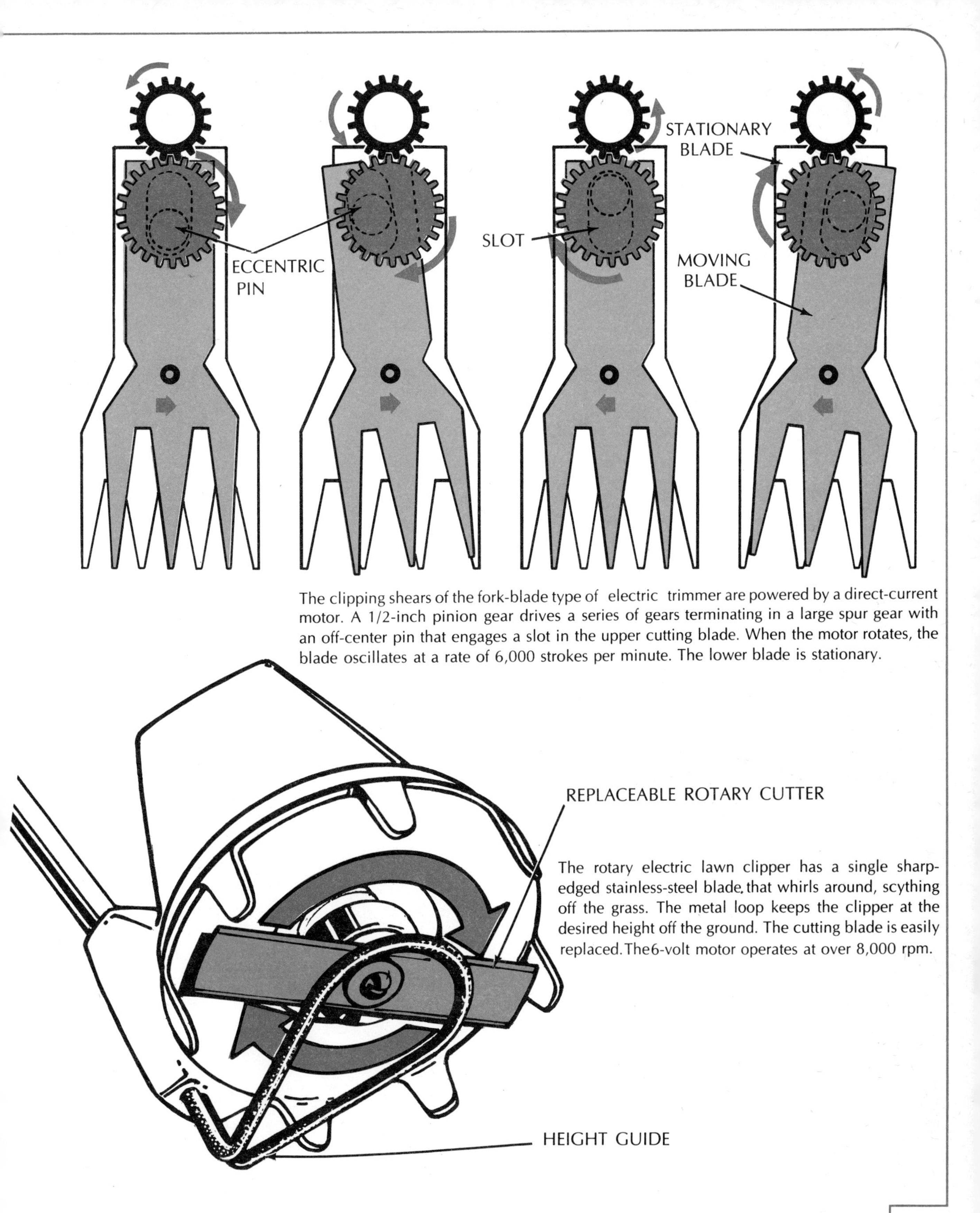

The clipping shears of the fork-blade type of electric trimmer are powered by a direct-current motor. A 1/2-inch pinion gear drives a series of gears terminating in a large spur gear with an off-center pin that engages a slot in the upper cutting blade. When the motor rotates, the blade oscillates at a rate of 6,000 strokes per minute. The lower blade is stationary.

The rotary electric lawn clipper has a single sharp-edged stainless-steel blade, that whirls around, scything off the grass. The metal loop keeps the clipper at the desired height off the ground. The cutting blade is easily replaced. The6-volt motor operates at over 8,000 rpm.

Chain Saw

A man with an ax looks small beside most mature trees. But with a gasoline-powered chain saw in his hands, he becomes a Paul Bunyan, able to clear a large stand of timber in a few hours. Chain saws develop their phenomenal cutting power by merging the brute force of a small, high speed, two-stroke cycle engine with a chain fitted with 40 or more hardened-steel cutting teeth. The chain rides the grooved track of an oval cutting bar that allows continuous chain drive. Either side of the bar can be employed for sawing. The assembly is carefully balanced between a pistol-grip and cross-bar handle to allow ease of operation. But since the blade has no shield, the operator is well advised to maintain a firm controlling grip to prevent the blade's glancing into his leg.

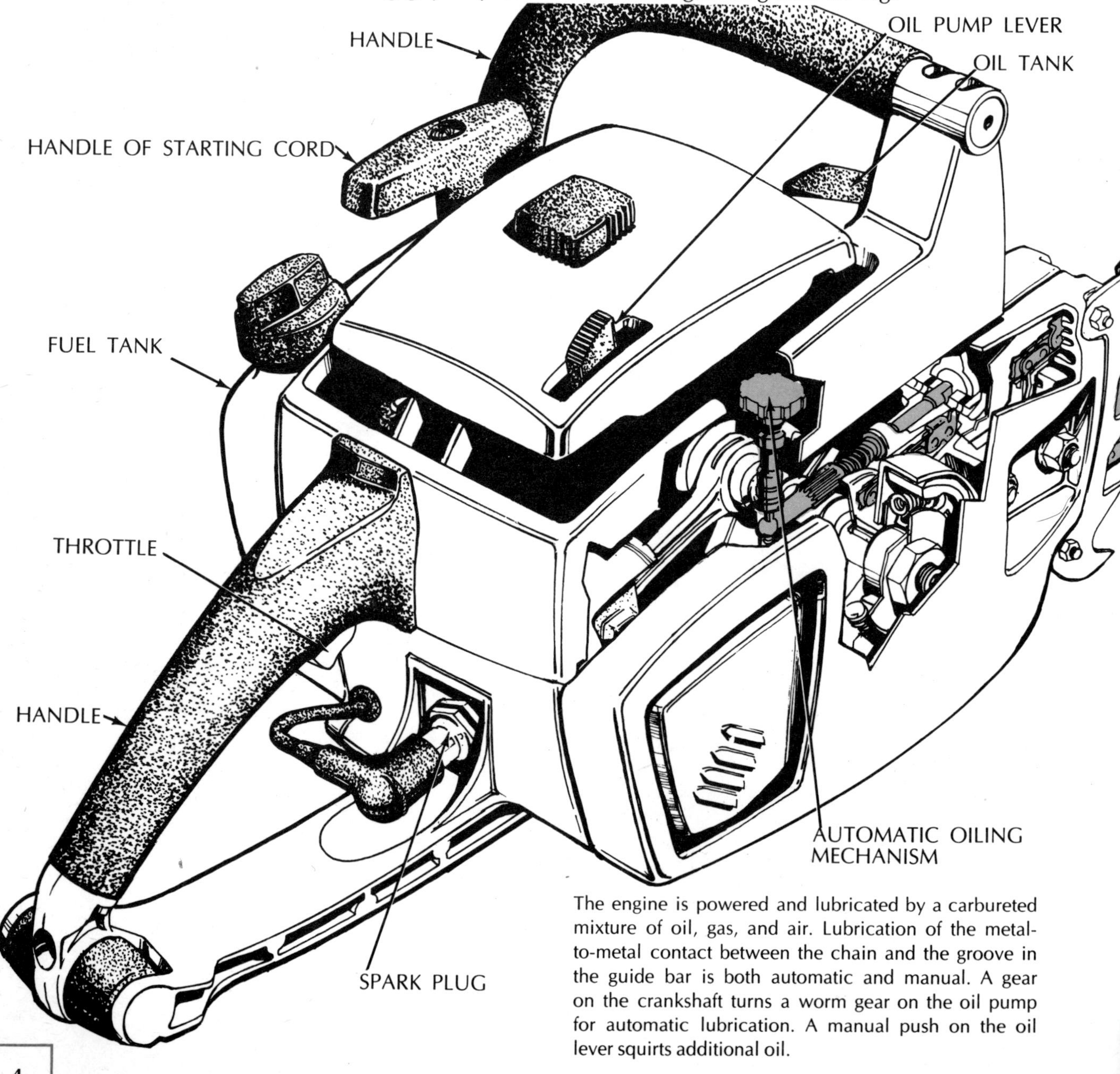

The engine is powered and lubricated by a carbureted mixture of oil, gas, and air. Lubrication of the metal-to-metal contact between the chain and the groove in the guide bar is both automatic and manual. A gear on the crankshaft turns a worm gear on the oil pump for automatic lubrication. A manual push on the oil lever squirts additional oil.

The crankshaft of the single-piston engine carries a centrifugal friction clutch (right) containing three spring-attached shoes which turn within the circumference of a short, shaft-mounted drum. The drum carries a sprocket that drives the saw chain.

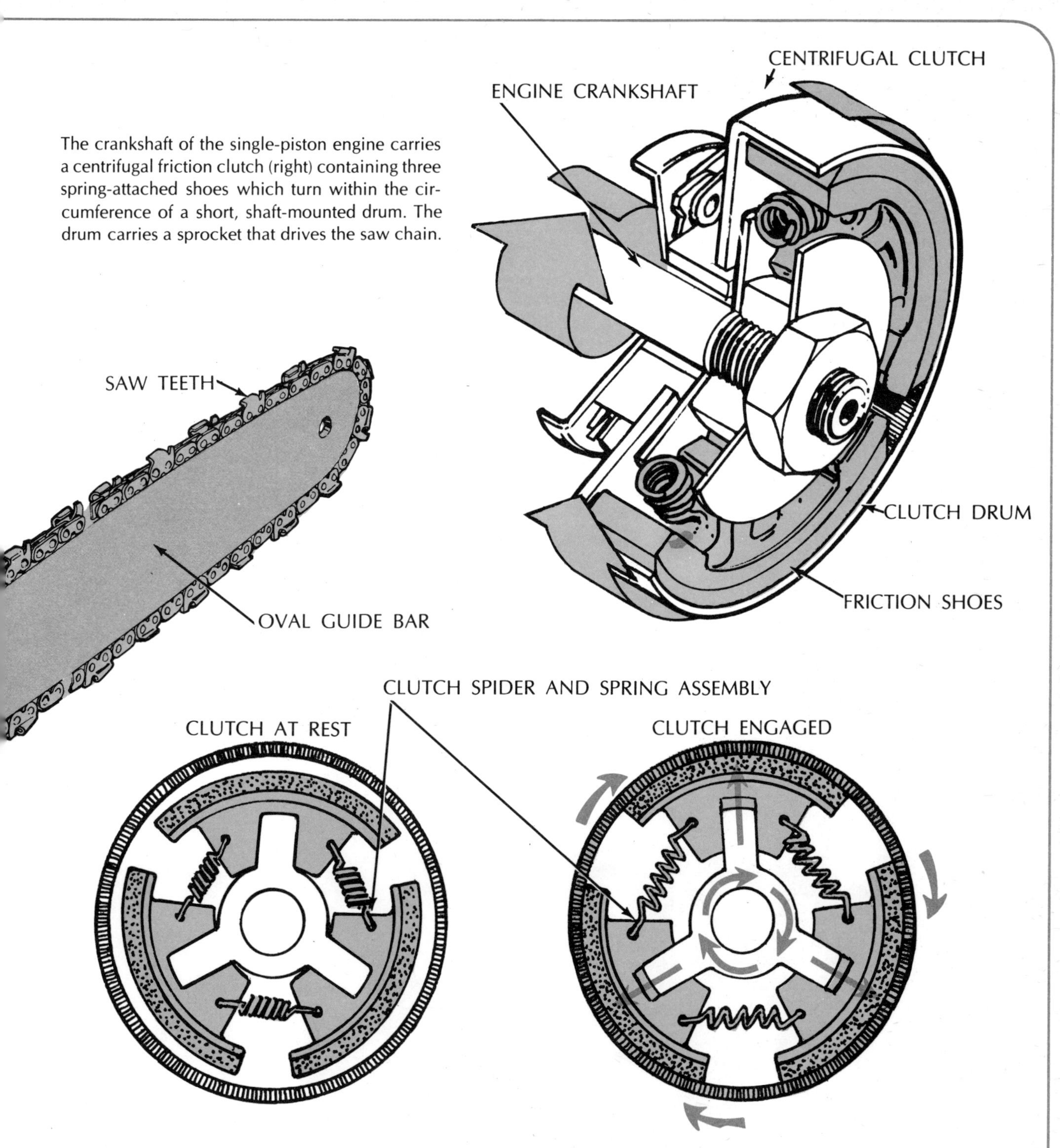

When the engine idles, tension springs hold the shoes close to the shaft. When the throttle trigger is pressed, the engine accelerates and the shoes fly outwards to engage the drum. This clutch now forms a "solid" connection between the engine and the saw chain, whipping the chain along the guide bar at high speed. The chain's sharp cutting teeth bite into wood with fiber-smashing force. If the chain binds and slows in the wood, the engine speed drops and the clutch disengages.

Rotary Engine

Some experts predict that the rotary internal combustion engine may eclipse the piston engine in automobiles during the 1980s. Why? The rotary engine promises to be less costly to produce. It is half the size and only one third the weight of piston engines of the same horsepower. It has only three internal moving parts that rub as compared to over 160 that rub and pound in a piston engine. Instead of pistons, rotary engines employ rotors that revolve on a shaft. Each rotor, shaped like a slightly bulging triangle, revolves in a combustion chamber shaped like a fat figure eight, called an epitrochoid.

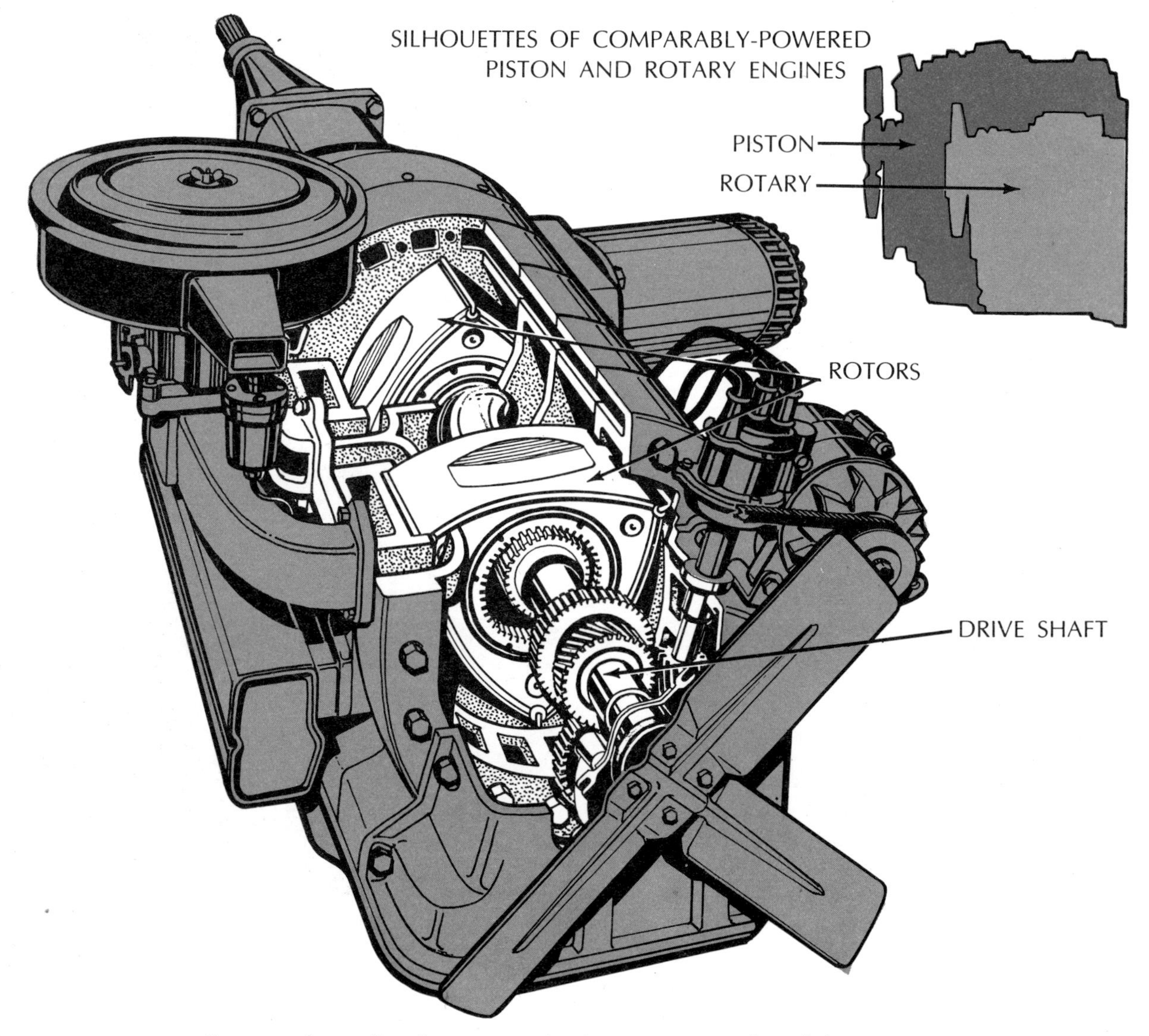

Key to understanding the rotary engine is an awareness that all three apexes of the eccentrically-revolving rotor are always in firm contact with the chamber walls. As the rotors revolve eccentrically, they create rhythmically changing spaces between themselves and the chamber walls. These volume changes accomplish the four essential phases of the internal combustion cycle: intake, compression, power, and exhaust.

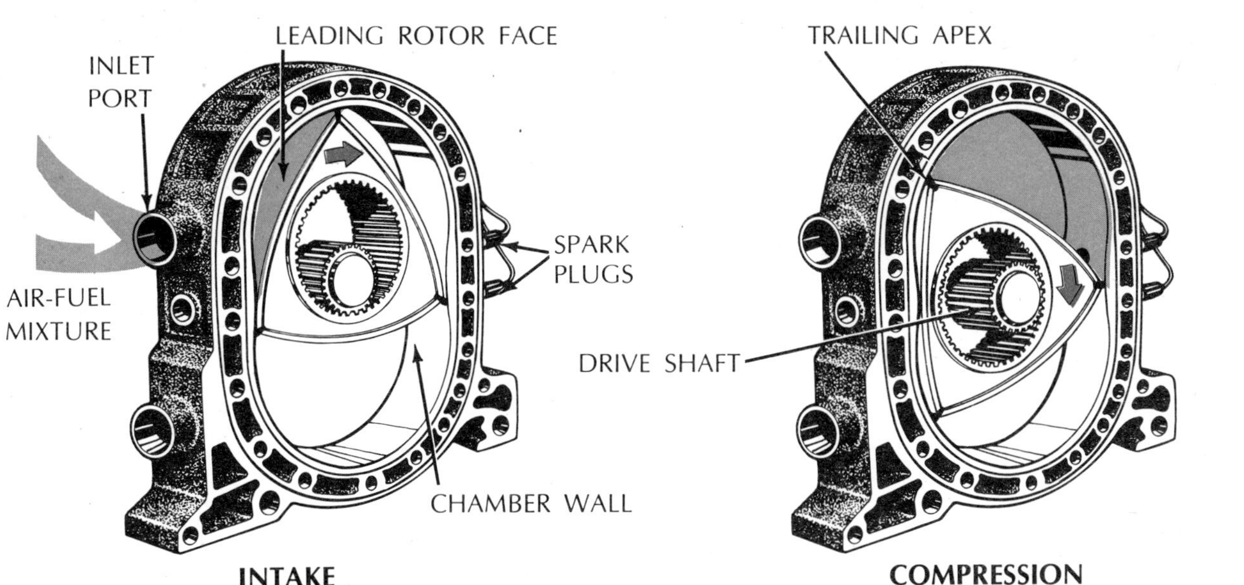

INTAKE

During the intake phase, the leading rotor face sweeps past and uncovers the inlet port. The gas-air mixture from the carburetor rushes into the combustion chamber vacuum created by the suddenly expanding space between the rotor's leading face and the chamber wall.

COMPRESSION

As the rotor continues to turn, the space between the rotor face and the chamber wall reaches its maximum. Meanwhile, the trailing apex of the rotor cuts off the inlet port. Then the space between the rotor face and the chamber wall diminishes, compressing the air-fuel mixture.

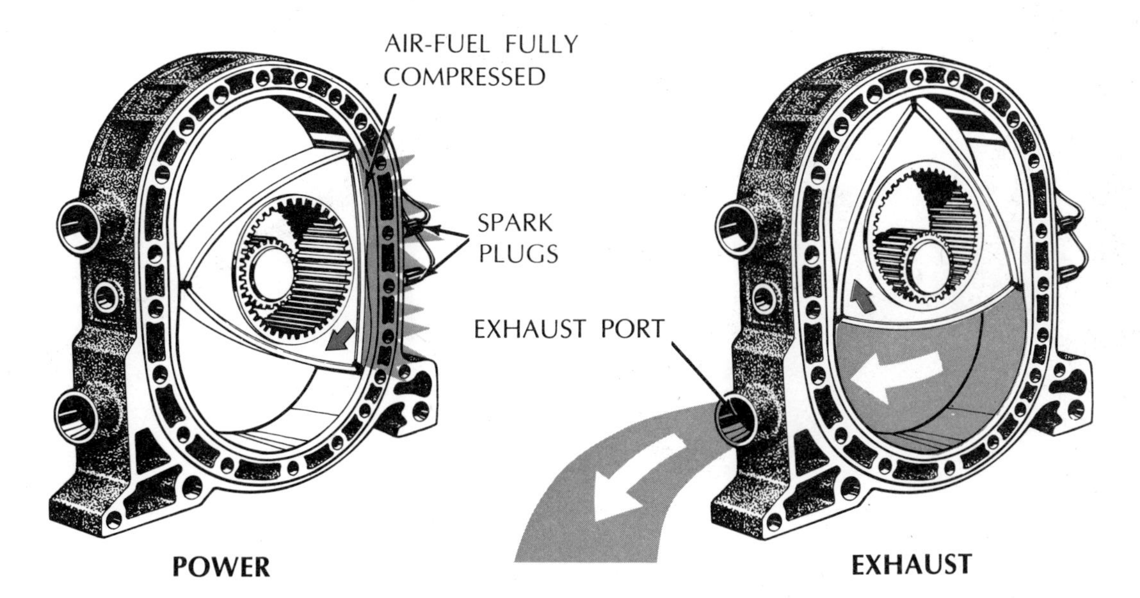

POWER

At maximum compression, the air-fuel mixture lies over the two spark plugs in the chamber wall. The plugs fire, igniting the mixture. Super-hot expanding gases pushwith great force against the rotor, adding impetus to the rotor's initial momentum and producing the power phase.

EXHAUST

As the rotor face yields to the force of the power phase, the leading apex of the rotor reaches and uncovers the exhaust port. When the space between the rotor and the chamber wall diminishes, exhaust gases are expelled. This exhaust phase completes one four-phase cycle.

Multi-speed Bicycle

Forerunners of the modern bicycle appeared in the early 1800s. In 1865 one model was known by spectators as a velocipede and by cyclists as a "boneshaker." It rolled on iron tires with wooden rims and was powered by foot pressure applied to direct-drive pedals—like those of modern tricycles. Hard rubber tires soon improved the ride, and by the 1880s the front wheel had grown to about 60 inches in diameter. This "high wheeler's" 16 feet of travel for every pedal revolution made it capable of goodly speeds downhill and on the level. But it left upgrades to only the robust. And its high center of gravity made it accident prone. First manufactured in 1885, a bicycle with equal-size wheels and the pedal sprocket and chain to the rear wheel allowed safer travel and a more favorable gear ratio for low speed and uphill travel.

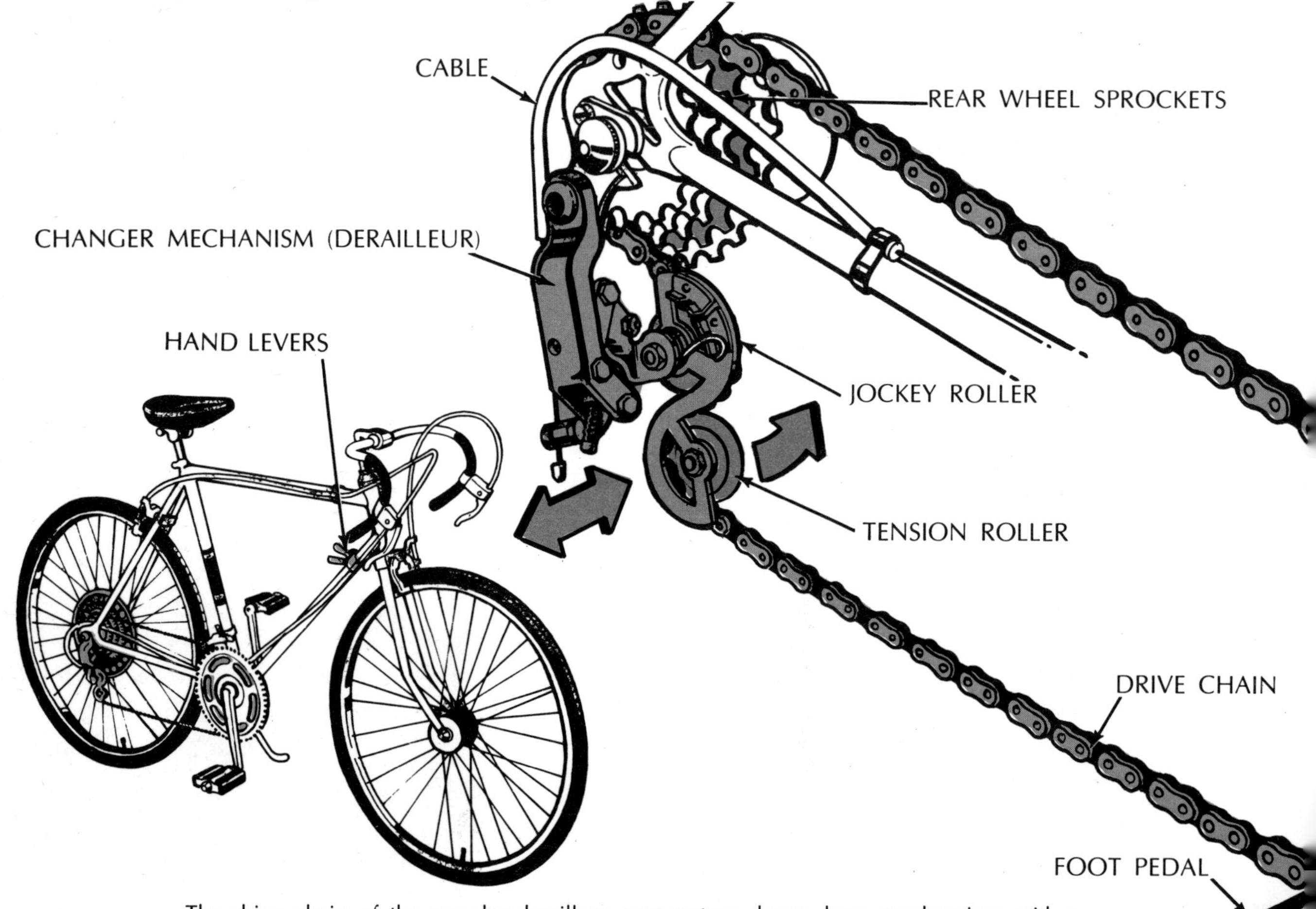

The drive chain of the popular derailleur gear system shown here synchronizes with a changer mechanism at the rear wheel. Here the drive chain meshes with individual sprockets mounted on a "free wheel" ratchet mechanism. Up front, the drive chain meshes with either of two or three sprockets. The combinations of two drive sprockets and five rear-wheel sprockets provide 10 gear options. Three drive sprockets would increase the options to 15. Chain tension is maintained in all gears by a spring-tensioned roller that works in conjunction with a jockey roller that alters the line of the chain, guiding it from one sprocket to another. The rear-wheel changer mechanism, governing the jockey roller and the tension roller, is actuated by a cable running to a hand lever on the bicycle frame. Up front, the chain is guided from one sprocket to another by means of a cage lever, connected through another lever assembly and cable to a hand lever.

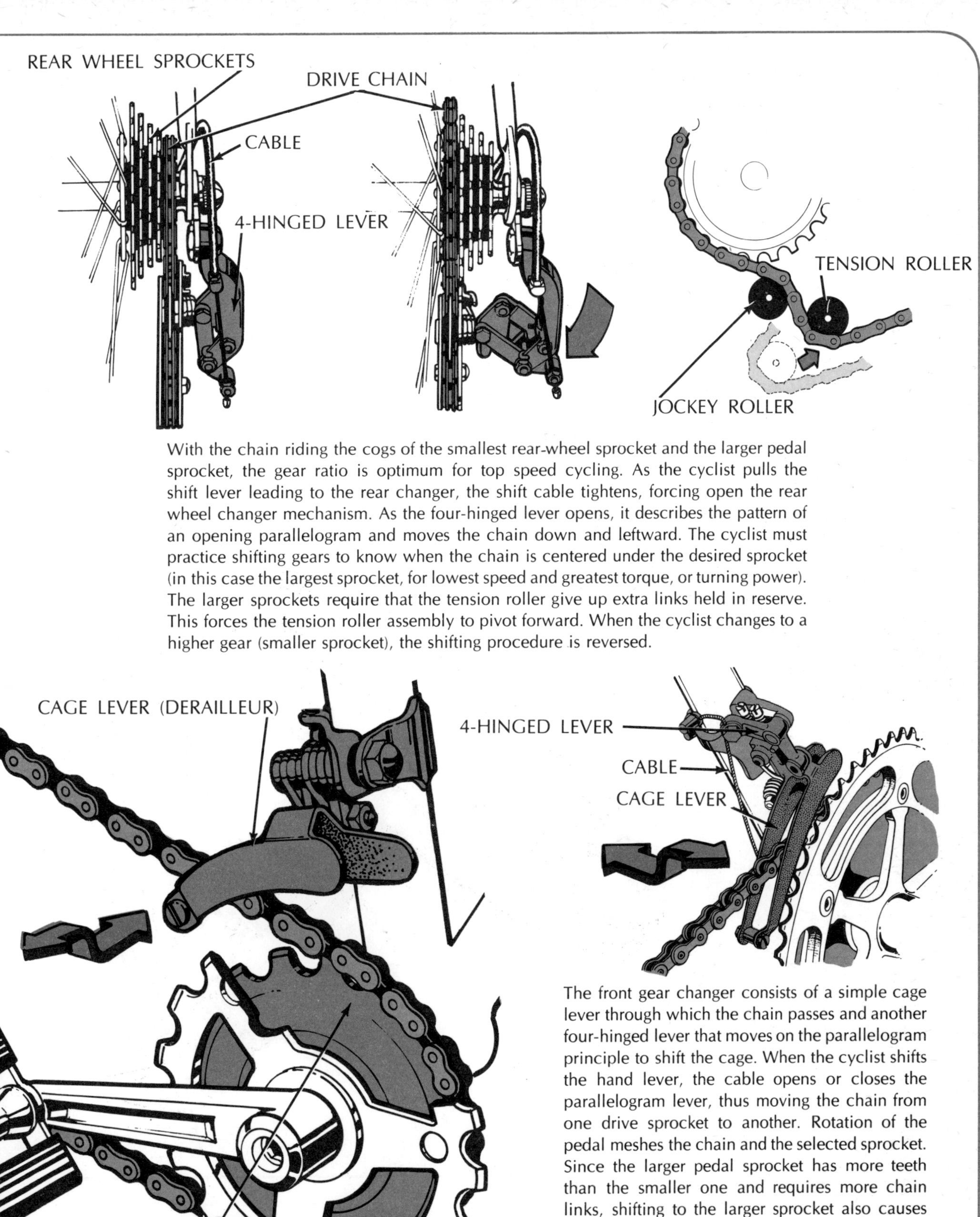

With the chain riding the cogs of the smallest rear-wheel sprocket and the larger pedal sprocket, the gear ratio is optimum for top speed cycling. As the cyclist pulls the shift lever leading to the rear changer, the shift cable tightens, forcing open the rear wheel changer mechanism. As the four-hinged lever opens, it describes the pattern of an opening parallelogram and moves the chain down and leftward. The cyclist must practice shifting gears to know when the chain is centered under the desired sprocket (in this case the largest sprocket, for lowest speed and greatest torque, or turning power). The larger sprockets require that the tension roller give up extra links held in reserve. This forces the tension roller assembly to pivot forward. When the cyclist changes to a higher gear (smaller sprocket), the shifting procedure is reversed.

The front gear changer consists of a simple cage lever through which the chain passes and another four-hinged lever that moves on the parallelogram principle to shift the cage. When the cyclist shifts the hand lever, the cable opens or closes the parallelogram lever, thus moving the chain from one drive sprocket to another. Rotation of the pedal meshes the chain and the selected sprocket. Since the larger pedal sprocket has more teeth than the smaller one and requires more chain links, shifting to the larger sprocket also causes the tension roller assembly near the rear wheel's cluster of sprockets to pivot forward. Shifting to the smaller pedal sprocket reverses the process.

Dial Telephone Mechanism

It is common knowledge that Alexander Graham Bell invented the first telephone. The year was 1876. But few people are aware of the legion of inventors and innovations that have brought about dial phones of today. One of the more interesting and, perhaps, most amusing stories goes back to 1889 in Kansas City. An undertaker named Almon B. Strowger began to suspect that he was losing "customers" to a competitor who had induced a switchboard operator to divert Strowger's phone calls to the competitor's parlor. As it turned out, Strowger's suspicion was unfounded, but it served as a catalyst in Strowger's invention of the dial telephone, first put into successful commercial use in LaPorte, Indiana, in 1895.

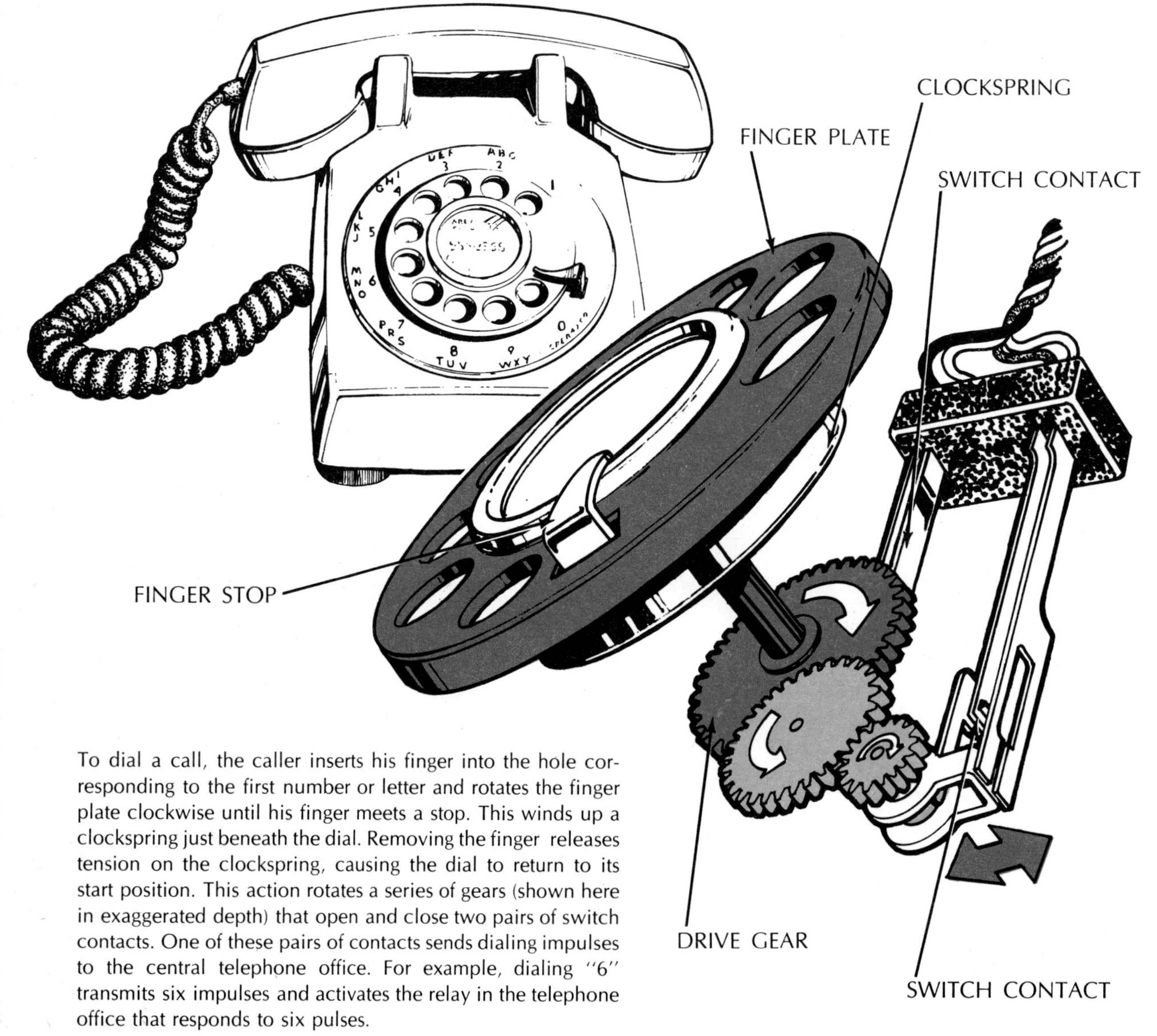

To dial a call, the caller inserts his finger into the hole corresponding to the first number or letter and rotates the finger plate clockwise until his finger meets a stop. This winds up a clockspring just beneath the dial. Removing the finger releases tension on the clockspring, causing the dial to return to its start position. This action rotates a series of gears (shown here in exaggerated depth) that open and close two pairs of switch contacts. One of these pairs of contacts sends dialing impulses to the central telephone office. For example, dialing "6" transmits six impulses and activates the relay in the telephone office that responds to six pulses.

When the mechanism is at rest, a nylon actuator arm resting on the main gear shaft holds a pair of switch contacts open. This pair of contacts are designed to short-circuit the telephone receiver so that no clicks are heard when the finger plate rotates in either direction. The actuator arm, itself, is held in place by a tab stop on the drive gear. The other pair of switch contacts remain closed completing the circuit to the phone line.

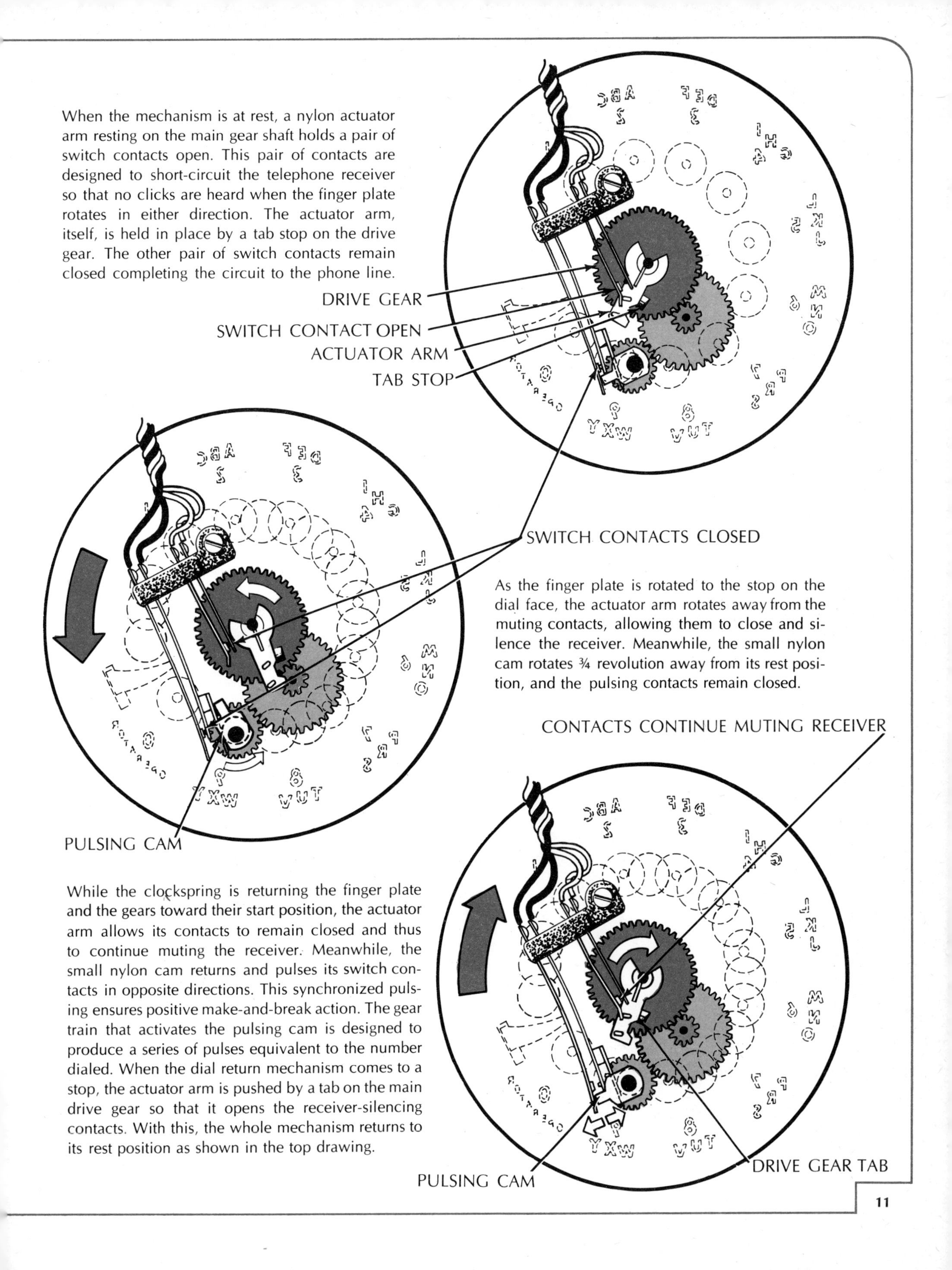

As the finger plate is rotated to the stop on the dial face, the actuator arm rotates away from the muting contacts, allowing them to close and silence the receiver. Meanwhile, the small nylon cam rotates ¾ revolution away from its rest position, and the pulsing contacts remain closed.

While the clockspring is returning the finger plate and the gears toward their start position, the actuator arm allows its contacts to remain closed and thus to continue muting the receiver. Meanwhile, the small nylon cam returns and pulses its switch contacts in opposite directions. This synchronized pulsing ensures positive make-and-break action. The gear train that activates the pulsing cam is designed to produce a series of pulses equivalent to the number dialed. When the dial return mechanism comes to a stop, the actuator arm is pushed by a tab on the main drive gear so that it opens the receiver-silencing contacts. With this, the whole mechanism returns to its rest position as shown in the top drawing.

Push-button Telephone Mechanism

In the push-button telephone, finger pressure operates electrical switches that replace the traditional rotary dial. The push-button telephone contains electronic oscillator circuits that produce groups of pleasing musical tones that fall either in the high frequency or the low frequency band. The switches are so connected that when a button is pushed, a *pair* of tones (different for each digit) are transmitted to the central telephone office where the tones are translated into pulses similar to those produced by the rotary telephone dial. For example, pressing "5" produces tones B and Y. Pressing "3" produces tones A and Z and so forth—always *two simultaneous* tones. This type of "dialing" is called DTMF (Dual Tone Multi-frequency).

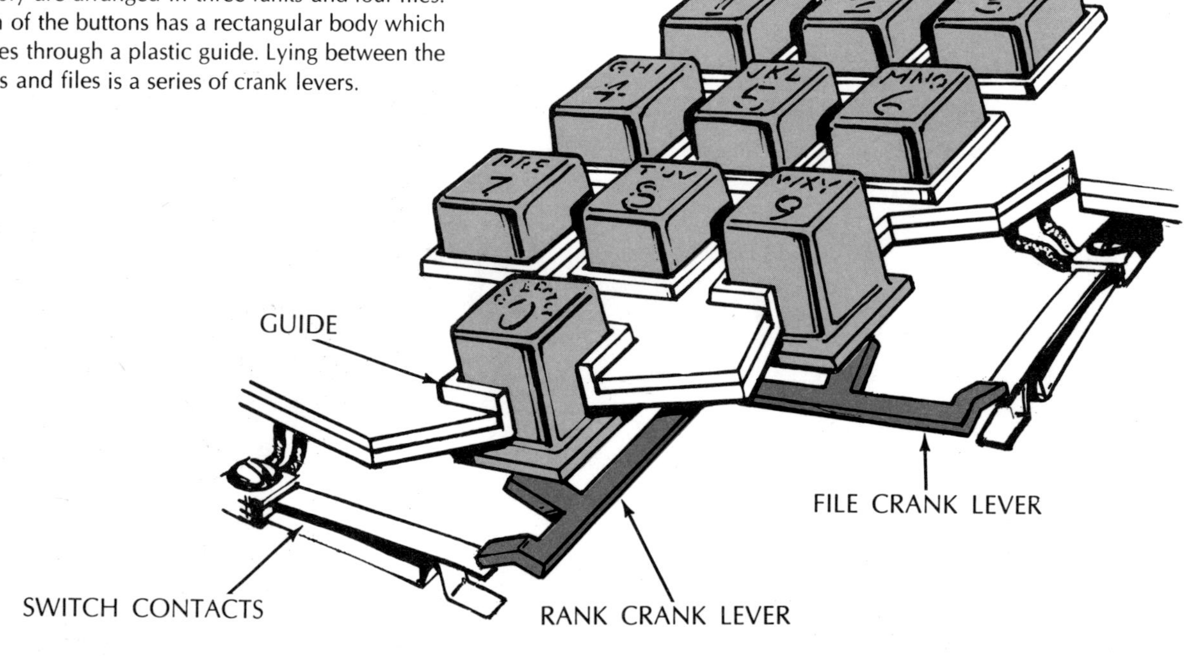

The ten push buttons of the tone generating assembly are arranged in three ranks and four files. Each of the buttons has a rectangular body which passes through a plastic guide. Lying between the ranks and files is a series of crank levers.

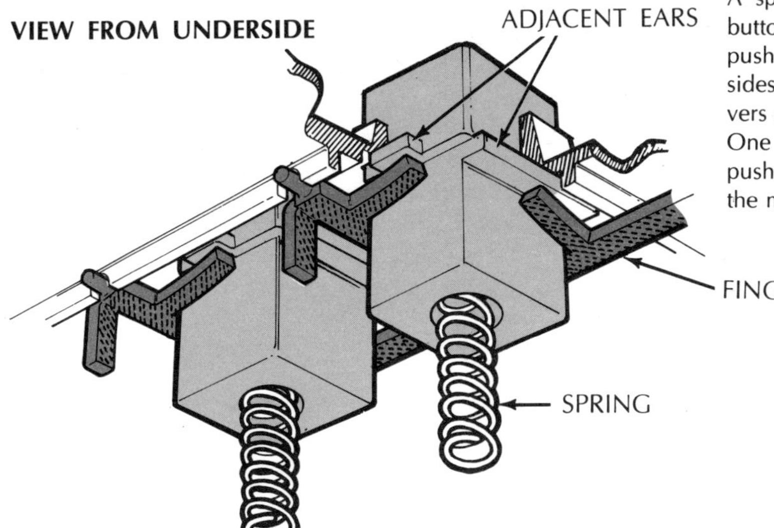

A spring passes from the center of each push button and bears against a backing plate. Each push button has two ears molded into adjacent sides of its body. Fingers on each of the crank levers contact the ears of the push-button switches. One finger on a rank lever is depressed by one push-button ear while the adjacent ear depresses the mating finger of a file lever.

The up and down motion of a push button is translated by rank and file fingers into a slight rotary motion at the tab ends of the crank levers. These tabs bear against switch contacts mounted on the sides of the tone generator assembly. When finger pressure is removed, spring tension causes the push button to return to its at-rest position. This also returns the spring-loaded crank levers to their start positions and thus opens the switch contacts of the tone generator circuits.

TAB

TAB

SWITCH CONTACT

REMAINS STATIONARY

SWITCH CONTACT

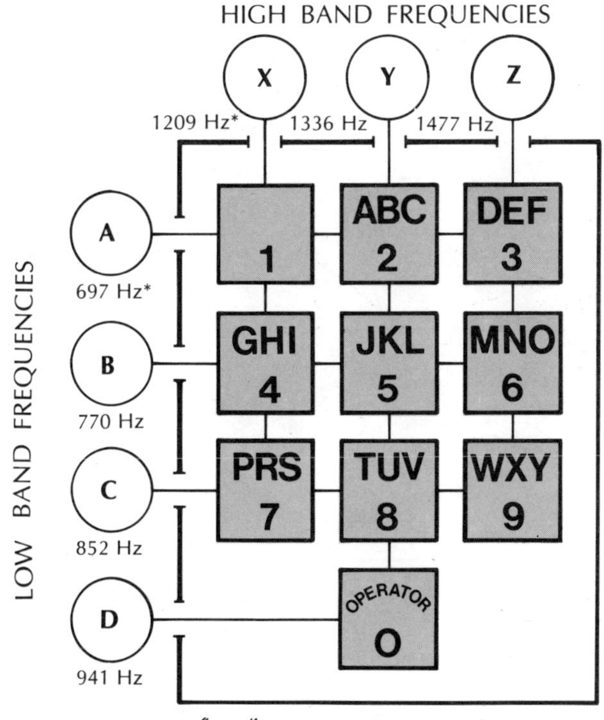

No matter which push button is depressed, a rank lever and a file lever will be operated, and each will close a switching contact. This causes tone generators to produce a unique two-tone signal for each push button depressed.

* Hz, the abbreviation for "hertz," meaning "cycles per second"

Polaroid SX-70 Land Camera

Closed, the SX-70 color camera is small enough to fit into a man's jacket pocket. Yet, amazingly, it houses the components for a fully automatic, single-lens reflex camera that does its own photo printing. The camera is simple to operate: unfold, focus through the lens, and press the red shutter button. After less than two seconds, a buzz, clunk, and whir produce a rectangular card from the front of the camera. Within the card's white border is a 3⅛-inch, blue-green square that is bone dry. In seconds a recognizable image appears within the square, and within minutes the image intensifies into a fully developed color photograph—all automatically outside the camera, with no goo or garbage.

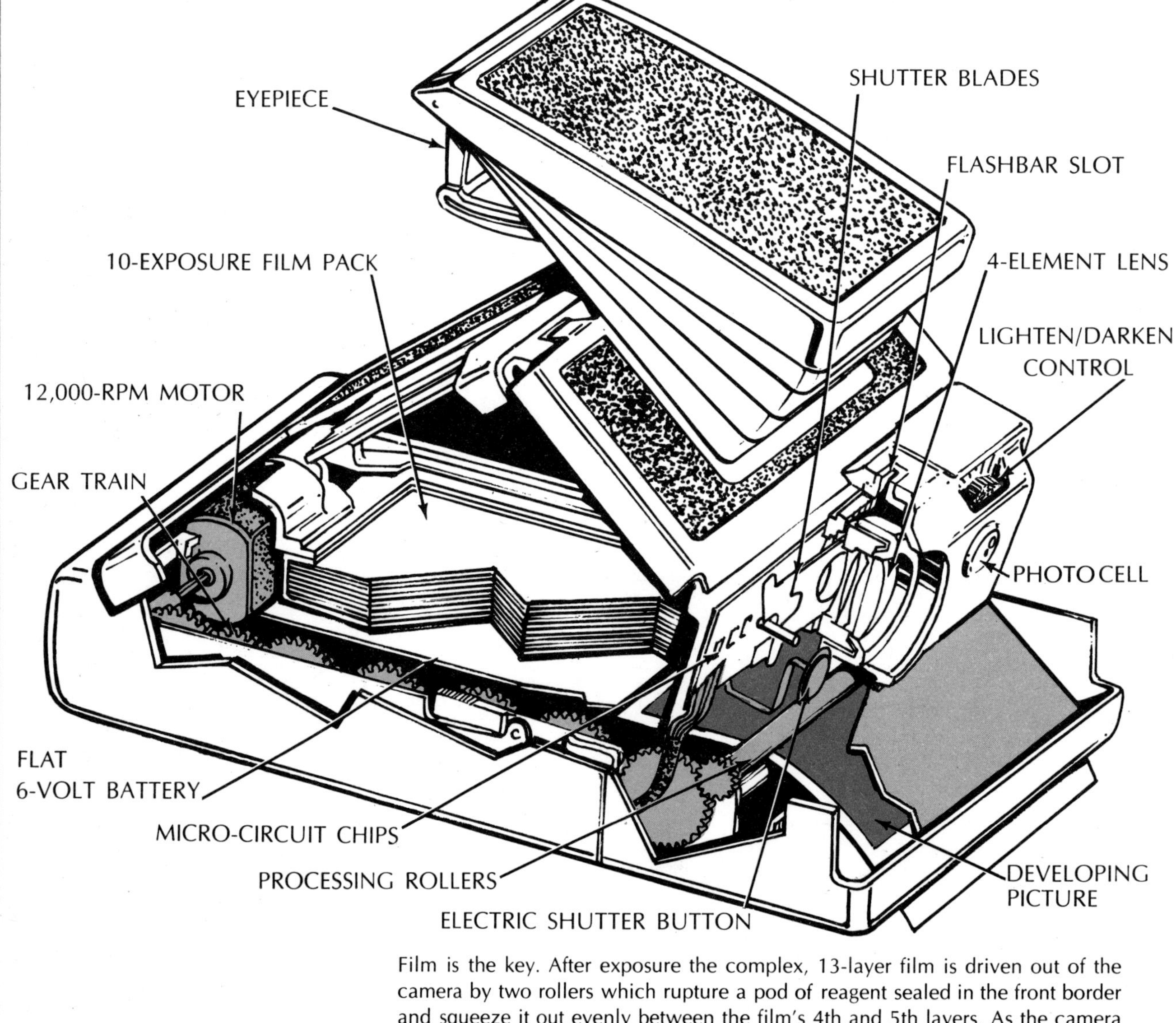

Film is the key. After exposure the complex, 13-layer film is driven out of the camera by two rollers which rupture a pod of reagent sealed in the front border and squeeze it out evenly between the film's 4th and 5th layers. As the camera ejects the film, the reagent starts the chemical developing process while opacifying the film so that no outside light can interfere with the developing process. As the image gains full intensity, the opacifying layer vanishes.

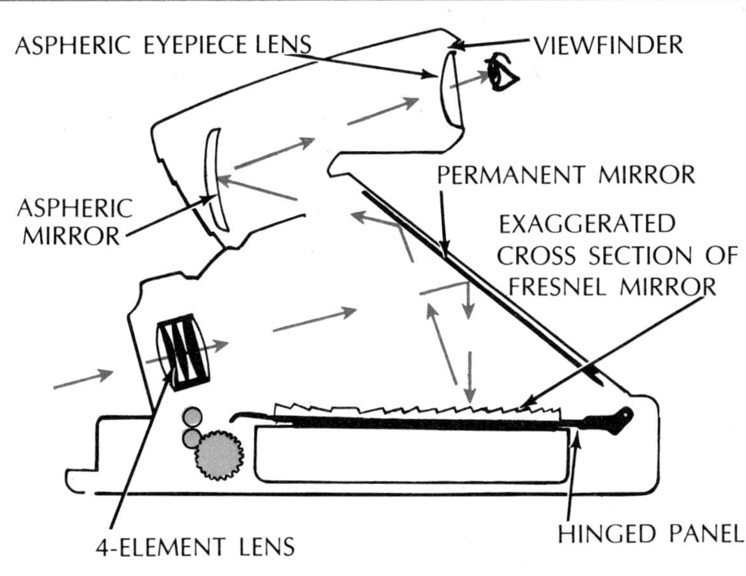

In single-lens reflex cameras, mirrors allow the photographer to see in the viewfinder the same image that will strike the film when the shutter is activated. In the SX-70, light from the subject enters the 4-element lens and then bounces off an angled permanent mirror down onto a special fresnel mirror atop a hinged panel. The special fresnel mirror bundles the light into a beam and projects it (at an increased angle of incidence) up to the permanent mirror. From there, the beam escapes through a small exit hole and strikes an aspheric mirror that reflects the beam through a magnifying aspheric eyepiece lens into the photographer's eye.

When the shutter button is pressed, the two blades that make up the camera shutter and iris diaphragm line up to seal the interior of the camera from light. The hinged mirror panel lying over the film flips up instantly, bringing the "picture-taking" mirror on its underside into the light path. A rubber flap on the panel's free end seals off light from the viewfinder. Backed by an electronic brain with over 200 transistors, a photocell senses the amount of light reflected by the subject, opens the shutter-diaphragm to the correct f-stop, and closes it. Then the mirror panel flops down to cover the film packet.

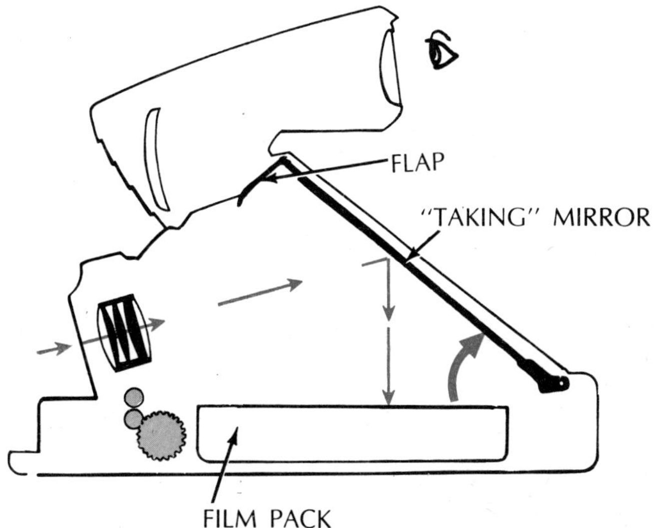

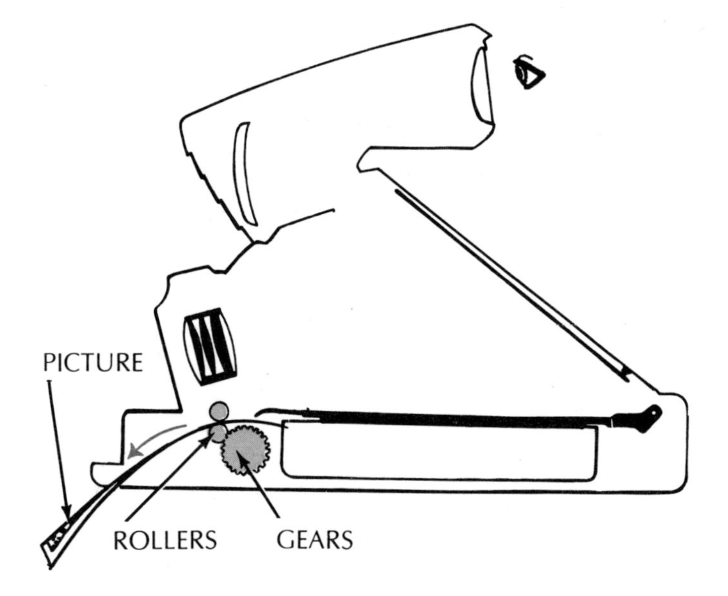

As the mirror panel flops down, a precision gear train, activated by the camera's motor, drives the exposed film through the rollers, as described on the preceding page. Power for the 12,000 rpm motor, flashcubes, and complex electronic micro-circuitry is furnished by a wafer-thin 3½-by-3¾-inch battery. To forestall battery failure, Polaroid supplies a fresh battery within every film pack. When speedy sequential shooting is desired, the film's entire 10-frame pack can be exposed in under 20 seconds, even with the flash attachment.

Hydraulic Car Lift

Electricity, acting through air or oil, can exert tremendous mechanical force. Car lifts are classed as fully hydraulic or semi-hydraulic. Fully hydraulic lifts rely on the dynamics of pressurized liquid, in this case oil. Semi-hydraulic lifts rely on a standing oil supply to seal in pressurized air. For both types of lifts, the supporting column is an elongated piston, or plunger, working in an equally long cylinder set into the ground. At the cylinder's lower end, a plunger bearing rides the cylinder wall to keep the column from wobbling under an off-center load. At the top of the cylinder, a wiper seal leaves only a thin film of oil on the rising plunger and keeps out water and dirt.

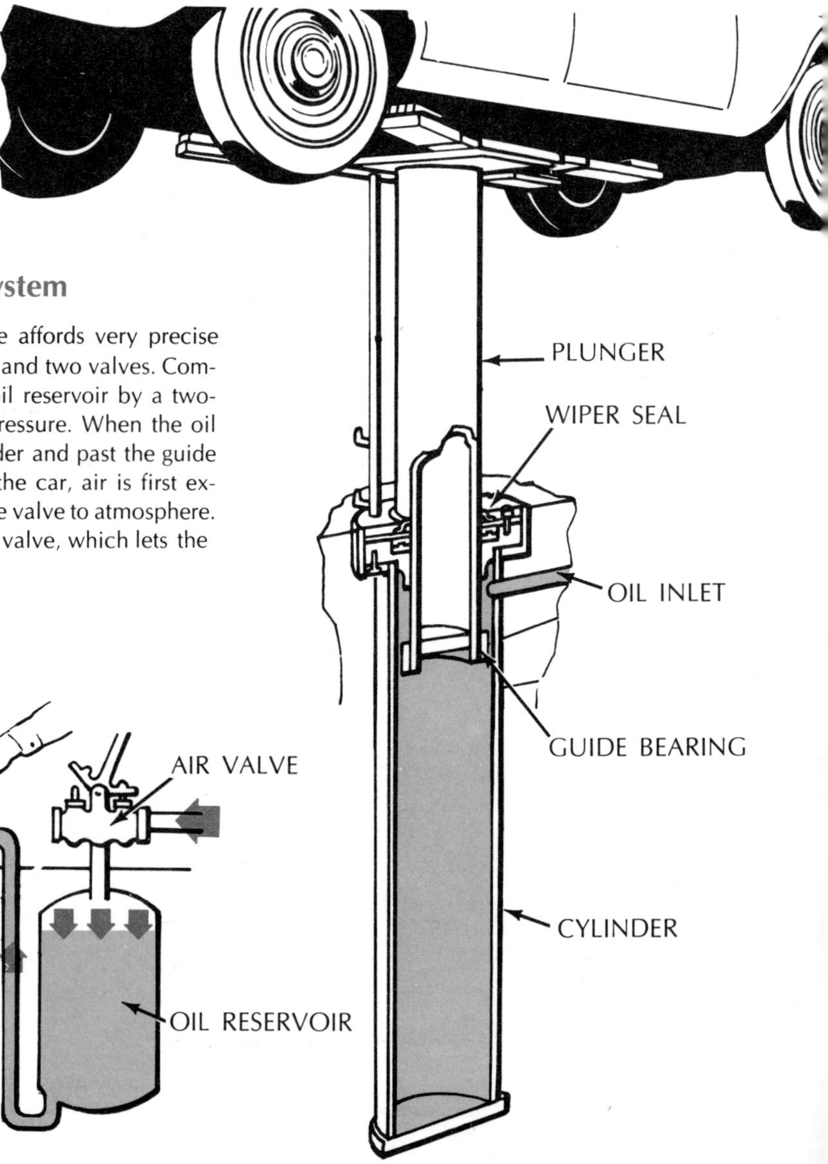

Fully Hydraulic System

The fully hydraulic system shown here affords very precise control but requires a separate reservoir and two valves. Compressed air is admitted to a separate oil reservoir by a two-position valve, putting the oil under pressure. When the oil valve is opened, oil goes into the cylinder and past the guide bearing, raising the plunger. To lower the car, air is first exhausted from the reservoir by opening the valve to atmosphere. Then the descent is controlledby the oil valve, which lets the oil flow back into the reservoir.

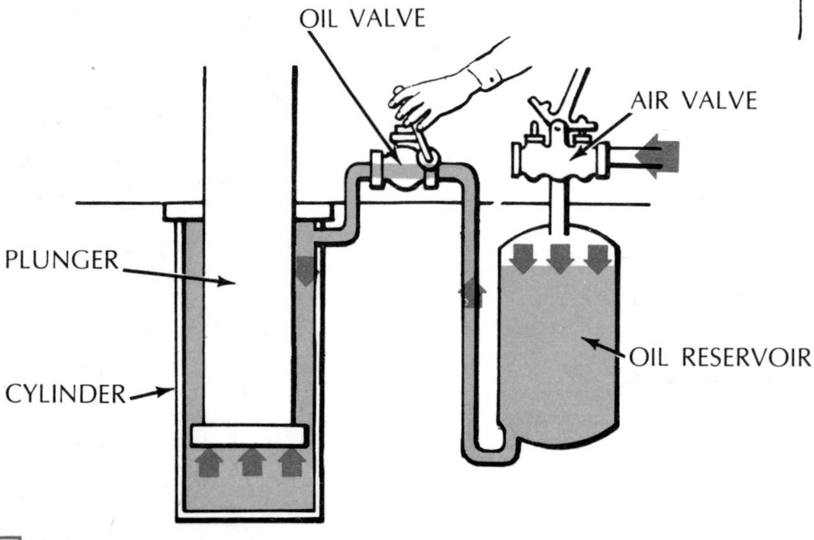

Semi-hydraulic System

In the semi-hydraulic lift, an air pipe runs through the bottom of the plunger up to the top of the cylinder. Oil flows freely into the plunger and between its outer wall and the cylinder proper. The only control is an air valve. When the valve is held open to admit air from the compressor line, pressure builds up between the closed upper end of the plunger and the oil beneath, raising the plunger. When the air is shut off, the plunger coasts slightly before stopping because air is more compressible than oil. To lower the plunger, the valve on the side leading to the atmosphere is opened, letting the pressurized air in the plunger bleed off slowly.

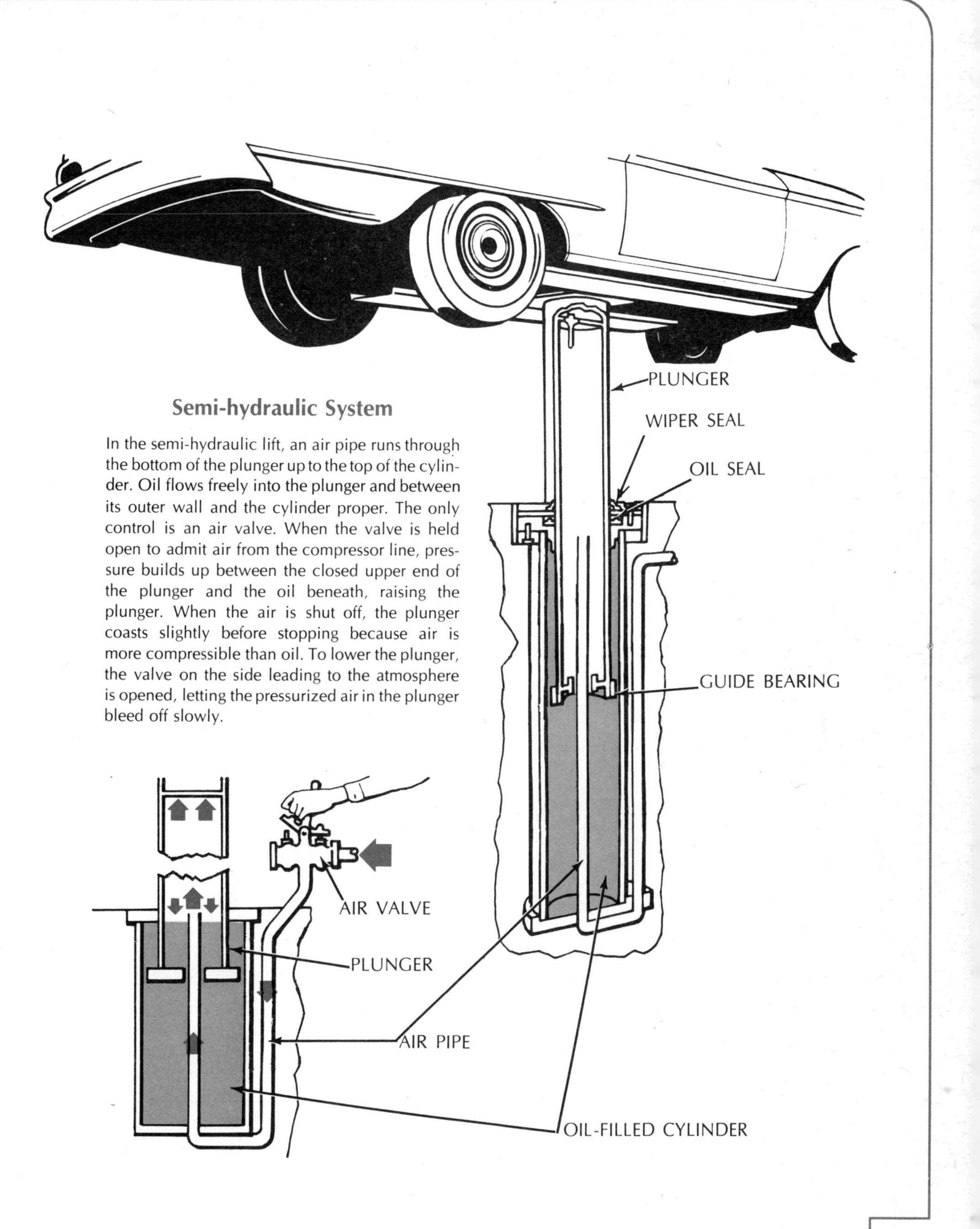

Pop Rivetool

Until recently, riveting was a metal-fastening job mainly reserved for professional ship-builders, steel-workers, and assembly-line riveters. Common rivets look like squat, flat-nosed nails and have varying head shapes. Generally a rivet is passed through aligned holes in two or more pieces of metal, hammered on the near end, and "bucked" on the far end against an anvil or hand tool. The two mushroom-like heads that result on each end of the rivet fasten the metal between them. This method works well when both rivet ends are accessible. But with access to only one end, "blind" riveting was impossible. The need for a single, easy-to-handle tool that could rivet blind led to the "Pop" Rivetool — as easy to use as a stapling gun and ideal for light riveting tasks in industry and at home.

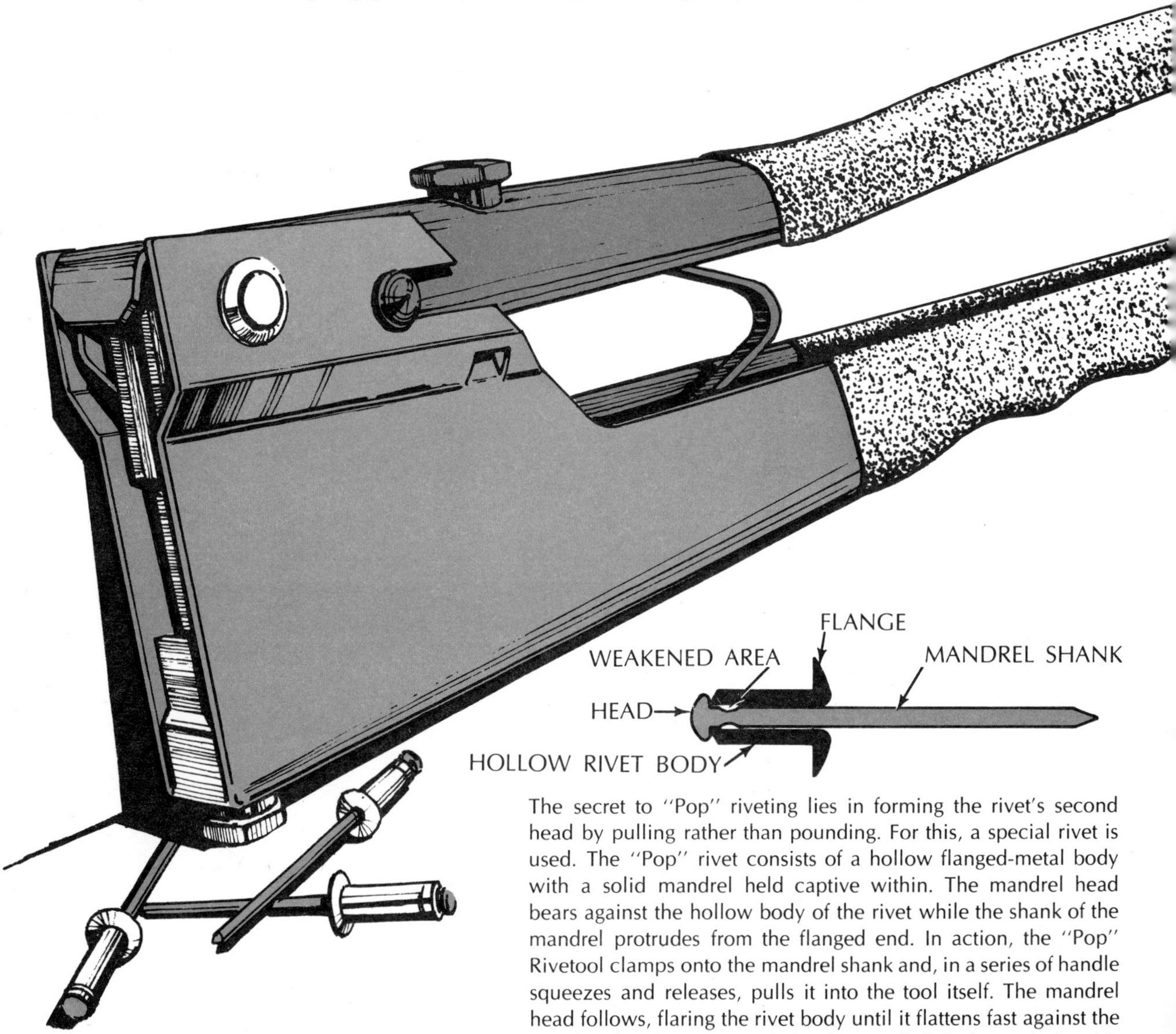

The secret to "Pop" riveting lies in forming the rivet's second head by pulling rather than pounding. For this, a special rivet is used. The "Pop" rivet consists of a hollow flanged-metal body with a solid mandrel held captive within. The mandrel head bears against the hollow body of the rivet while the shank of the mandrel protrudes from the flanged end. In action, the "Pop" Rivetool clamps onto the mandrel shank and, in a series of handle squeezes and releases, pulls it into the tool itself. The mandrel head follows, flaring the rivet body until it flattens fast against the metal surface being riveted. On the mandrel shaft near its head is a purposely-weakened breakage area designed to shear when subjected to a specific amount of pressure. As the mandrel shears, it makes a "pop" sound, giving the "Pop" Rivetool its name.

Mandrel shank is slipped into Rivetool's hollow nose piece, and mandrel head is inserted into work surface. Shank is clenched between clamp and teeth of gripper jaw. As handle is compressed, clamp engages mandrel, wedging it against gripper teeth. Further compression of lever forces clamp and gripper upward.

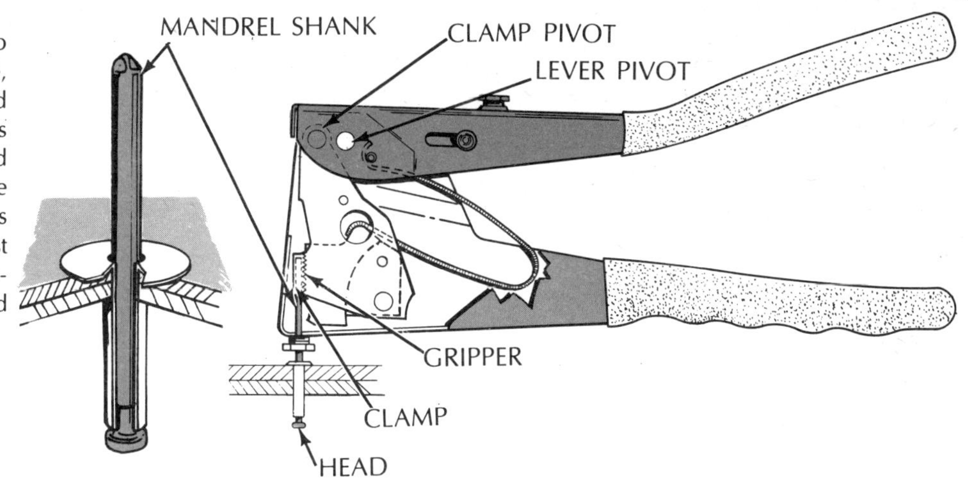

As clamp and gripper draw mandrel into tool, mandrel head balloons rivet body outward. Handle lever reaches maximum travel. When pressure on handle is released, U-spring returns lever to start position while forcing clamp and gripper farther down shank to re-engage it. Process continues, inching mandrel head up, until rivet body is flattened to form "blind side" fastening.

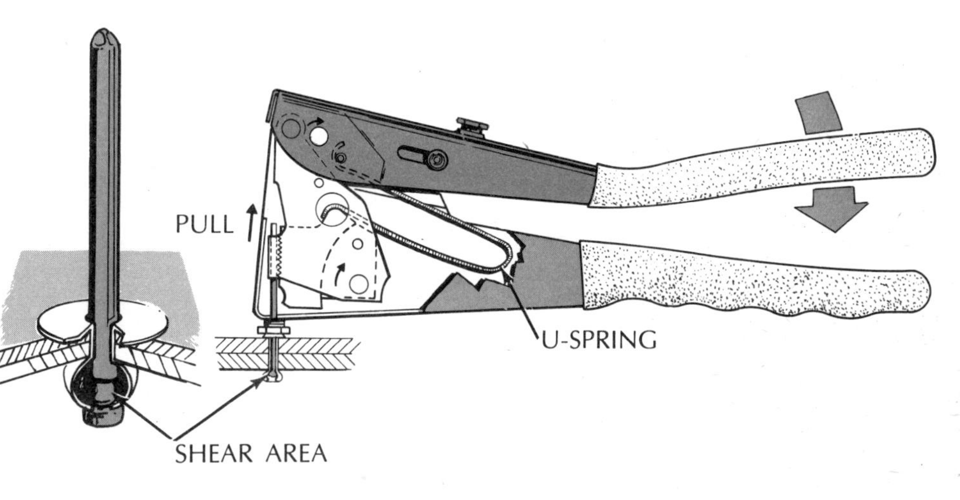

Once rivet body is flattened, clamp and gripper wedge mandrel head snugly into rivet body. With increased hand pressure, shank breaks away from its head, creating "pop" sound. With release of pressure, U-spring returns mechanism to starting position. Unused part of mandrel shank is shaken out of tool and discarded.

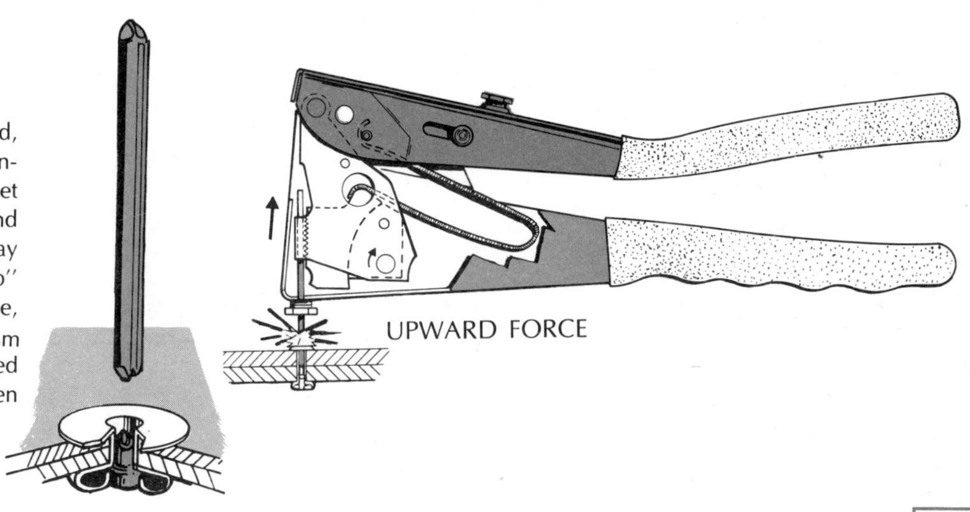

Tire Pressure Gauge

Correct tire pressure for varying vehicle loads in all seasons can improve tire performance and decrease wear. To assure this, a pencil-size gauge accurately measures tire pressures in pounds per square inch. And the gauge is designed to fit all standard tire valves. Its operation principle is simple: A hollow tube in the gauge head depresses the spring-loaded valve pin in the tire. This taps pressurized air from the tire into the gauge. Within the gauge, a plunger is driven by the inrushing air much the way a piston is driven during the power stroke in an internal combustion engine. This thrusts a bar indicator out into view and provides a reading. The "magic" factor about the tire gauge is that the indicator stays in the "out" position until it is manually pushed back inside the gauge.

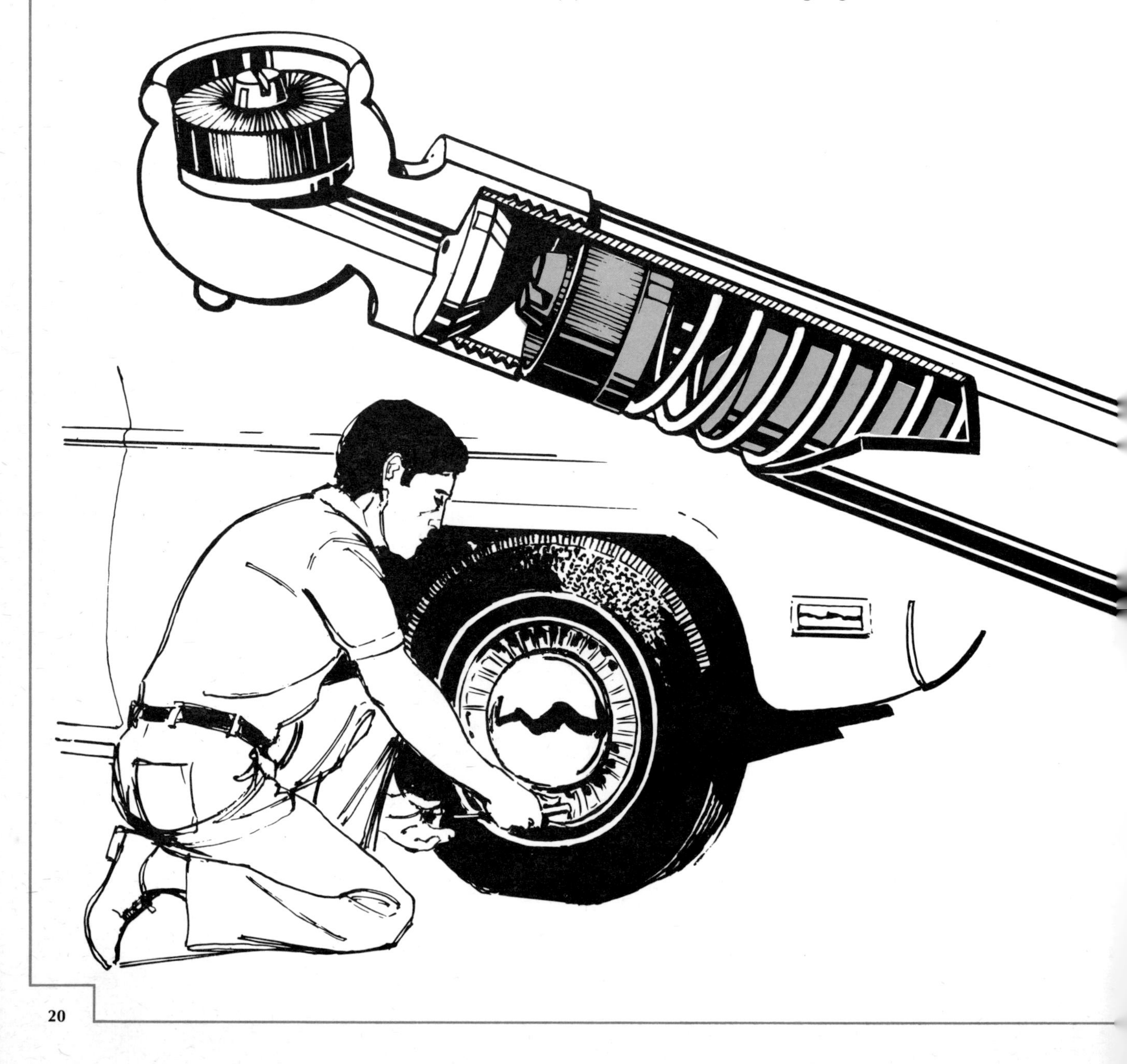

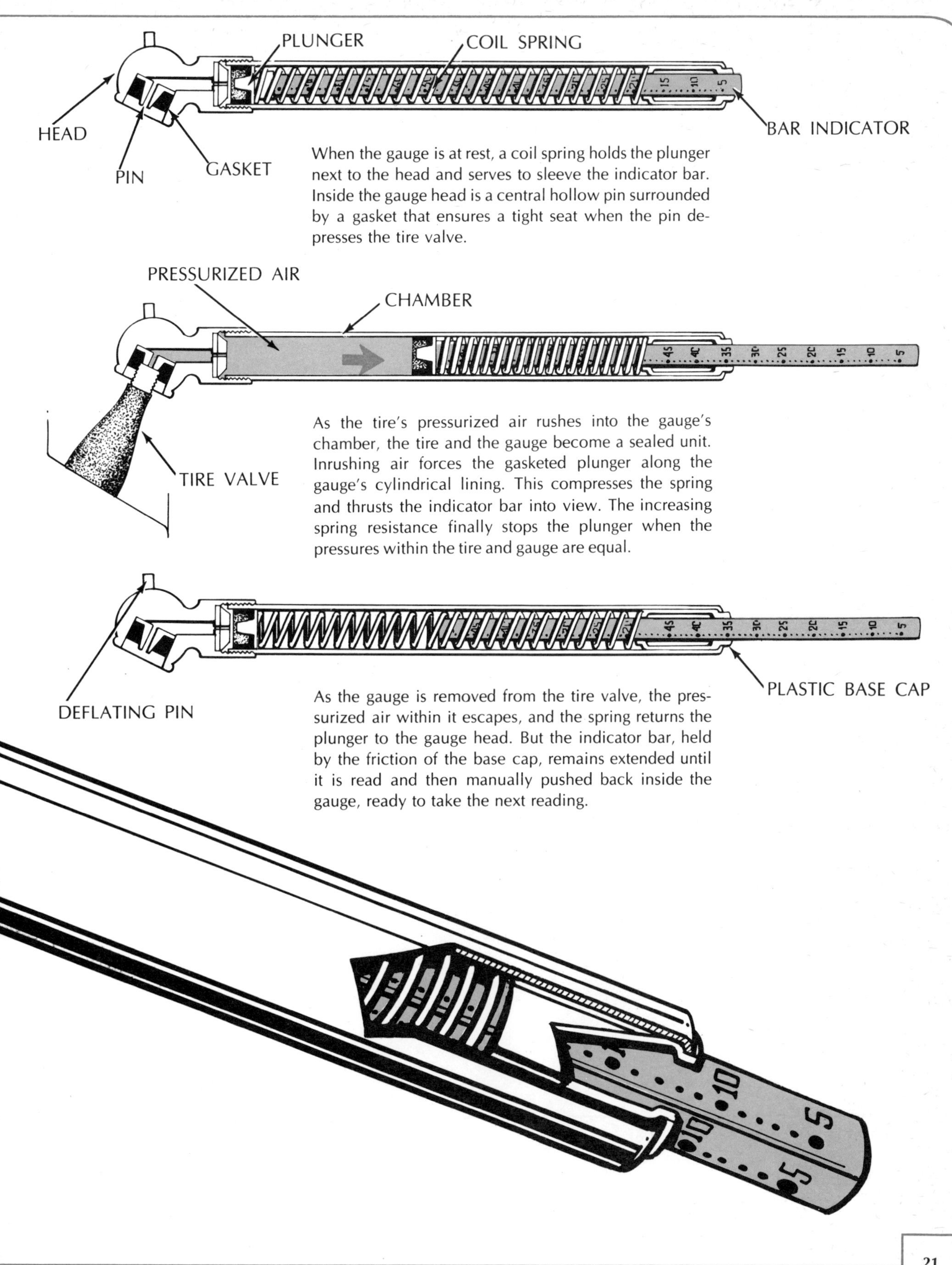

When the gauge is at rest, a coil spring holds the plunger next to the head and serves to sleeve the indicator bar. Inside the gauge head is a central hollow pin surrounded by a gasket that ensures a tight seat when the pin depresses the tire valve.

As the tire's pressurized air rushes into the gauge's chamber, the tire and the gauge become a sealed unit. Inrushing air forces the gasketed plunger along the gauge's cylindrical lining. This compresses the spring and thrusts the indicator bar into view. The increasing spring resistance finally stops the plunger when the pressures within the tire and gauge are equal.

As the gauge is removed from the tire valve, the pressurized air within it escapes, and the spring returns the plunger to the gauge head. But the indicator bar, held by the friction of the base cap, remains extended until it is read and then manually pushed back inside the gauge, ready to take the next reading.

Automatic-focus Slide Projector

Only a bright, crisply focused picture can bring out the subtleties of a good slide. But before automatic focus machines, the projectionist had to constantly fiddle with the lens to maintain clarity. All projectors use high-intensity lamps for bright images. These lamps generate considerable heat, much of which is trapped within the machine. When a slide is snapped from room temperature into the hot housing, it tends to buckle slightly. This temporary distortion, which varies from slide to slide, is enough to throw off the initial fine focus adjustment. The ingenious automatic-focus mechanism on many modern projectors teams light with electronic circuitry to make the small focusing corrections needed for consistently clear pictures.

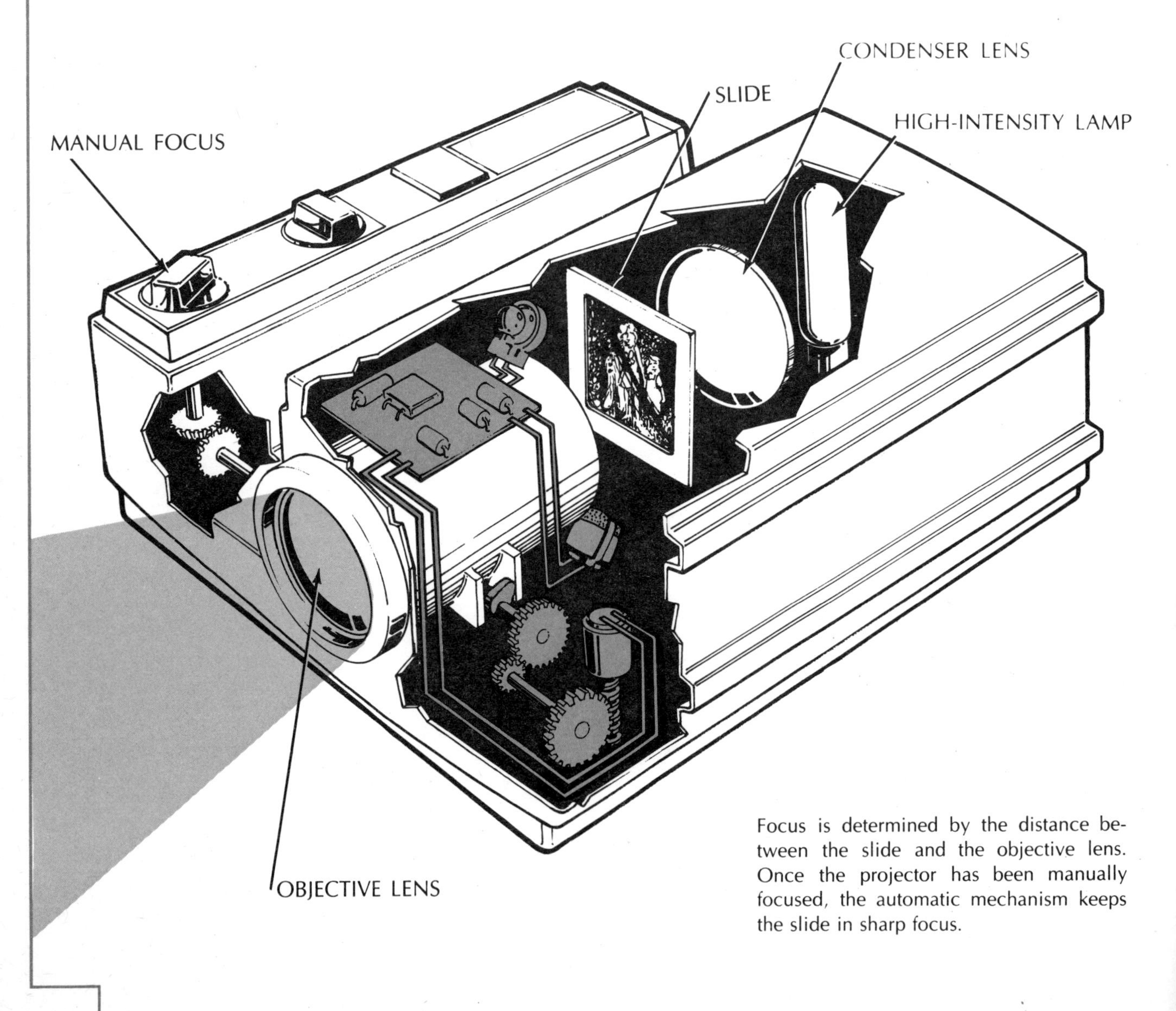

Focus is determined by the distance between the slide and the objective lens. Once the projector has been manually focused, the automatic mechanism keeps the slide in sharp focus.

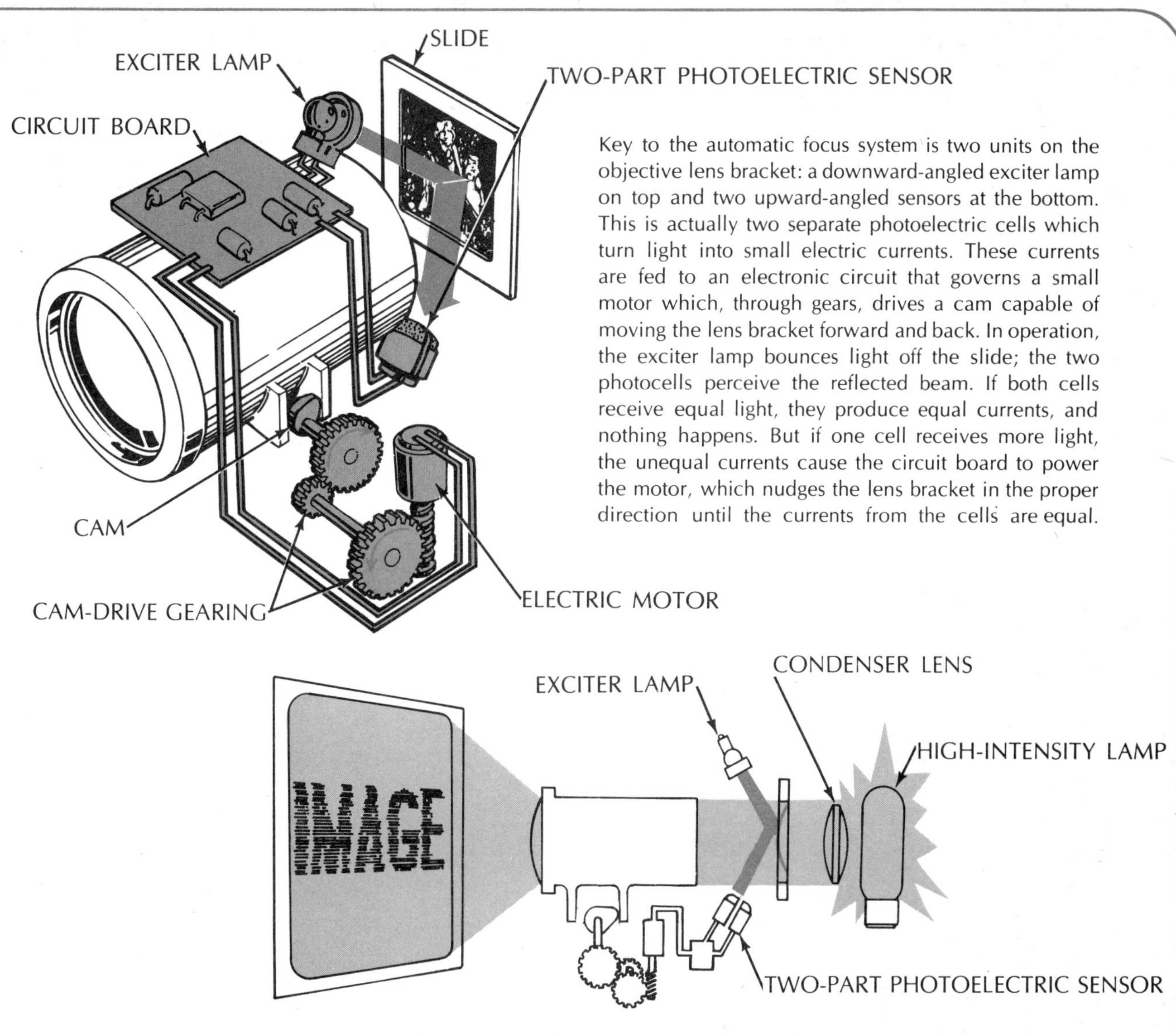

Key to the automatic focus system is two units on the objective lens bracket: a downward-angled exciter lamp on top and two upward-angled sensors at the bottom. This is actually two separate photoelectric cells which turn light into small electric currents. These currents are fed to an electronic circuit that governs a small motor which, through gears, drives a cam capable of moving the lens bracket forward and back. In operation, the exciter lamp bounces light off the slide; the two photocells perceive the reflected beam. If both cells receive equal light, they produce equal currents, and nothing happens. But if one cell receives more light, the unequal currents cause the circuit board to power the motor, which nudges the lens bracket in the proper direction until the currents from the cells are equal.

The fuzzy image shown is caused by a warped slide. The distortion causes the exciter lamp's beam to reflect mostly onto the upper photoelectric cell. The circuit board then activates the motor-driven cam, which pushes the entire lens assembly forward until both cells receive an equal amount of light, at which point the slide is again in sharp focus.

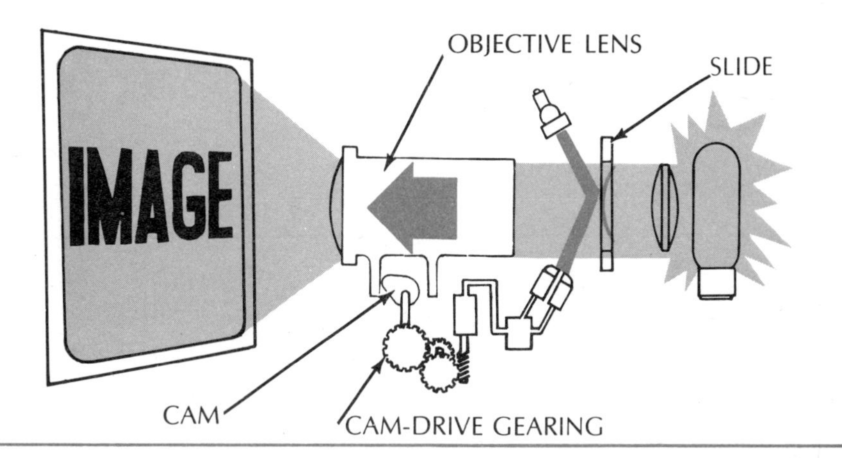

Barometer and Altimeter

The barometer and the altimeter are two similar instruments that measure atmospheric (air) pressure. At sea level, air pressure averages 14.6 pounds per square inch. At 3.5 miles up, pressure is about half that, and at an altitude of 11 miles the pressure of the atmosphere is down to about 0.146 pounds per square inch. Measuring air pressure is thus one way to measure altitude. Barometers are calibrated in inches of mercury because the first, invented by Evangelista Torricelli, used a vertical column of the liquid metal which rose or fell to indicate air pressure changes. Good weather is generally accompanied by high atmospheric pressure—29.9 inches, or more. A sudden drop in atmospheric pressure indicates a change in the weather—usually for the worse. Because mercury barometers are clumsy to transport, the aneroid, or dry, barometer (below and top of opposite page) is the more common weather watching instrument.

Barometer

The aneroid barometer's pressure sensor is a flexible metal can-like bellows, from which some of the air has been partially evacuated. One side is fixed, the other is attached to a lever, crankshaft and arm that moves a bell crank which terminates in a sector gear meshed to a pinion gear on the instrument's pointer shaft. Increasing or decreasing outside air pressure causes the bellows to contract or expand. Bellows movement is magnified by the levers and gears. A slight change in air pressure thus produces a wide swing of the pointer over the barometer's dial.

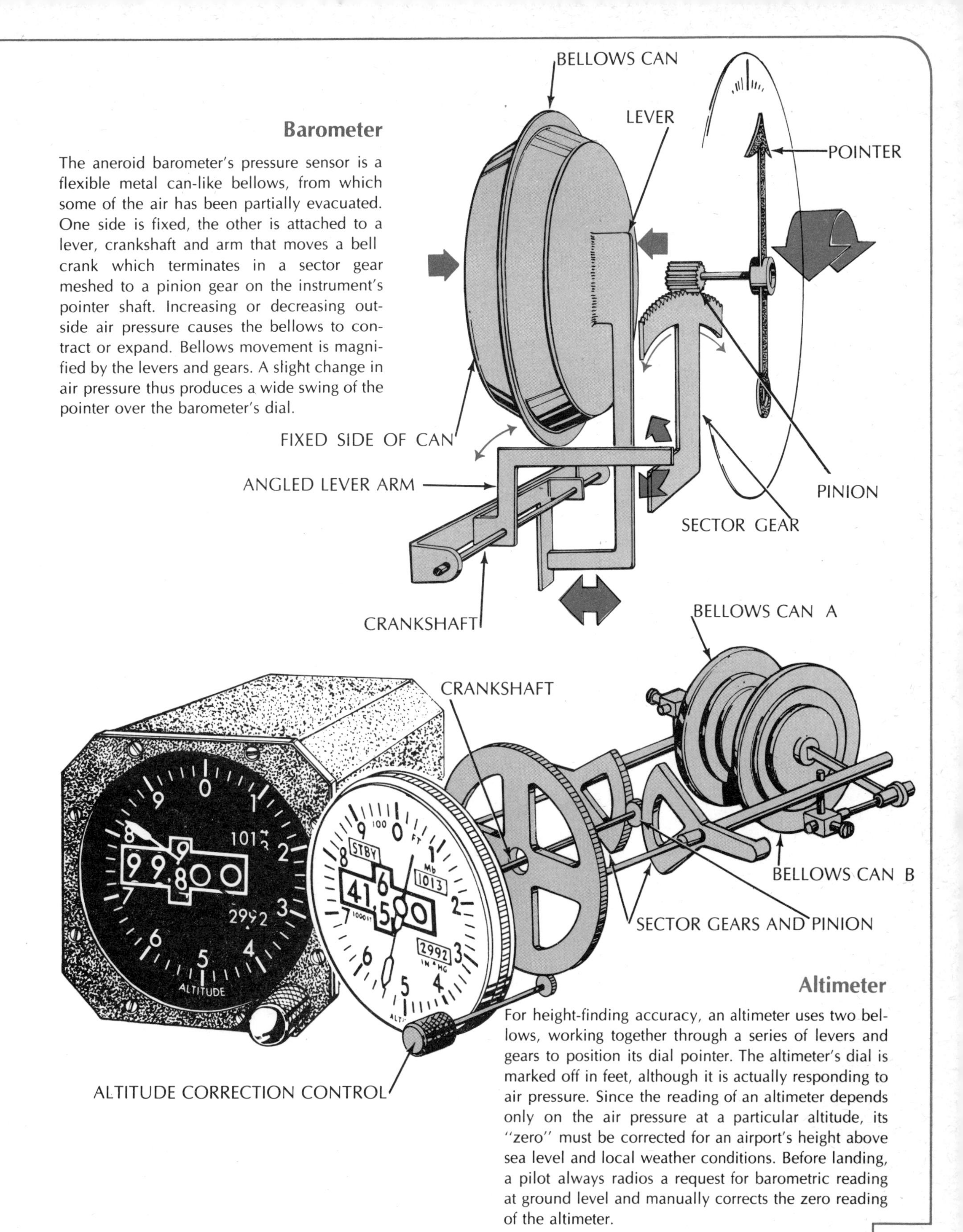

Altimeter

For height-finding accuracy, an altimeter uses two bellows, working together through a series of levers and gears to position its dial pointer. The altimeter's dial is marked off in feet, although it is actually responding to air pressure. Since the reading of an altimeter depends only on the air pressure at a particular altitude, its "zero" must be corrected for an airport's height above sea level and local weather conditions. Before landing, a pilot always radios a request for barometric reading at ground level and manually corrects the zero reading of the altimeter.

Depth Finder

From Biblical times to World War II, the sole instrument available to mariners for gauging depths was a weighted and marked "lead line." A ship's leadman would cast the line overboard, wait for it to bottom and then read off the depth. But in the early war years, the advent of submarine fleets triggered a frantic search by scientists of several nations for a way to "see" underwater. The result: **so**und **na**vigation **r**anging, or sonar. After the war, sonar was quickly adapted for commercial use. Today, whether it is called a depth sounder, depth finder or fathometer, it is one of the boatsman's most useful pieces of electronic equipment.

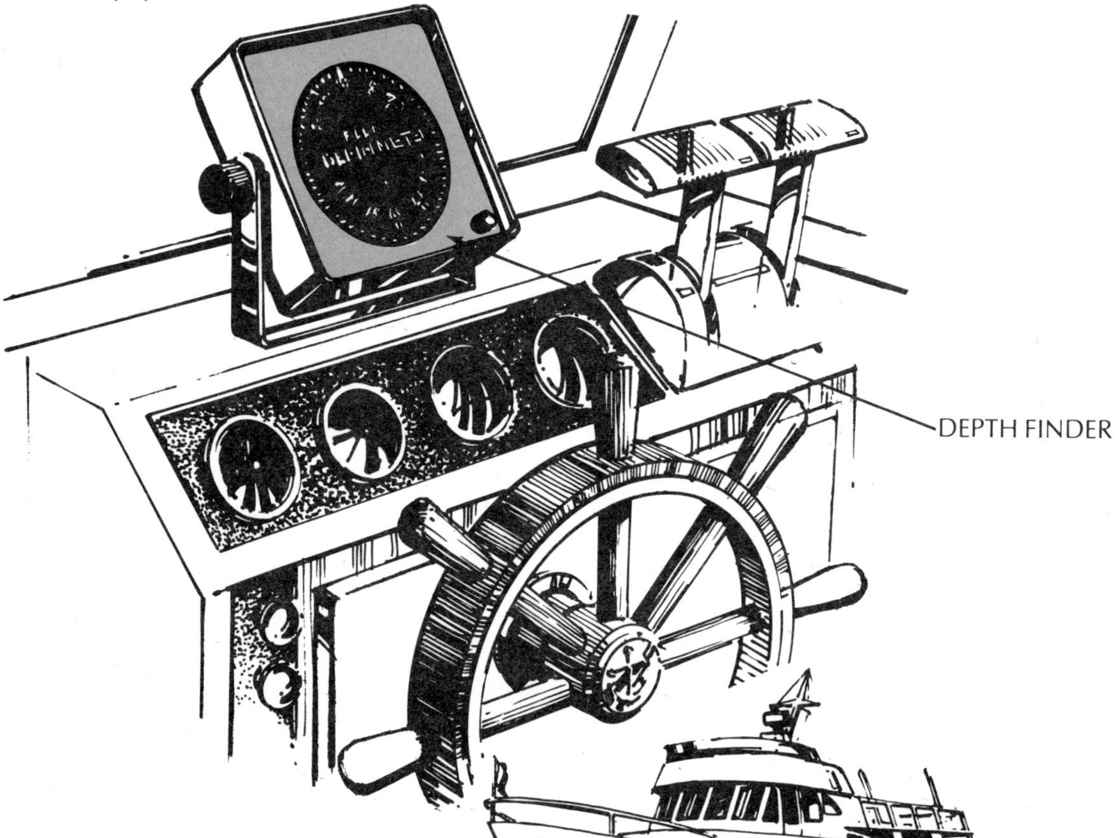

DEPTH FINDER

TRANSDUCER

TRANSMITTED SIGNAL

The main components of the depth finder include a transmitter, receiver and gauge, usually in a single unit as shown above. The transmitter sends electrical impulses to the hull-mounted transducer, right, which converts them into high-frequency ultrasound waves. These waves travel through water at 4,800 feet per second—almost five times the speed of sound in air. When they strike an underwater object, the ultrasound waves are reflected. The echoes that reach the transducer are converted into electrical signals which are amplified by the receiver and then fed to the gauge.

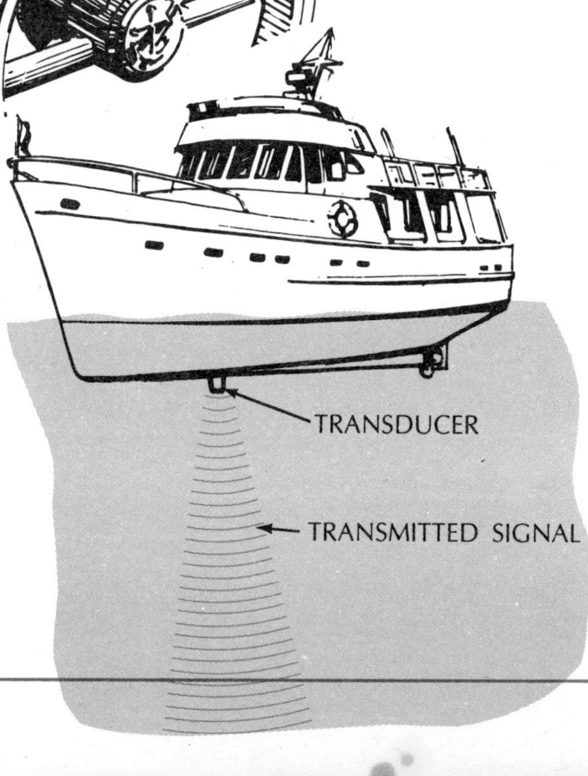

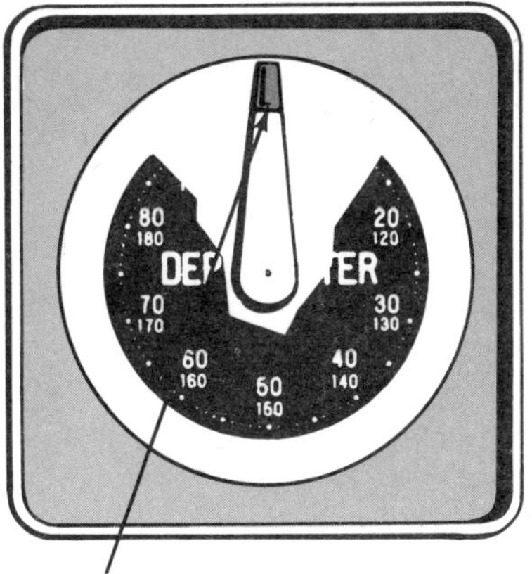

The clock-like gauge, left, features a sweeping hand with a neon bulb at the end that sits behind a scale calibrated in feet or fathoms. The hand can be set to revolve at speeds ranging from 200 to 3,600 rpm—slow speeds for great depths, high speeds for shallow depths. In operation, right, the neon bulb automatically flashes every time the hand passes the "O" setting. In addition, it flashes every time a signal echoes into the receiver.

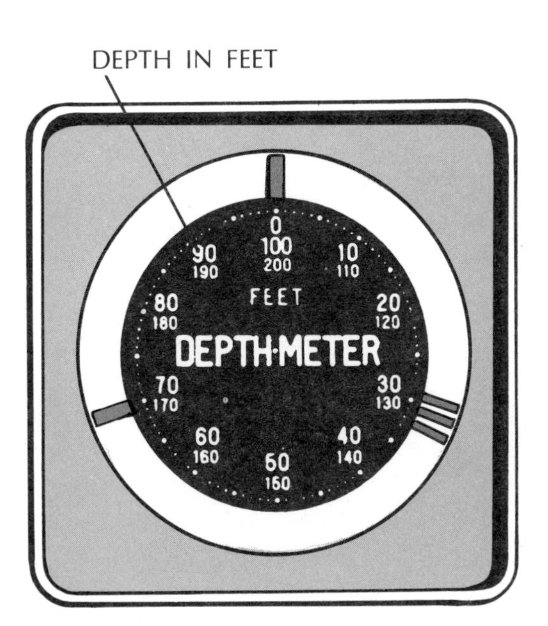

The brightness of the flash on the gauge depends on the strength of the echo picked up by the transducer. Thus, weaker lines usually indicate schools of fish, while the strongest flash usually marks the bottom.

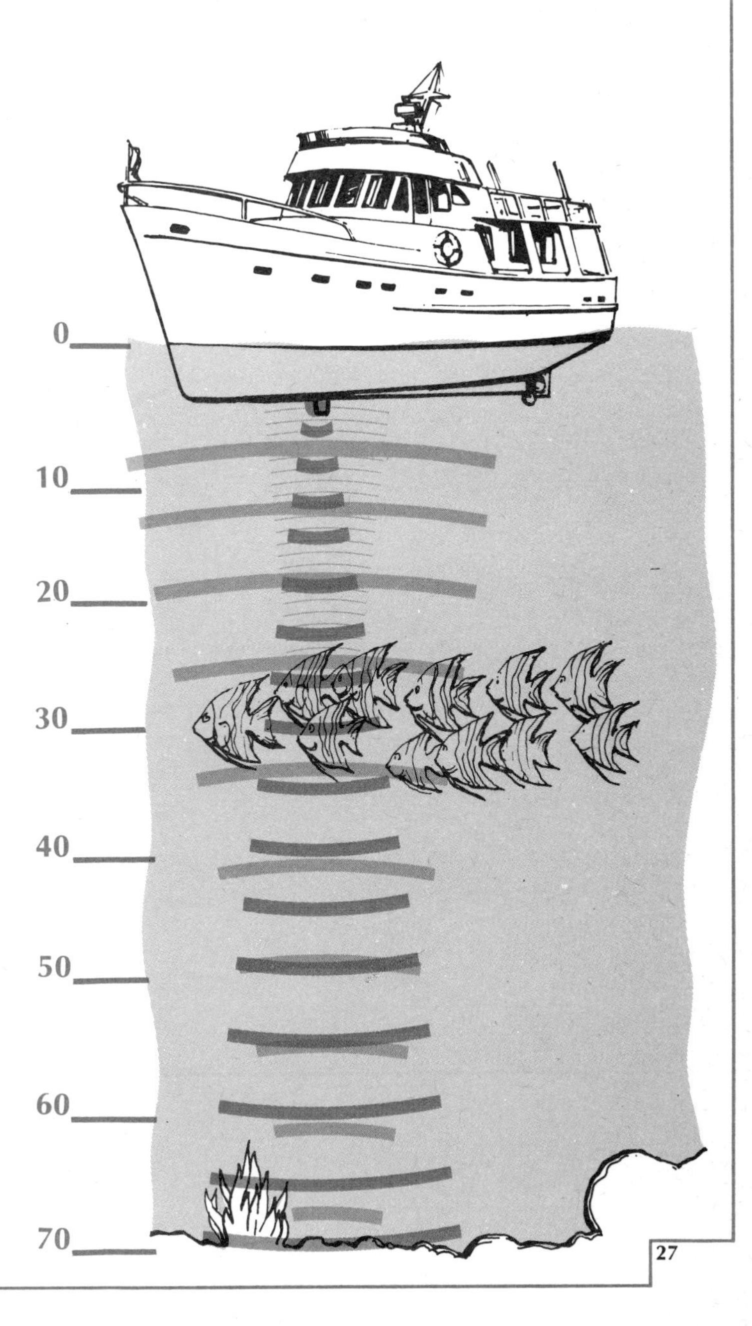

Vise-grip Pliers

Neither pliers, wrench, clamp, nor wire cutters, the vise-grip pliers is a hybrid that successfully blends the best features of all four. This clever mechanism puts more than one ton of gripping power into the user's hand. It can be adjusted so that its jaws will self-lock onto any firm object up to 1½ inches across. The vise-grip's tooth pattern and tremendous gripping force make it capable of grasping uneven objects and turning sheared nut heads. And the vise grip's self-lock mechanism holds the jaws on the work surface even when the user's hands are removed. In operation, the squeezing of the handles produces a lever-magnified force in the jaws.

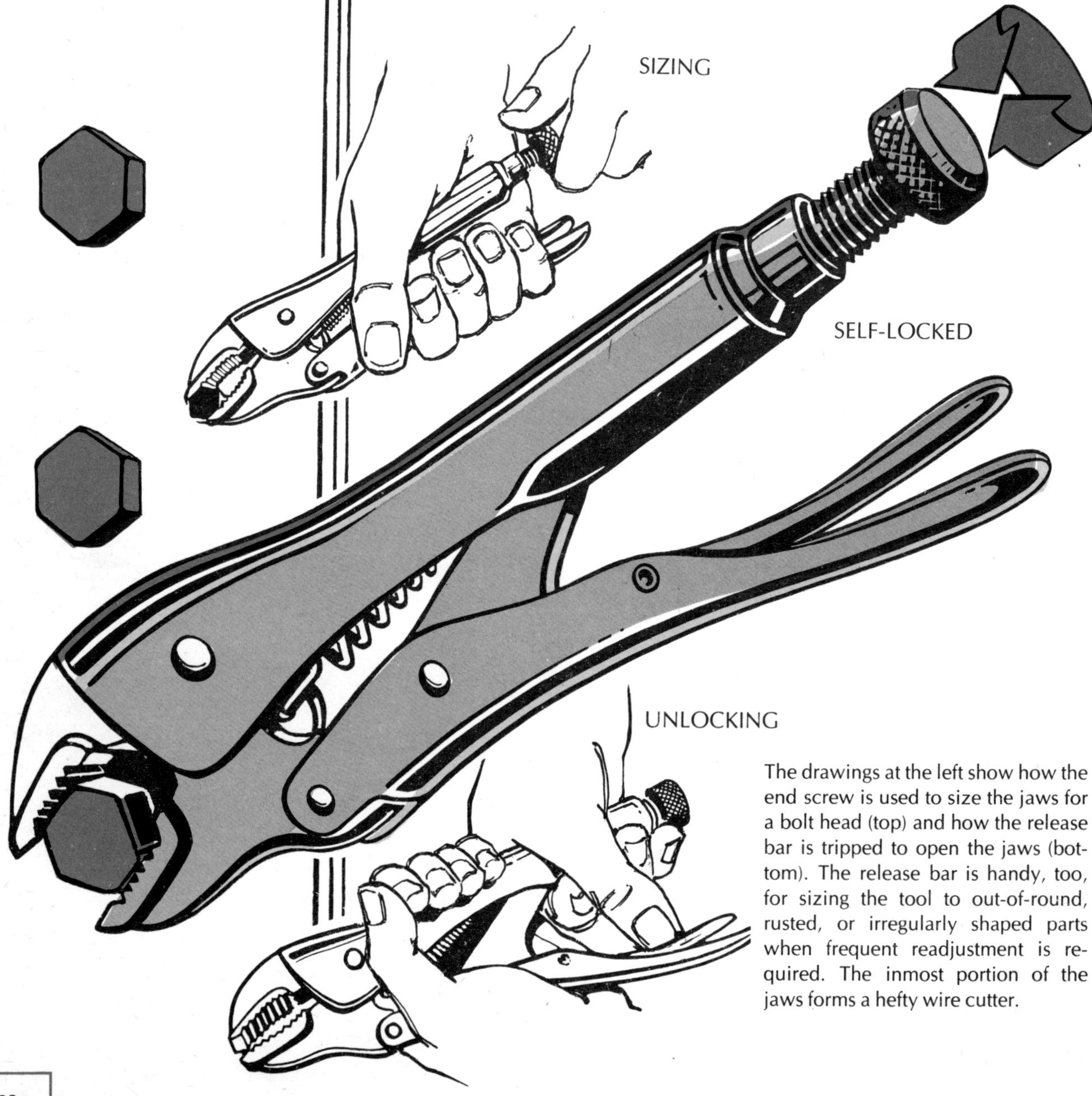

The drawings at the left show how the end screw is used to size the jaws for a bolt head (top) and how the release bar is tripped to open the jaws (bottom). The release bar is handy, too, for sizing the tool to out-of-round, rusted, or irregularly shaped parts when frequent readjustment is required. The inmost portion of the jaws forms a hefty wire cutter.

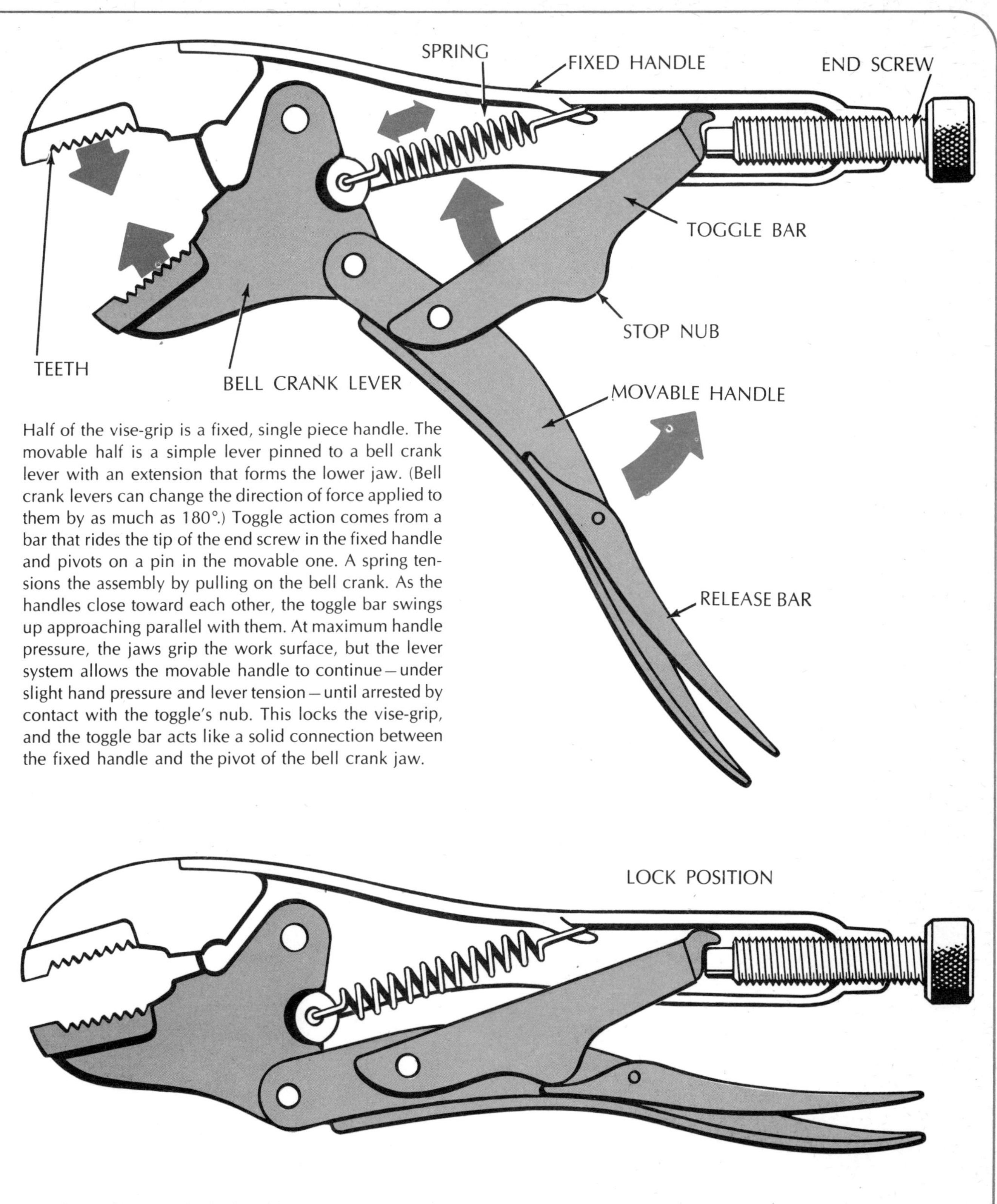

Half of the vise-grip is a fixed, single piece handle. The movable half is a simple lever pinned to a bell crank lever with an extension that forms the lower jaw. (Bell crank levers can change the direction of force applied to them by as much as 180°.) Toggle action comes from a bar that rides the tip of the end screw in the fixed handle and pivots on a pin in the movable one. A spring tensions the assembly by pulling on the bell crank. As the handles close toward each other, the toggle bar swings up approaching parallel with them. At maximum handle pressure, the jaws grip the work surface, but the lever system allows the movable handle to continue—under slight hand pressure and lever tension—until arrested by contact with the toggle's nub. This locks the vise-grip, and the toggle bar acts like a solid connection between the fixed handle and the pivot of the bell crank jaw.

Once the jaws lock, hand pressure on the movable handle couples the force to the jaws. The greater the pressure, the tighter the jaws grip.

Electric Hedge Trimmer

Clipping a hedge with hand shears can be a tedious exercise. With an electric hedge trimmer, though, the chore of chopping through a mass of pulpy stems while maintaining a firm hand becomes much easier. The lightweight trimmer runs on household current. Its muscle comes from a universal motor that rotates at 18,000 rpm. This high-speed energy is converted, by gears, into slower but stronger reciprocating strokes. Of the machine's two toothed blades, one is fixed. The other, which contains the sharp cutting edges, slides back and forth to give 3,600 cutting strokes per minute.

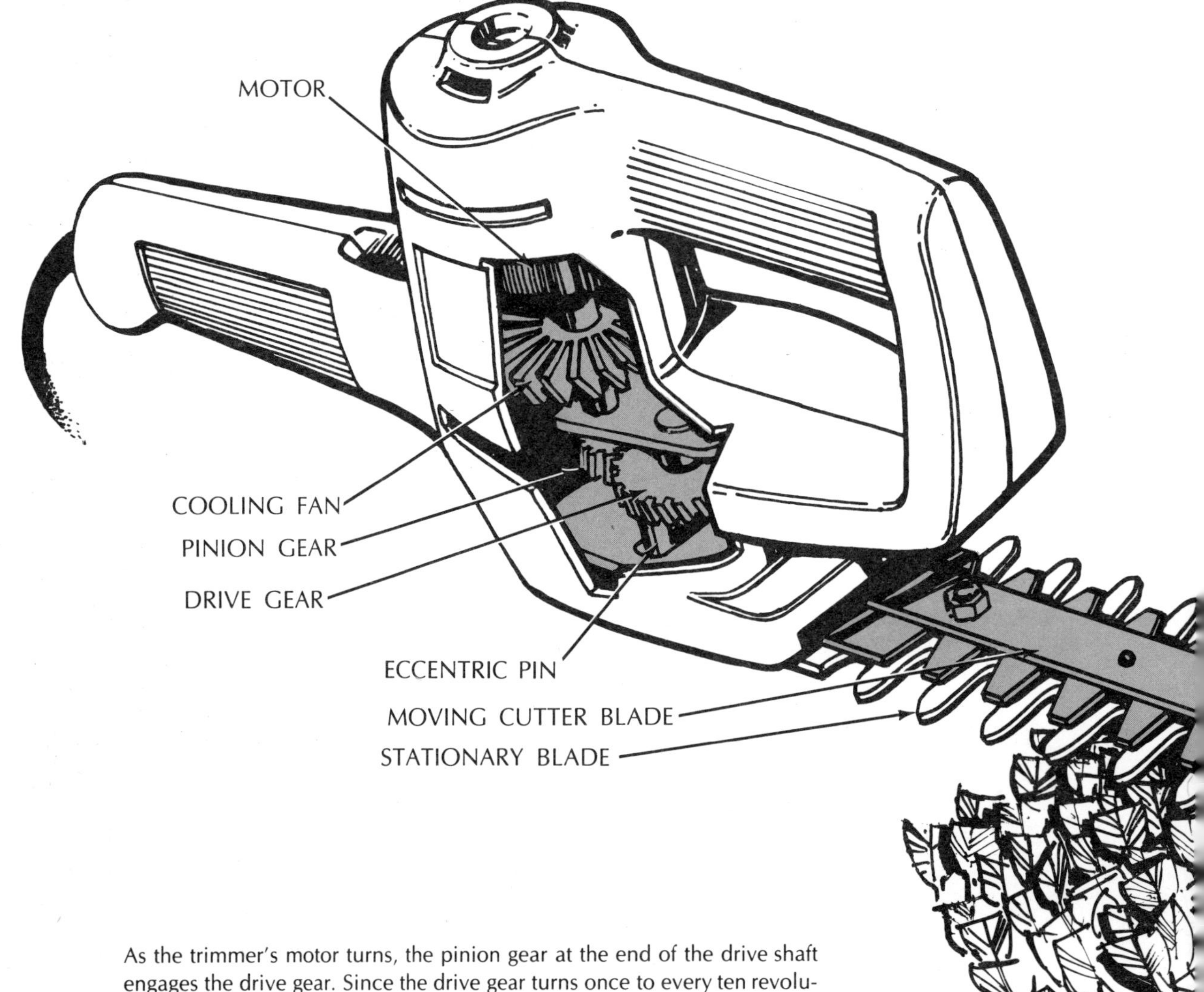

As the trimmer's motor turns, the pinion gear at the end of the drive shaft engages the drive gear. Since the drive gear turns once to every ten revolutions of the smaller pinion, the motor's speed is transformed into power. The drive gear has an off-center pin on its underside that engages a slot in the movable cutting blade. Each revolution of the drive gear produces one forward and one backward cutting stroke.

As the eccentric pin moves past position *A*, the movable blade begins its forward cutting stroke. When the eccentric pin reaches position *B*, the blades are closed. As the eccentric pin moves past position *C*, the movable blade begins its backward cutting stroke. When the eccentric pin reaches position *D*, the blades are again closed, completing one cutting cycle.

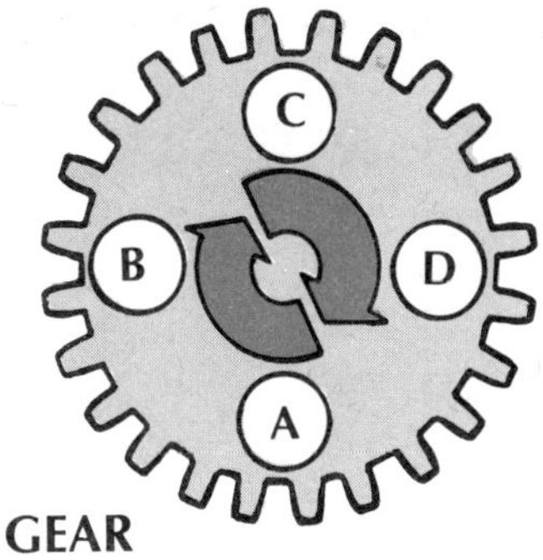

C
B
D
A

DRIVE GEAR

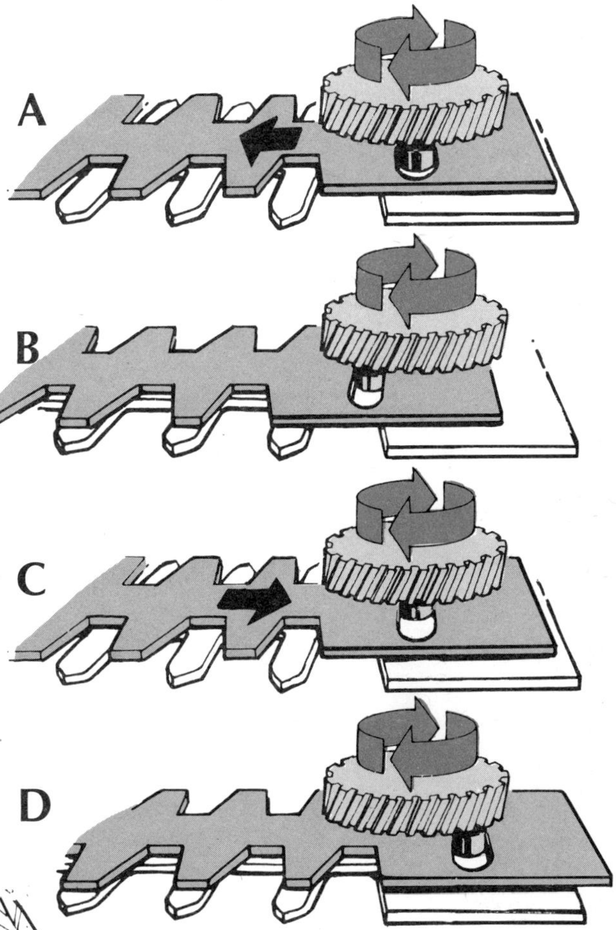

A

B

C

D

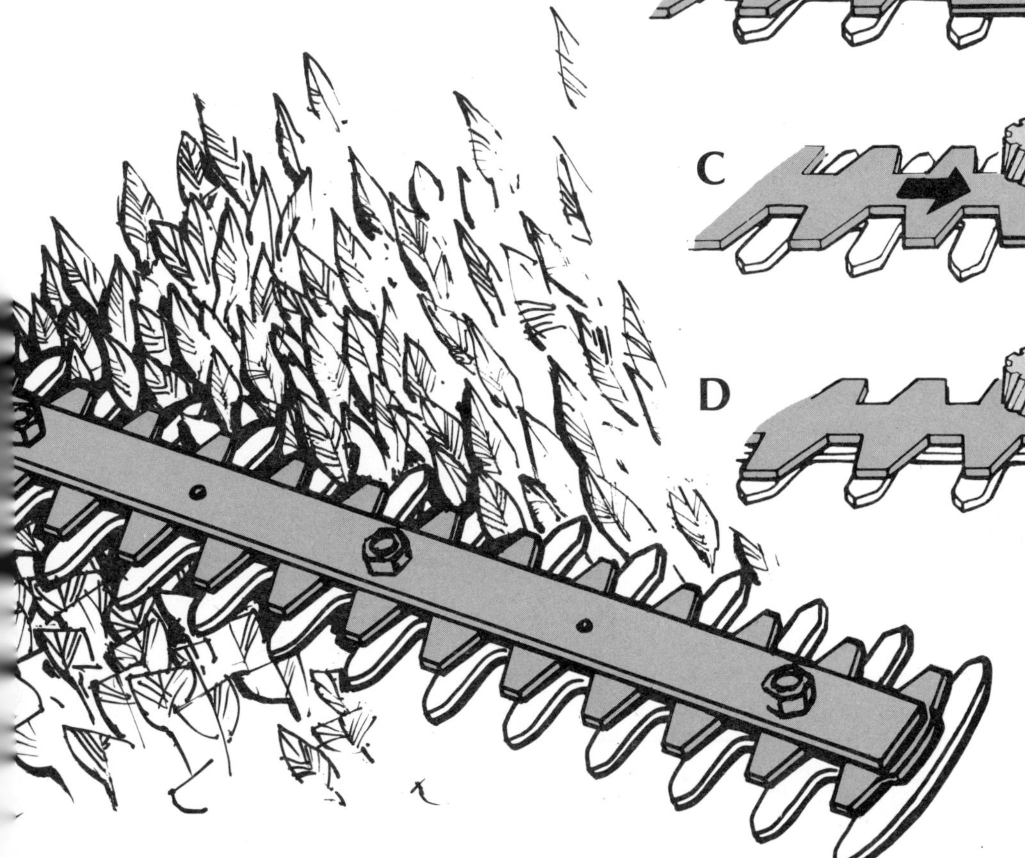

Hand Mixer

Lightweight hand mixers eliminate the drudgery of beating batter with a ladle. Inside the mixer are a hefty universal electric motor and a speed reducing gear train. A thumb switch activates the motor, which produces strong starting torque (rotational force) that can overcome the resistance of a thick batter. All mixers employ two coordinated beaters whirling in opposite directions. This beater action lifts the batter mixture from the bottom of the bowl and propels it upward between the blades. Most manufacturers offer a variety of blade styles for different mixing jobs. Standard are the rectangular blades shown here and those that have a teardrop shape. Other styles include hook-wire blades for mixing sauces and shallow, propeller-like blades for mixing drinks, or even paint.

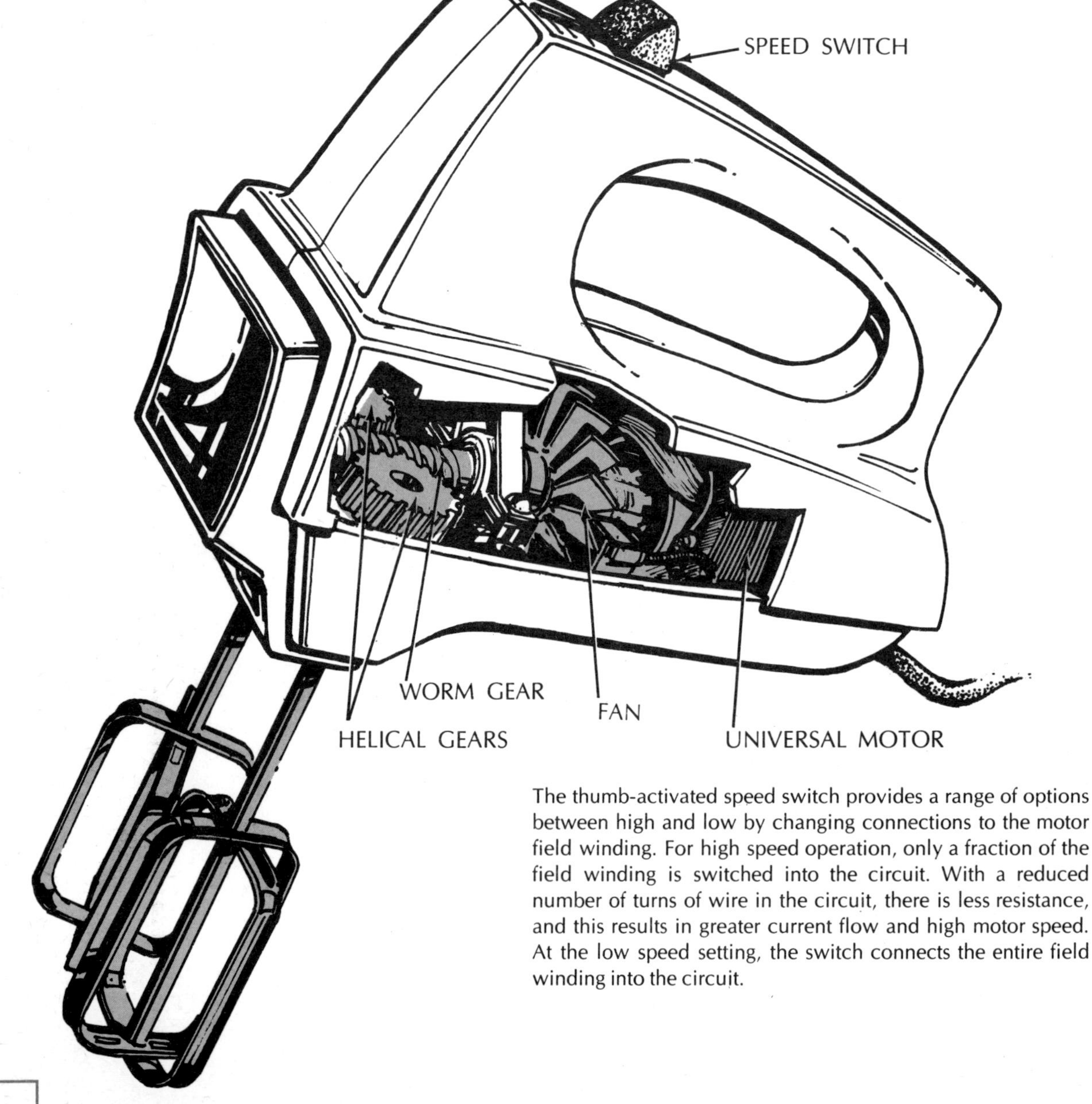

The thumb-activated speed switch provides a range of options between high and low by changing connections to the motor field winding. For high speed operation, only a fraction of the field winding is switched into the circuit. With a reduced number of turns of wire in the circuit, there is less resistance, and this results in greater current flow and high motor speed. At the low speed setting, the switch connects the entire field winding into the circuit.

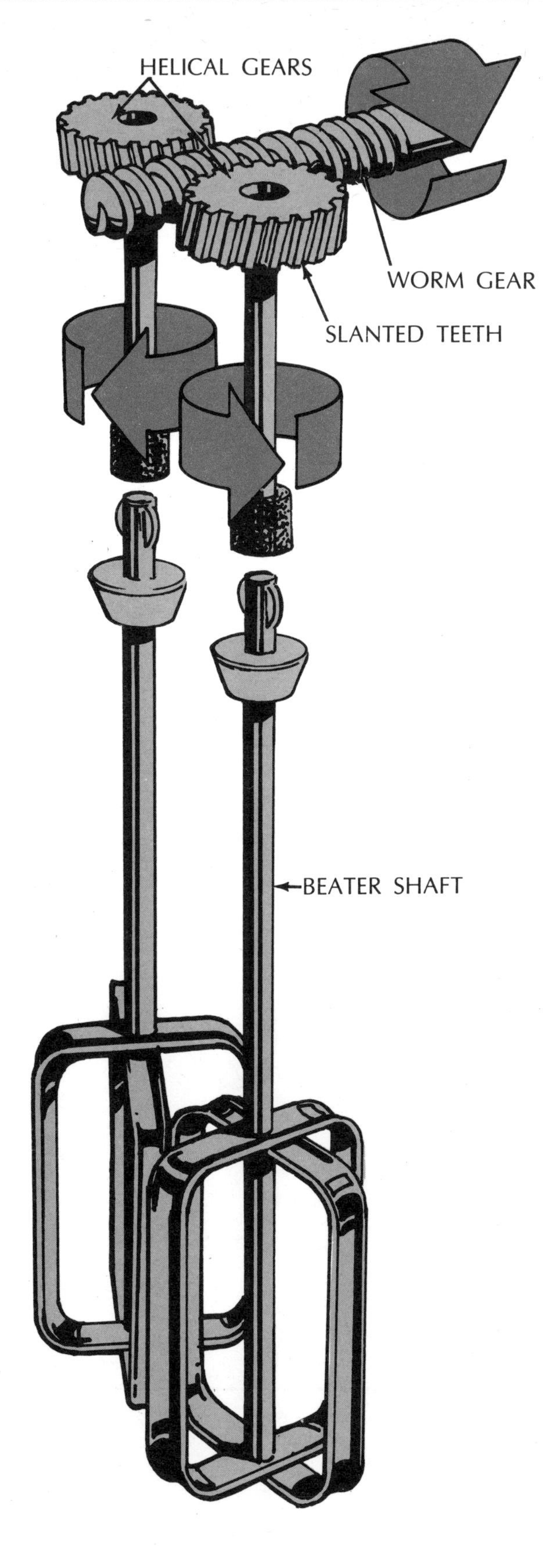

In the mixer the helical gears and worm gear act as *reduction gears,* converting the high-speed/moderate-torque shaft input from the motor to moderate-speed/high torque output at the beater blades. The teeth of the helical gear slant across the face of the gear, providing a larger contact surface than would an ordinary straight-cut, spur gear. Thus, a helical gear has a greater load-carrying capacity than other common gears of the same size. In some mixers, the helical gears are made of nylon or plastic, rather than metal, to hush the sound of operation. Heads of beater shafts (left) are designed to snap into keywayed holes in the bearing-mounted center supports of the helical gears. Once locked in place, the beaters rotate with the helical gears, converting gear motion into the whipping or mixing action needed.

Electric Scissors

For the home seamstress, cutting out clothing patterns with hand shears can be a difficult and tedious chore. For this reason the electric scissors has become a welcome addition to the sewing basket. Electric scissors generally wield short blades that project from a plastic housing. The upper blade is stationary and is set at an angle to the housing so that it can serve as a stabilizing foot as the scissors is guided though fabric lying on a flat work surface. The upper blade oscillates at a rate of 120 times per second but doesn't close with the lower blade completely. This incomplete closure, together with the restricted blade angle, prevents the accidental cutting of a stray finger.

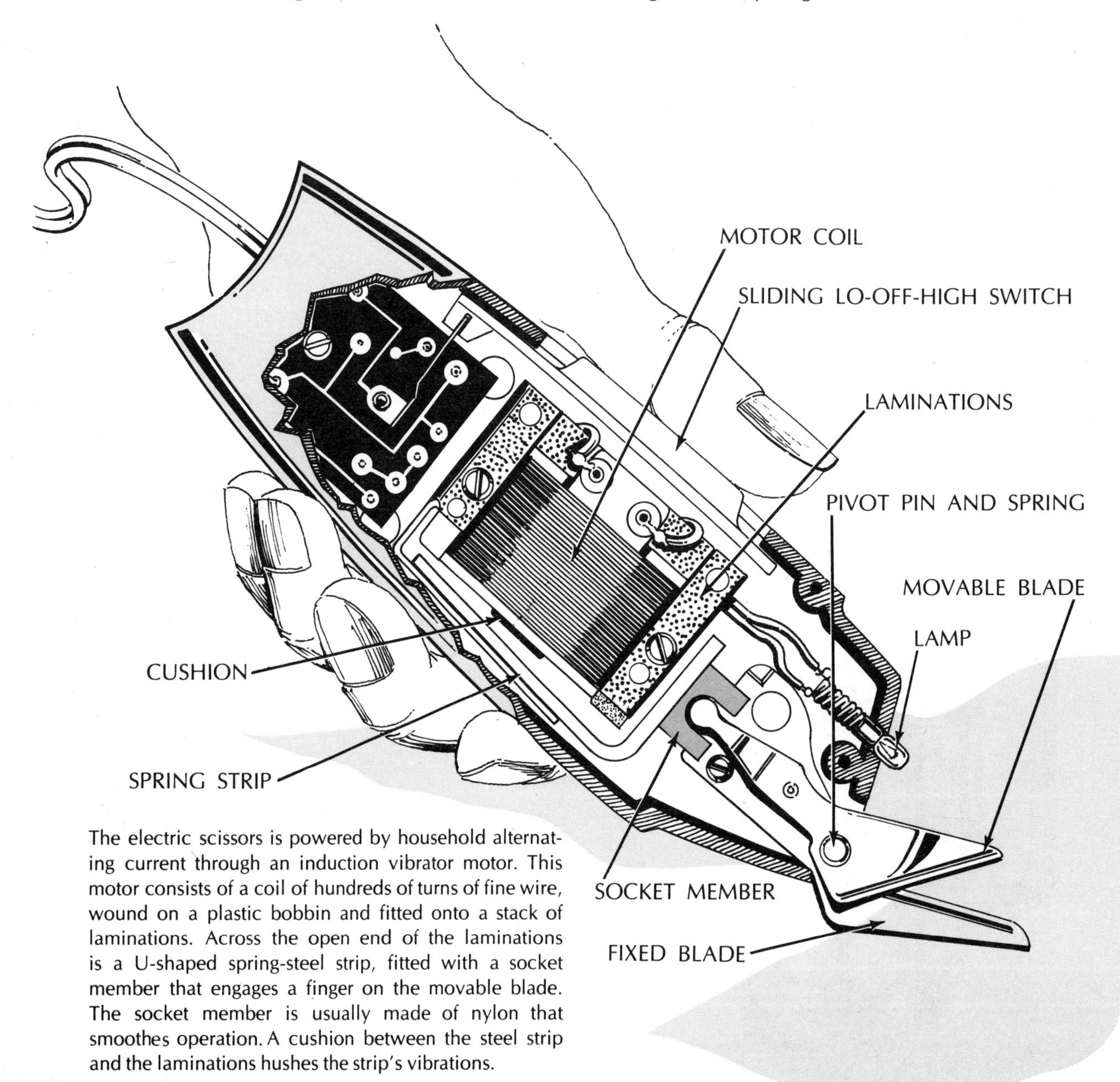

The electric scissors is powered by household alternating current through an induction vibrator motor. This motor consists of a coil of hundreds of turns of fine wire, wound on a plastic bobbin and fitted onto a stack of laminations. Across the open end of the laminations is a U-shaped spring-steel strip, fitted with a socket member that engages a finger on the movable blade. The socket member is usually made of nylon that smoothes operation. A cushion between the steel strip and the laminations hushes the strip's vibrations.

CURRENT FLOW

STRIP PULLED IN BY ELECTROMAGNETIC FIELD

BLADES CLOSE

Alternating Current

Household alternating current (ac) regularly rises and falls in value, first flowing in one direction and then in the other. This occurs 120 times per second in most U.S. power lines. A complete flow in both directions is called a *cycle*. Thus, household power is referred to as 60 cycle ac.

1st Half of AC Cycle

In the motor, the initial flow of current through the coil instantly magnetizes the laminations so that their ends exert a strong pull on the spring strip. As the strip moves toward the laminations, its movement is imparted to the upper blade.

STRIP SPRINGS BACK

BLADES OPEN

Midpoint of AC Cycle

Meanwhile, current flow begins to reverse through the motor coil, and the weakening electromagnetic field releases its pull on the strip. As current passes through zero, the spring strip moves back to its normal position, opening the blades.

2d Half of AC Cycle

As current rises through the coil in the reverse direction, the electromagnetic field again pulls the strip in. The movement of the spring strip thus follows the rise and fall of current through the motor coil, causing the upper blade to open and close at a rate of 60 times per second.

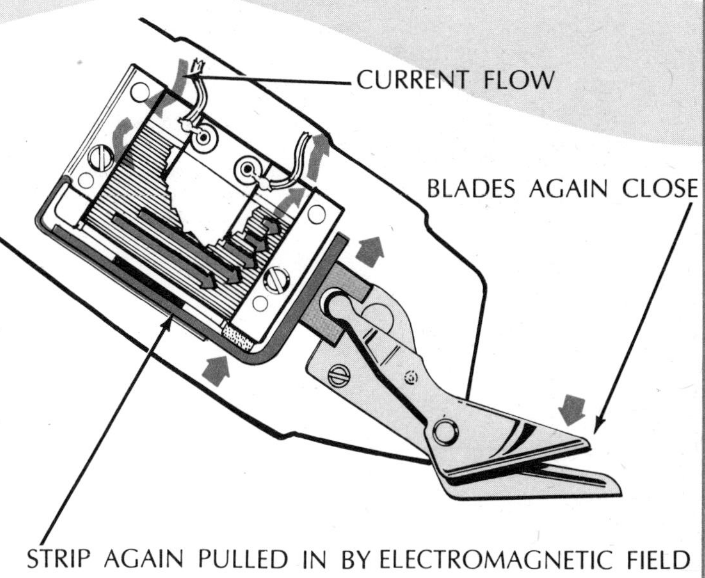

Straight-Line/Orbital Sander

Until recently, most electric sanders offered only straight-line action. Back-and-forth thrusts are ideal for fine, grain-exposing finish work. But when the home craftsman has to dress a rough surface, he usually switches to an orbital sander that produces orbital (circular) strokes—especially well suited to heavy-duty sanding. Combination sanders offer the user a choice of straight-line or orbital sanding at the flick of a toggle lever.

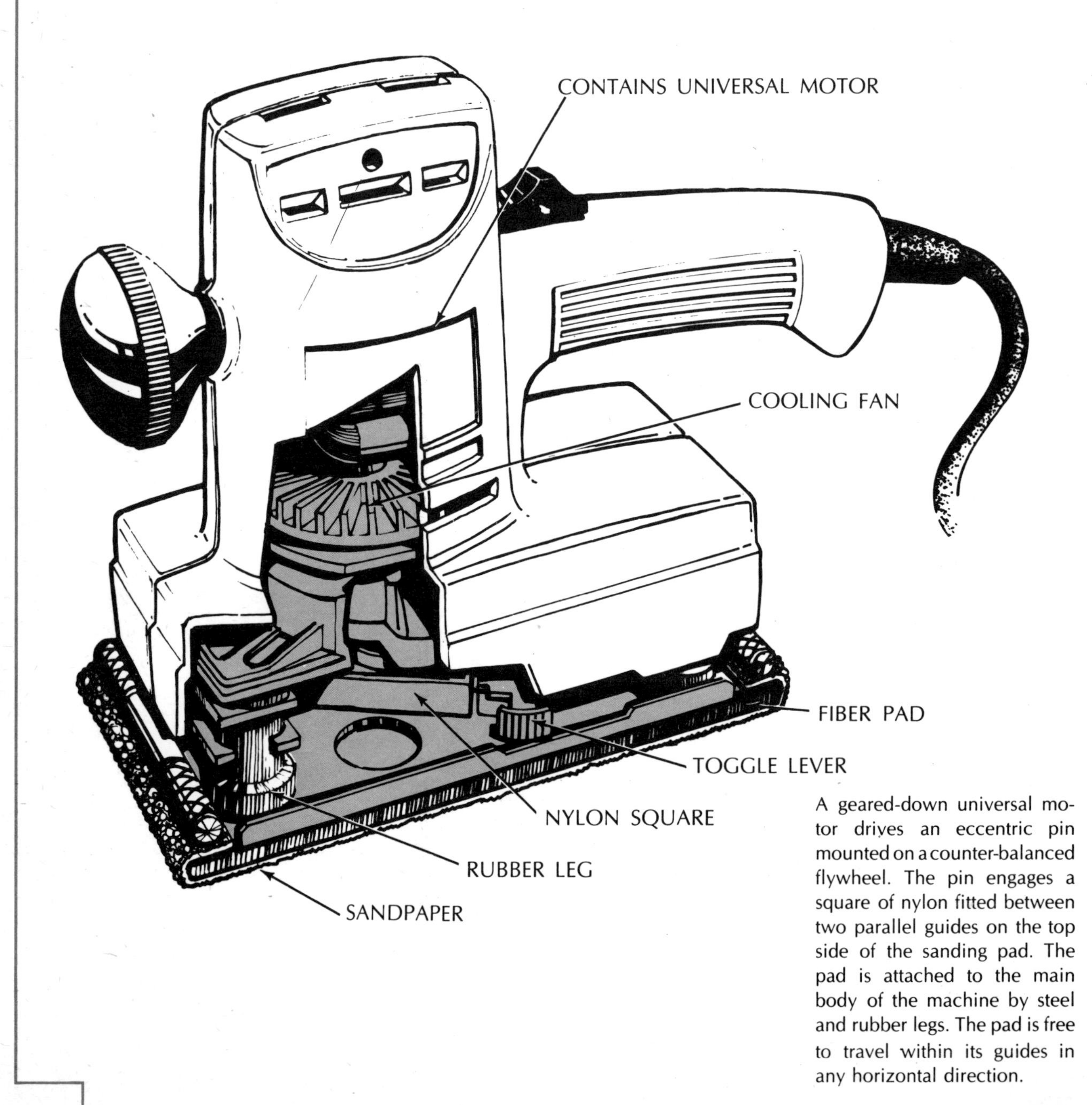

A geared-down universal motor drives an eccentric pin mounted on a counter-balanced flywheel. The pin engages a square of nylon fitted between two parallel guides on the top side of the sanding pad. The pad is attached to the main body of the machine by steel and rubber legs. The pad is free to travel within its guides in any horizontal direction.

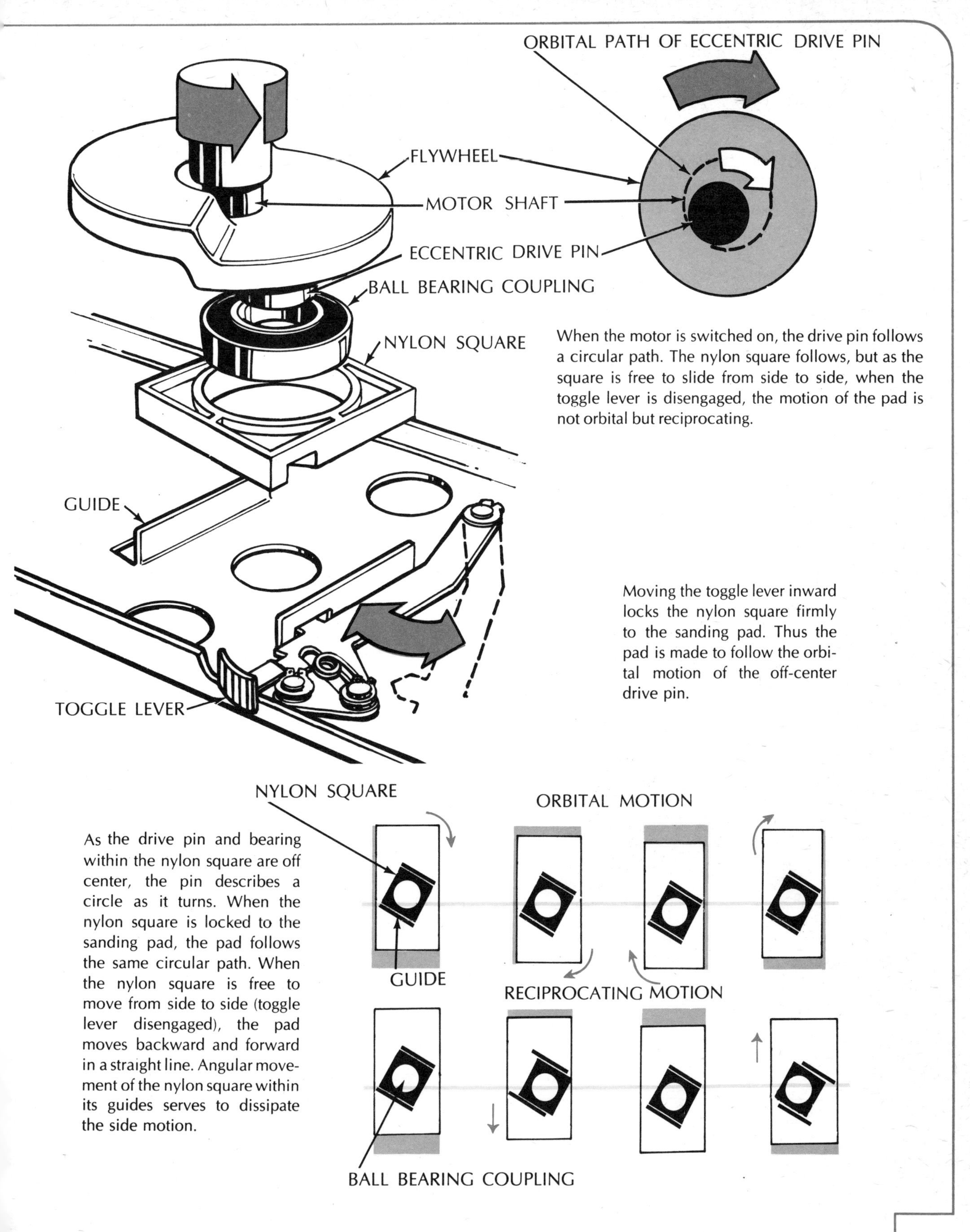

When the motor is switched on, the drive pin follows a circular path. The nylon square follows, but as the square is free to slide from side to side, when the toggle lever is disengaged, the motion of the pad is not orbital but reciprocating.

Moving the toggle lever inward locks the nylon square firmly to the sanding pad. Thus the pad is made to follow the orbital motion of the off-center drive pin.

As the drive pin and bearing within the nylon square are off center, the pin describes a circle as it turns. When the nylon square is locked to the sanding pad, the pad follows the same circular path. When the nylon square is free to move from side to side (toggle lever disengaged), the pad moves backward and forward in a straight line. Angular movement of the nylon square within its guides serves to dissipate the side motion.

Electric Can Opener

With increasing tens of billions of cans used, the day of the automatic can opener was inevitable. Early can openers had a fearsome hook-shaped blade that stabbed through the can lid. Then, while the operator turned the can with one hand, he applied cutting leverage to the blade. Sharp can edges and human error caused many accidents. In time, a hand opener with a gear drive appeared. Clamping the device over the can rim forced the opener's cutting blade through the lid. Squeezing two handles together pressed a serrated drive wheel against the can rim. And rotating a T-handle turned a toothed drive wheel, forcing the lid's perimeter along the cutting blade. Today, these same cutting principles—with a few refinements—are employed in electric can openers.

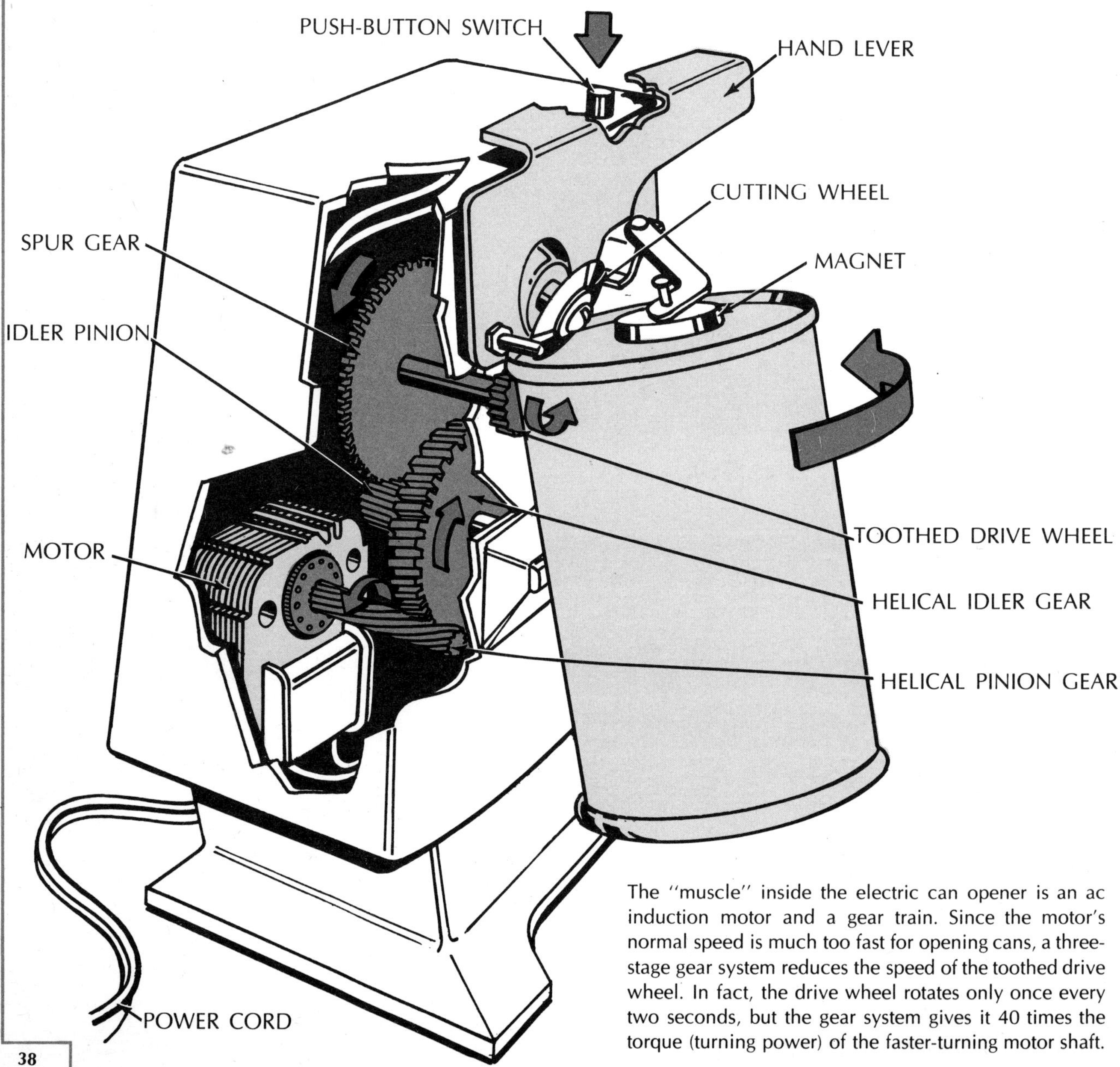

The "muscle" inside the electric can opener is an ac induction motor and a gear train. Since the motor's normal speed is much too fast for opening cans, a three-stage gear system reduces the speed of the toothed drive wheel. In fact, the drive wheel rotates only once every two seconds, but the gear system gives it 40 times the torque (turning power) of the faster-turning motor shaft.

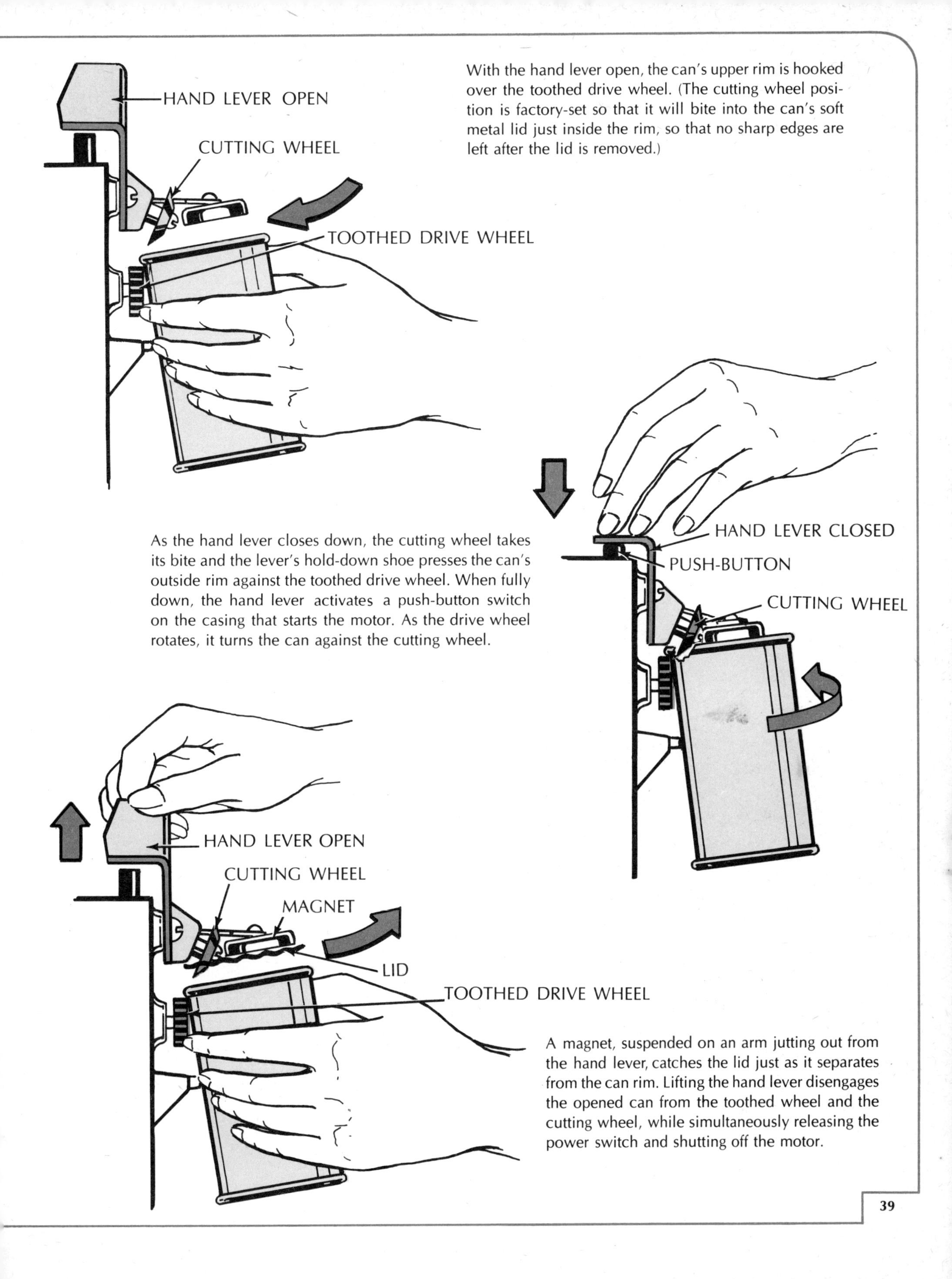

With the hand lever open, the can's upper rim is hooked over the toothed drive wheel. (The cutting wheel position is factory-set so that it will bite into the can's soft metal lid just inside the rim, so that no sharp edges are left after the lid is removed.)

As the hand lever closes down, the cutting wheel takes its bite and the lever's hold-down shoe presses the can's outside rim against the toothed drive wheel. When fully down, the hand lever activates a push-button switch on the casing that starts the motor. As the drive wheel rotates, it turns the can against the cutting wheel.

A magnet, suspended on an arm jutting out from the hand lever, catches the lid just as it separates from the can rim. Lifting the hand lever disengages the opened can from the toothed wheel and the cutting wheel, while simultaneously releasing the power switch and shutting off the motor.

Digital Clock

There is good reason for the popularity of the digital clock. With its bold-numeraled flip cards telling time, the digital clock satisfies current demand for simplicity of design while eliminating the need for sharp eye focus on small clock-face increments. In developing the digital clock, designers borrowed a mechanism once used in penny arcade Mutoscope machines that provided "motion" pictures long before filmstrips. In response to a hand crank the Mutoscope produced the illusion of motion by rapidly flipping slightly different pictures under a stationary finger. The digital clock employs two slow-motion card flip mechanisms—one that flops forward a card each minute, the other that flops hour cards.

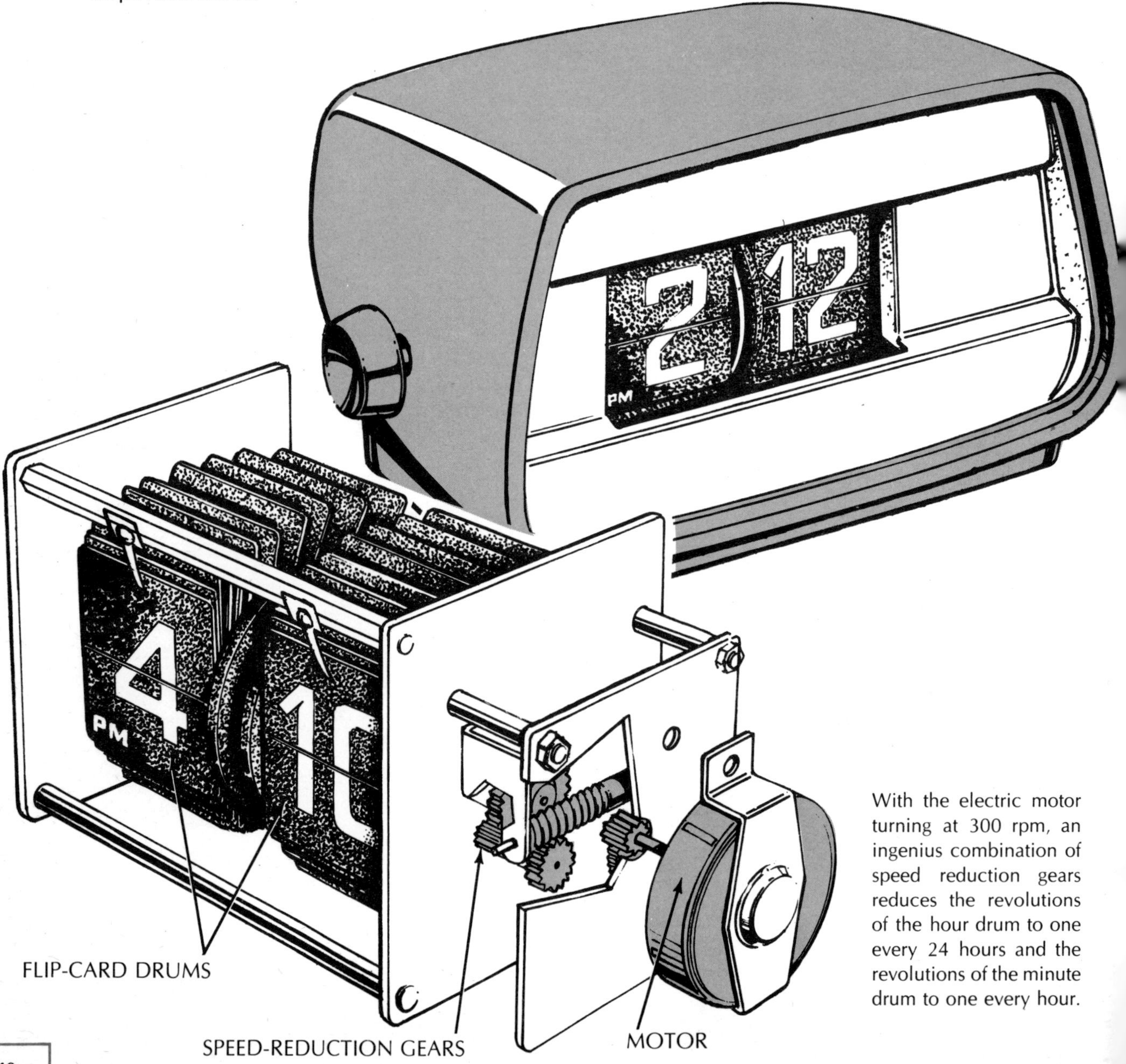

FLIP-CARD DRUMS

SPEED-REDUCTION GEARS

MOTOR

With the electric motor turning at 300 rpm, an ingenius combination of speed reduction gears reduces the revolutions of the hour drum to one every 24 hours and the revolutions of the minute drum to one every hour.

The hour drum contains a set of cards marked "1 PM" through "12 PM" and a similar set of cards for the hours 1 AM through 12 AM, a pair of cards for each hour—for a total of 48 hour cards. During the first half of each hour, the first of each pair of hour cards is kept from flopping forward by a finger on the clock frame. On the half hour, the rotation of the hour drum causes the first card of the pair to overcome the finger's control, flopping the card forward. When the hour is nearly up, the second of the pair of hour cards is kept from flopping forward by a small arm protruding from the 59-minute card, so that the hour card drops only when the 59-minute card, itself, flops forward.

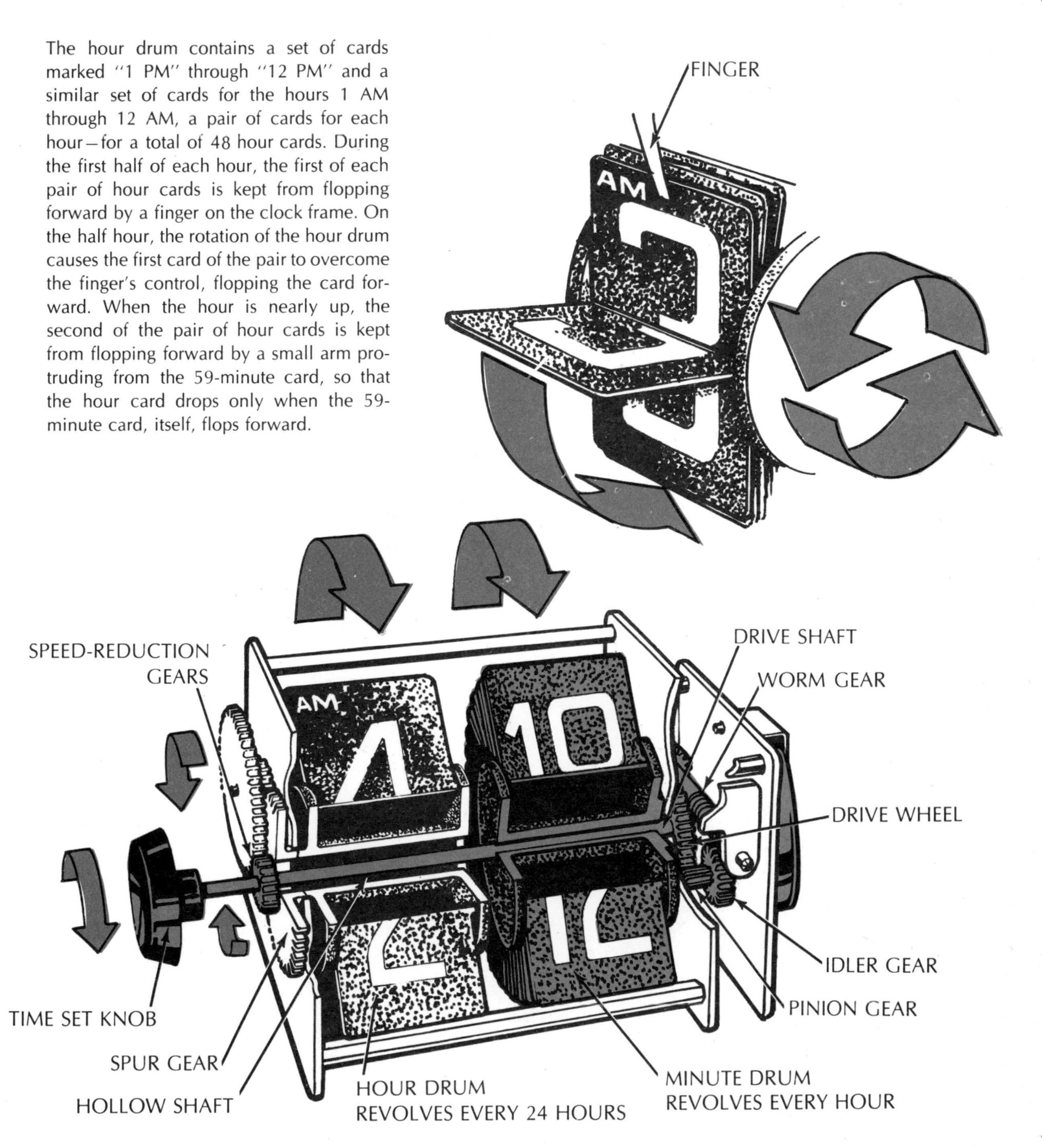

The complex combination of idler, pinion, and worm gears driven by the motor reduces the drive wheel revolutions to one per hour. Thus, the shaft extending from the drive wheel rotates all cards on the attached minute drum once every hour. The shaft also drives a second set of speed-reducing gears near the time set knob. Here, a large-diameter spur gear is connected to a hollow shaft that fits coaxially, like a sleeve, about the minute drum's drive shaft. The hour drum is fitted to this hollow shaft and revolves with it once every 24 hours.

Electronic Calculator

The electronic calculator can't really "think." Yet, with lightning speed it can always out-compute anyone with paper and pencil. Because early man based his counting on ten fingers, a universally accepted *decimal* number system evolved based on the numerals 1 through 10. But since an electronic device can be "on" or "off," the electronic calculator employs a *binary* (two-state) method of computation represented by the numerals 0 and 1, and by their positions in a *set,* or series of numerals. The calculator "reads" in binary form the decimal number pressed on its keyboard. Then the calculator converts the result of its binary computation through a switching circuit called a "decoder," to illuminated decimals that appear on the display panel.

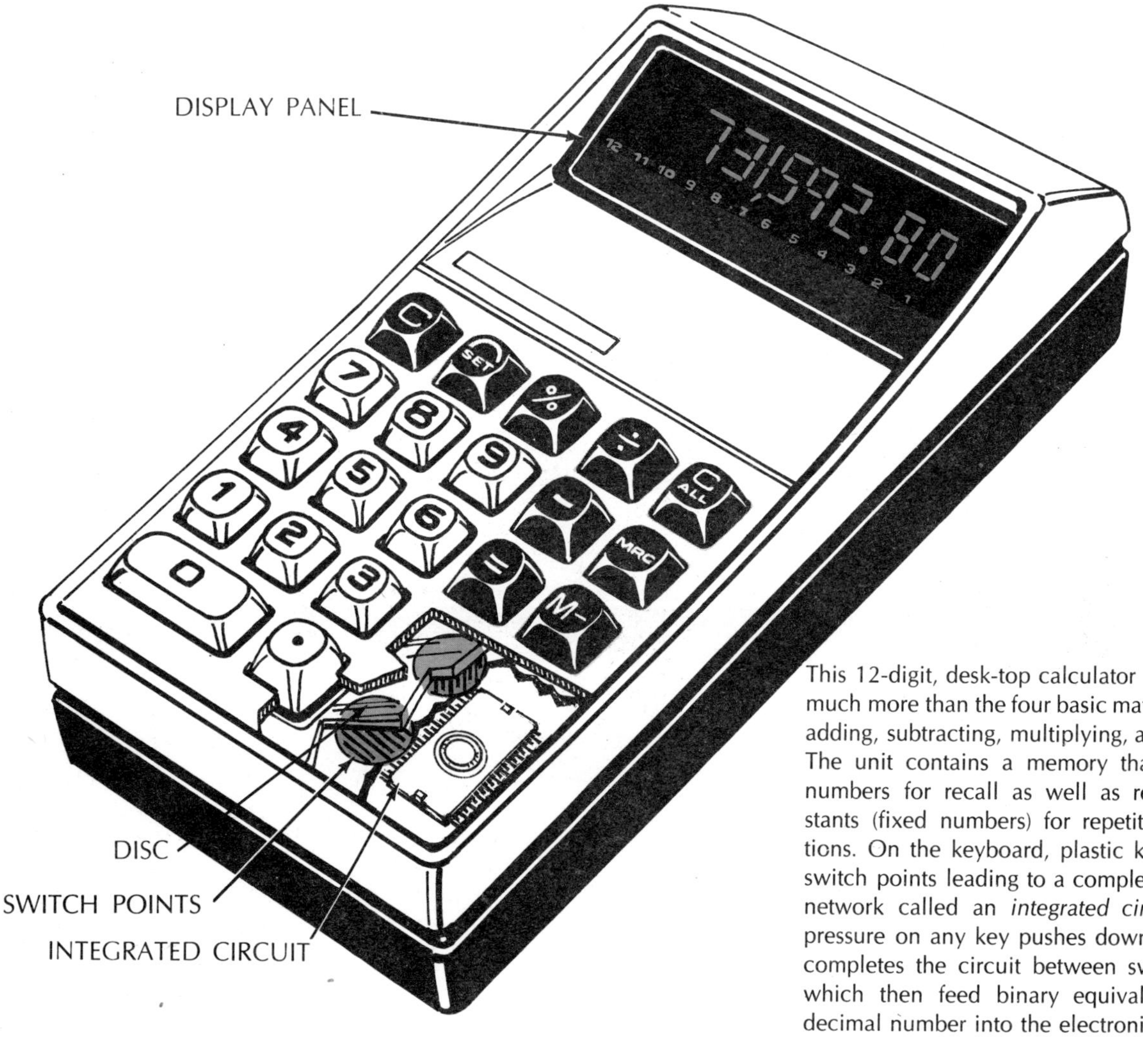

DISPLAY PANEL

DISC

SWITCH POINTS

INTEGRATED CIRCUIT

This 12-digit, desk-top calculator can perform much more than the four basic math functions: adding, subtracting, multiplying, and dividing. The unit contains a memory that can store numbers for recall as well as reapply constants (fixed numbers) for repetitive calculations. On the keyboard, plastic keys lie over switch points leading to a complex electronic network called an *integrated circuit.* Finger pressure on any key pushes down a pad that completes the circuit between switch points which then feed binary equivalents of the decimal number into the electronics.

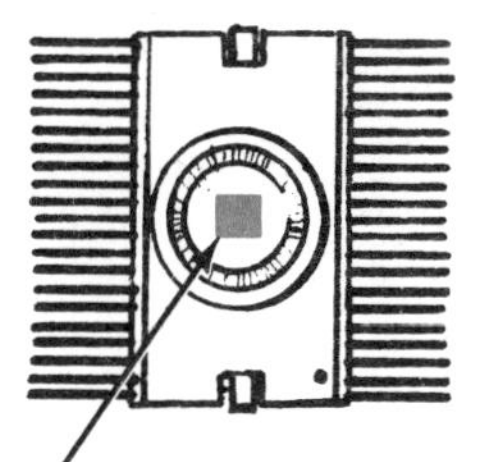

ACTUAL SIZE CHIP

The heart of the calculator is a postage-stamp-size, mult-lead electronic package that contains a tiny slab of silicon called an integrated circuit "chip." Incredibly, this tiny chip contains the equivalent of 1600 or more transistors, diodes, and other electronic components that are microscopically grown. (Hundreds of chips at one time are manufactured this way on a thin slice of silicon.) In response to pulses from the keyboard, all of the binary math operations are carried out at the speed of light on this amazing little chip.

SEVEN LED SEGMENTS THAT FORM ANY NUMERAL

Outputs of the chip are routed to the display, consisting of an in-line array of numeric units, each having seven segments plus a decimal point. Each segment uses a tiny LED (light emitting diode), second cousin to the transistor. An LED gives off a reddish glow when power is applied from the chip outputs. As the drawings illustrate, specific energized combinations of the diode segments can form any decimal number.

HOW TO FIND THE BINARY EQUIVALENT OF ANY DECIMAL NUMBER

Digital placement from right	**11th**	**10th**	**9th**	**8th**	**7th**	**6th**	**5th**	**4th**	**3d**	**2d**	**1st**
Value of 1 in that digit	*1024*	*512*	*256*	*128*	*64*	*32*	*16*	*8*	*4*	*2*	*1*
Decimal 1 equals binary.											1
Decimal 2 equals binary.										1	0
Decimal 3 equals binary.										1	1
Decimal 9 equals binary.								1	0	0	1
Decimal 137 equals binary. . .				1	0	0	0	1	0	0	1
Decimal 1,739 equals binary. .	1	1	0	1	1	0	0	1	0	1	1

In the binary system, each left-wise digit has twice the decimal value of the digit to its right. But only those digits occupied by "1" have decimal value. The "0" occupies a position but is not counted. In the chart above, the decimal numbers converted to binary form were chosen at random. To convert *any* decimal number, one simply refers to the chart's second line, headed *"Value of 1 in that digit."* To convert 137, for example, one finds the largest value on the second line that "fits" into 137. That would be 128. The next step is to find the largest second-line value that fits into the difference between 137 and 128 – a difference of 9. This would be 8 on the second line, leaving 1 remaining. On the line showing the binary equivalent of 137, the numeral 1 appears three times, under the columns headed 128, 8, and 1 – for a total of 137, represented in binary form as 10001001.

THE FOUR MATH FUNCTIONS

Simply stated, the functions are performed by computer circuits that switch off and on, changing the placements of binary 1 and 0.

Addition of decimal number 2 (binary 10) and 2 gives 4 (binary 100).

Subtraction works like addition except that binary number positions shift to right instead of to left.

Multiplication is treated like a series of addition steps. To multiply 23 by 9, the computer's built-in "clock" adds and re-adds 23 nine times. The answer seems to appear instantly.

Division is a series of subtractions carried out in sequence much like that for multiplication.

Camera Flash Cube

The "Instamatic" approach to photography put cameras into the hands of millions seeking good snapshots without technical know-how or bother. But early models used a battery-powered flash with unreliable electrical connections. Then, engineers achieved a breakthrough with a new *battery-less* flash and called it the Type X Magicube. This reliable, four-bulb cube is easily inserted into its mated camera pedestal. The cube allows the photographer to take as many as four shots in rapid succession—as rapidly as he can 1) depress the camera shutter, which exposes the film *and* fires the bulb and 2) rotate the film advance lever, which rotates the flash cube a quarter turn and positions a fresh bulb.

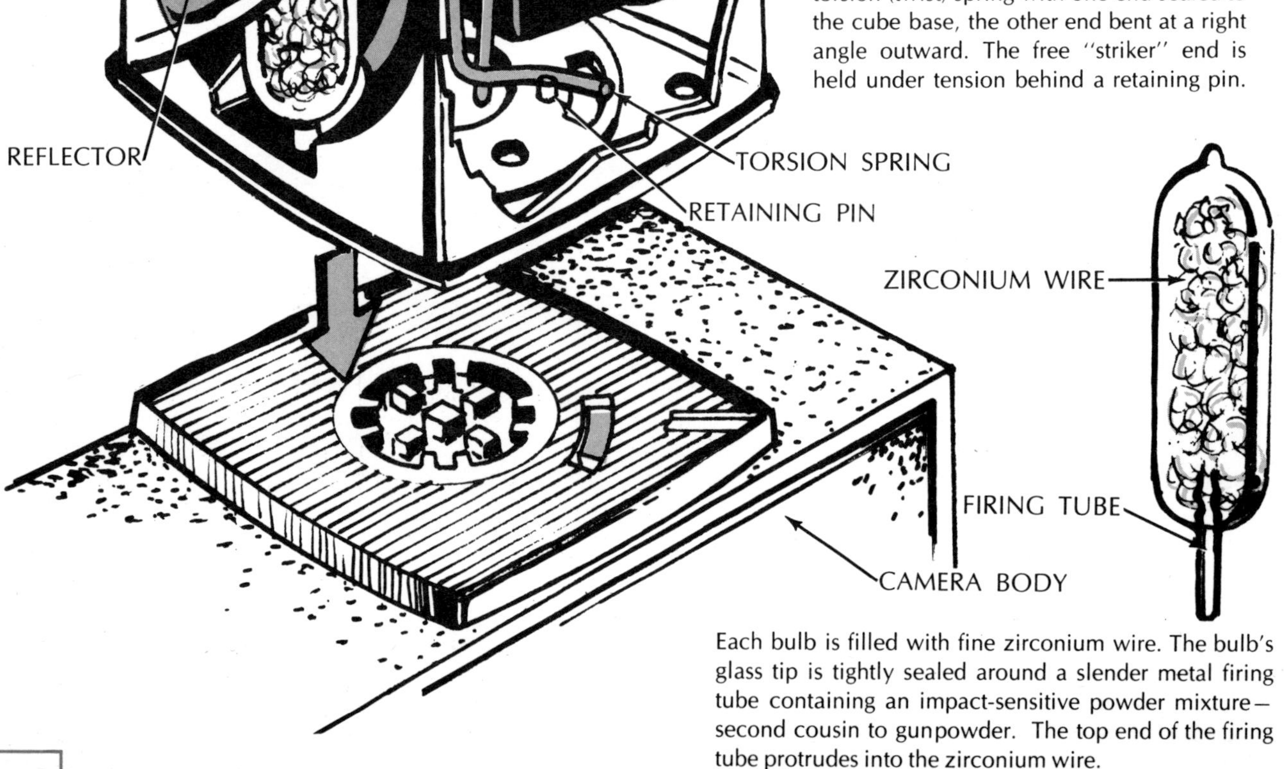

Each of the four bulbs sits vertically in its own reflector on the cube's plastic base. Behind each reflector is a hair-pin shaped, torsion (twist) spring with one end sealed to the cube base, the other end bent at a right angle outward. The free "striker" end is held under tension behind a retaining pin.

Each bulb is filled with fine zirconium wire. The bulb's glass tip is tightly sealed around a slender metal firing tube containing an impact-sensitive powder mixture—second cousin to gunpowder. The top end of the firing tube protrudes into the zirconium wire.

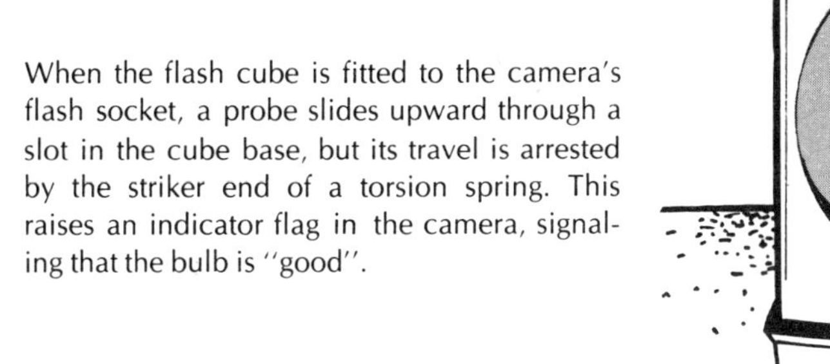

When the flash cube is fitted to the camera's flash socket, a probe slides upward through a slot in the cube base, but its travel is arrested by the striker end of a torsion spring. This raises an indicator flag in the camera, signaling that the bulb is "good".

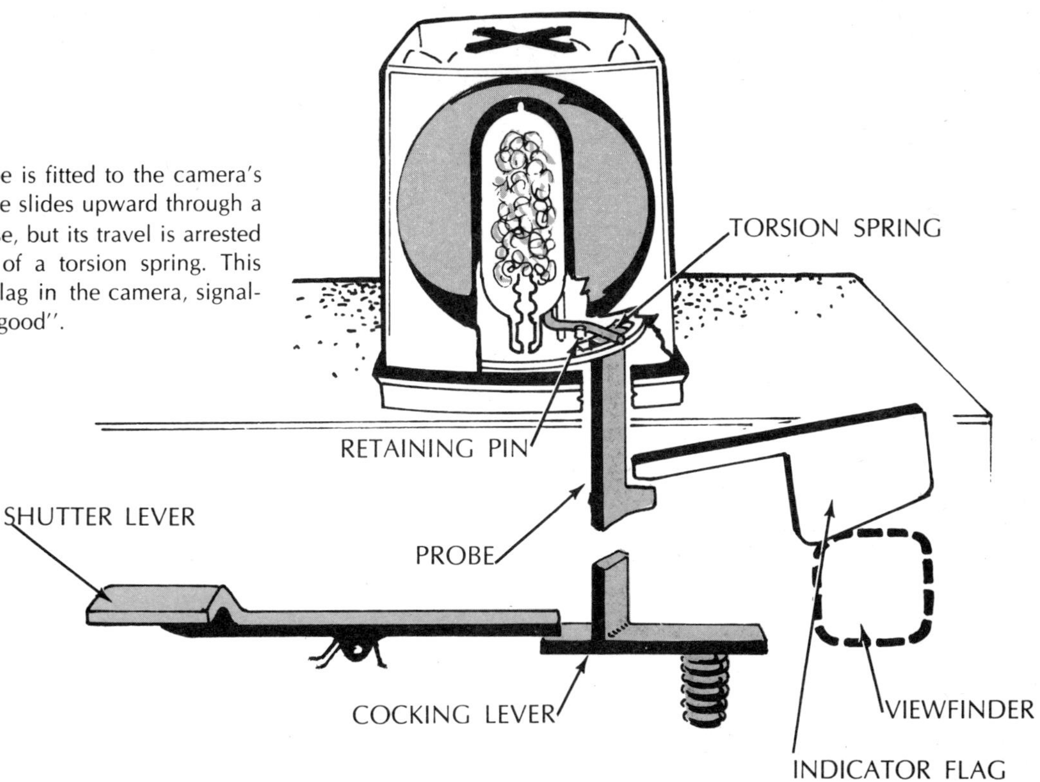

When the shutter lever is depressed, a spring-loaded cocking lever strikes the probe from below, ramming it up through the cube base and lifting the striker end of the torsion spring above the retaining pin. Instantly, the torsion spring whips the striker end sideways, so that it impacts sharply against the bulb's firing tube. At impact, molecules of oxygen, bound up in the firing mixture, are shaken loose to mix with the "fuel" content (or reductant, as chemists call it). The reaction is like the striking of a match but is far more violent. The firing tube erupts hot shards upwards into the zirconium wire, causing it to ignite in a brilliant 2,000-candlepower burst of light lasting only 30 millionths of a second. As the lamp fires and the picture is recorded, the indicator flag appears in the viewfinder to warn: "STOP—BAD BULB".

INDICATOR FLAG
FIRING TUBE
TORSION SPRING
PROBE
SHUTTER LEVER
COCKING LEVER
VIEWFINDER

Antenna Rotator

Not even the finest TV set can perform well if it receives a weak signal at its antenna terminals. The best antennas—those that provide strongest signals with least interference from other sources— are "directional," meaning they are designed to be aimed directly at individual signal sources. Antennas with directional pickup patterns are especially important for long-distance color reception but also prove valuable to black-and-white viewers in TV "fringe country." The more directional the antenna, the more necessary it is to aim it exactly at the signal source. And since the locations of these sources may be many, a mechanical antenna rotator is recommended.

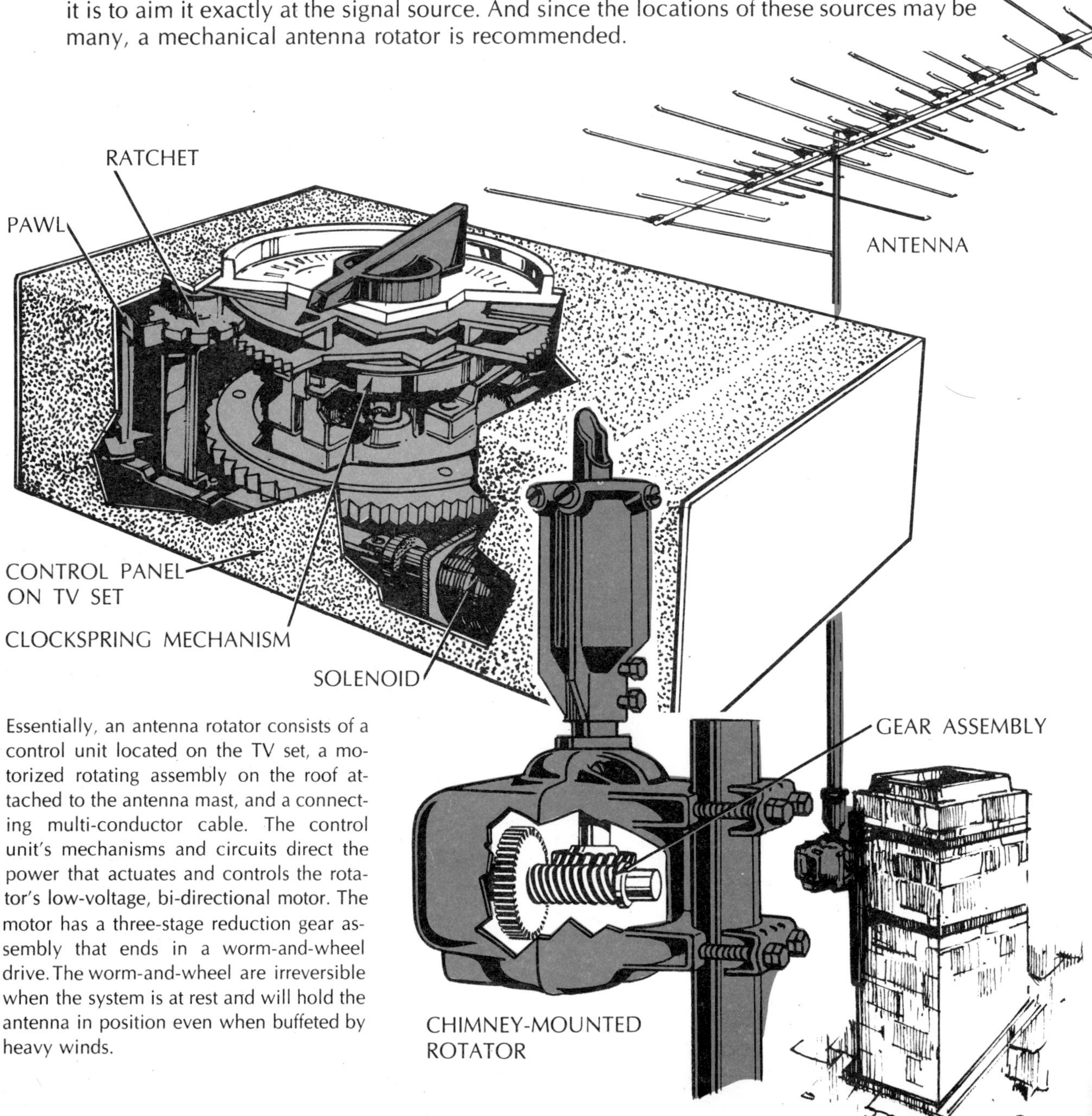

Essentially, an antenna rotator consists of a control unit located on the TV set, a motorized rotating assembly on the roof attached to the antenna mast, and a connecting multi-conductor cable. The control unit's mechanisms and circuits direct the power that actuates and controls the rotator's low-voltage, bi-directional motor. The motor has a three-stage reduction gear assembly that ends in a worm-and-wheel drive. The worm-and-wheel are irreversible when the system is at rest and will hold the antenna in position even when buffeted by heavy winds.

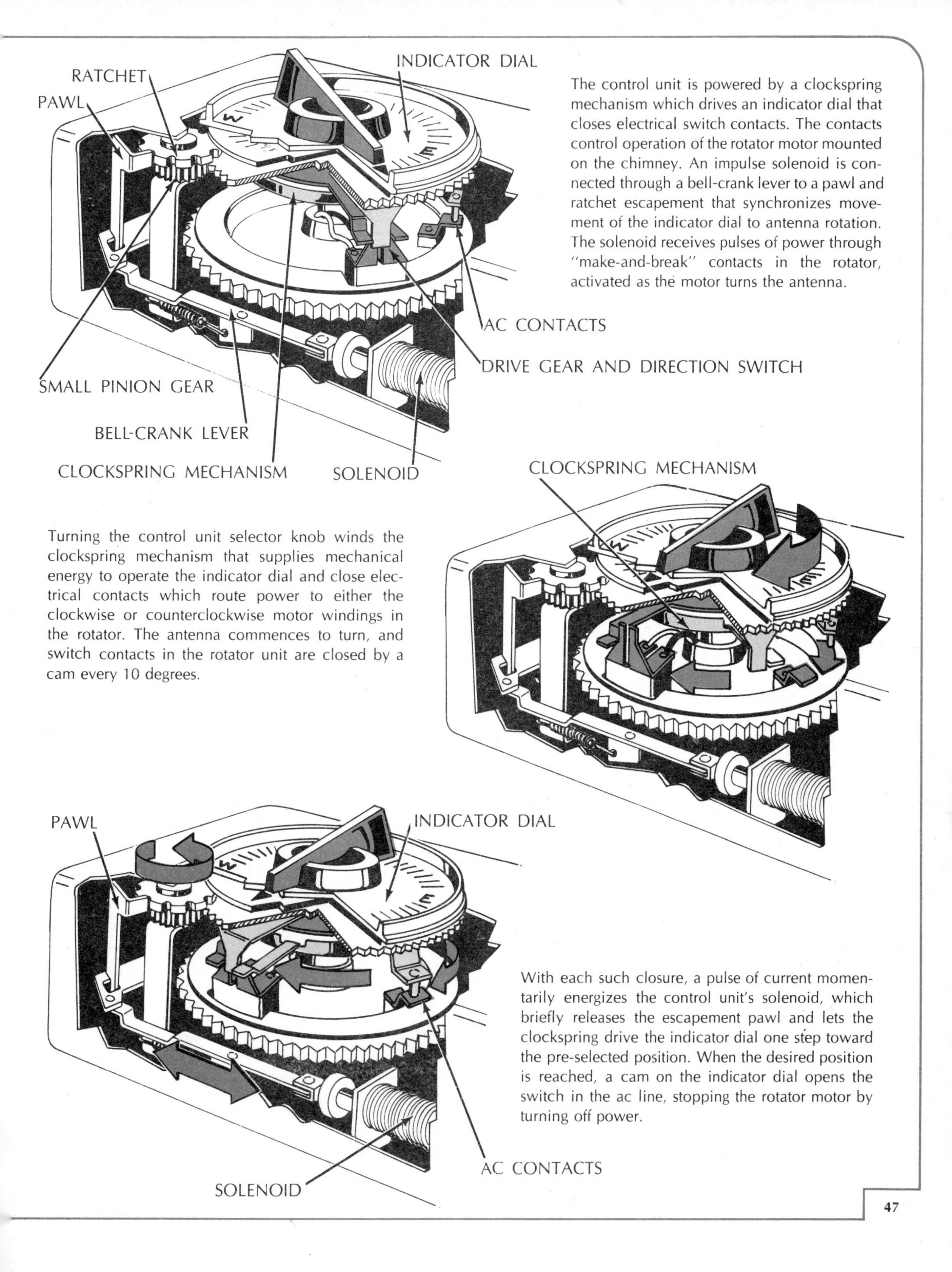

The control unit is powered by a clockspring mechanism which drives an indicator dial that closes electrical switch contacts. The contacts control operation of the rotator motor mounted on the chimney. An impulse solenoid is connected through a bell-crank lever to a pawl and ratchet escapement that synchronizes movement of the indicator dial to antenna rotation. The solenoid receives pulses of power through "make-and-break" contacts in the rotator, activated as the motor turns the antenna.

Turning the control unit selector knob winds the clockspring mechanism that supplies mechanical energy to operate the indicator dial and close electrical contacts which route power to either the clockwise or counterclockwise motor windings in the rotator. The antenna commences to turn, and switch contacts in the rotator unit are closed by a cam every 10 degrees.

With each such closure, a pulse of current momentarily energizes the control unit's solenoid, which briefly releases the escapement pawl and lets the clockspring drive the indicator dial one step toward the pre-selected position. When the desired position is reached, a cam on the indicator dial opens the switch in the ac line, stopping the rotator motor by turning off power.

Label Maker

Since the advent of the label maker, embossed tape has become a highly popular means of showing ownership and identifying objects. This popularity is warranted, for alternative means of producing legible and durable labels require a variety of materials and tools as well as drafting skill. With the label maker, the user simply turns the character selector wheel to the desired character and squeezes a hand trigger, repeating the process to print any combination of letters and numbers with speed and precision. The toughness of the tape and its adhesive make it capable of indefinite wear once affixed to a surface.

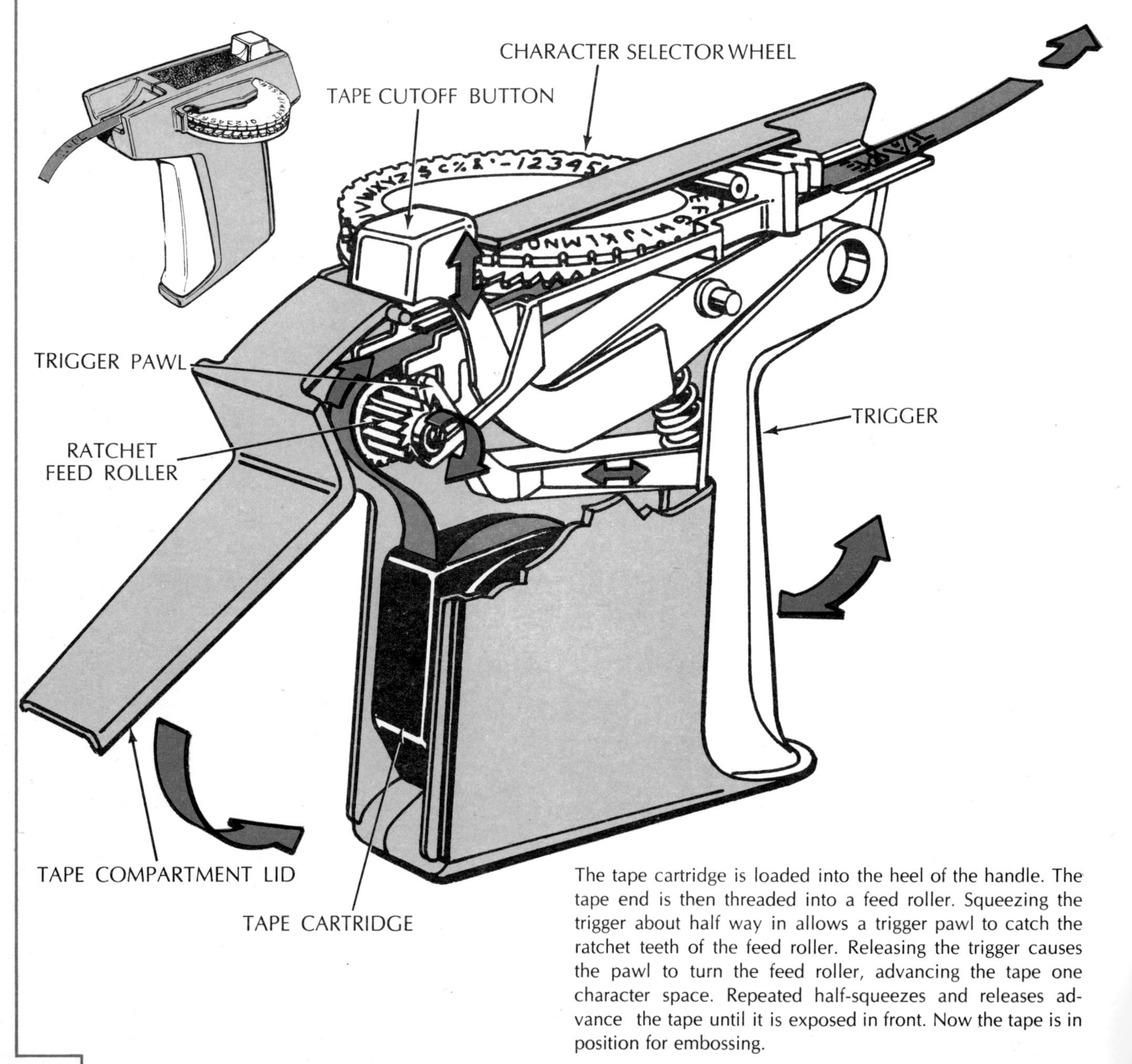

The tape cartridge is loaded into the heel of the handle. The tape end is then threaded into a feed roller. Squeezing the trigger about half way in allows a trigger pawl to catch the ratchet teeth of the feed roller. Releasing the trigger causes the pawl to turn the feed roller, advancing the tape one character space. Repeated half-squeezes and releases advance the tape until it is exposed in front. Now the tape is in position for embossing.

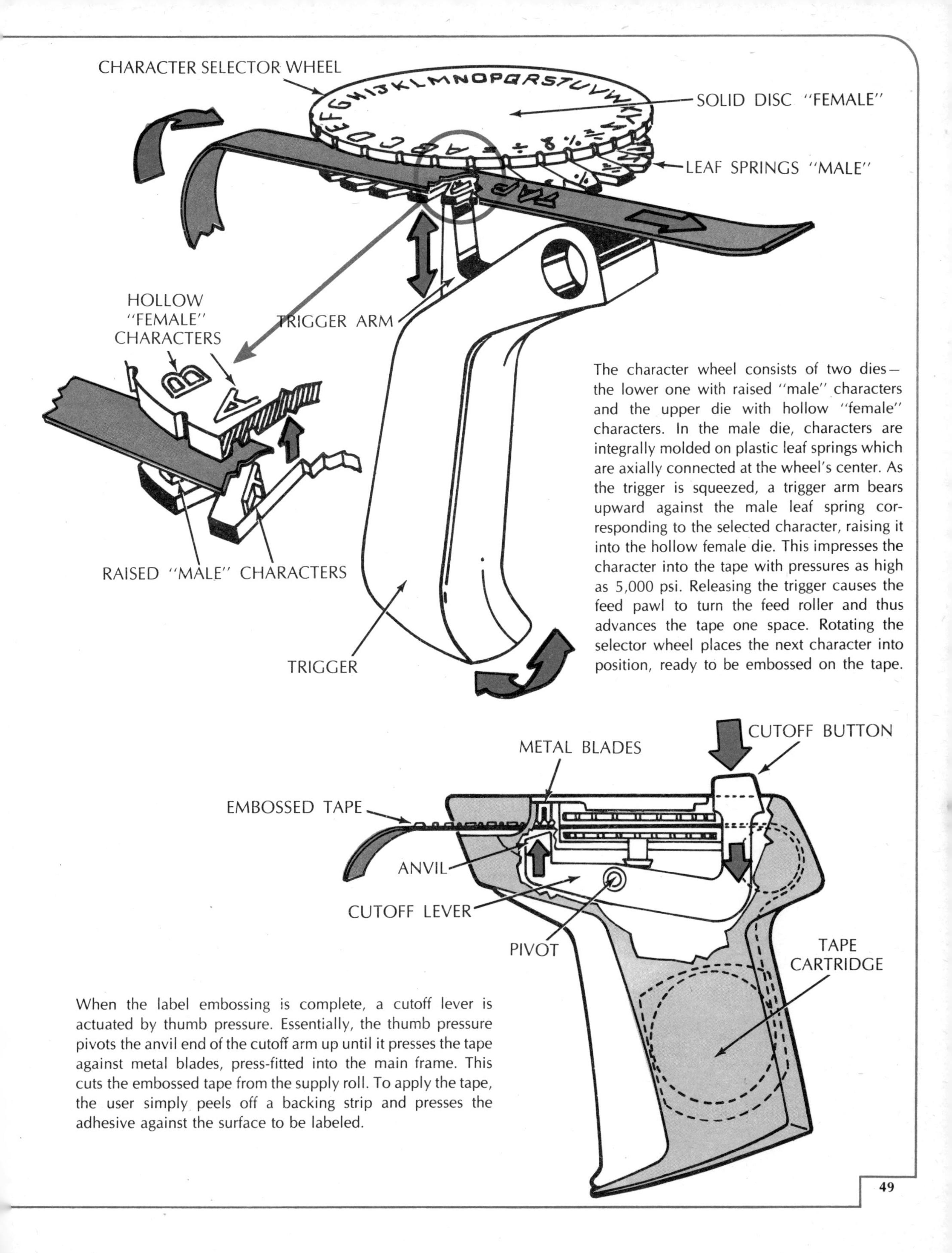

The character wheel consists of two dies—the lower one with raised "male" characters and the upper die with hollow "female" characters. In the male die, characters are integrally molded on plastic leaf springs which are axially connected at the wheel's center. As the trigger is squeezed, a trigger arm bears upward against the male leaf spring corresponding to the selected character, raising it into the hollow female die. This impresses the character into the tape with pressures as high as 5,000 psi. Releasing the trigger causes the feed pawl to turn the feed roller and thus advances the tape one space. Rotating the selector wheel places the next character into position, ready to be embossed on the tape.

When the label embossing is complete, a cutoff lever is actuated by thumb pressure. Essentially, the thumb pressure pivots the anvil end of the cutoff arm up until it presses the tape against metal blades, press-fitted into the main frame. This cuts the embossed tape from the supply roll. To apply the tape, the user simply peels off a backing strip and presses the adhesive against the surface to be labeled.

Automatic Garage Door Opener

The first garage door that swung open at the touch of a remote control button probably belonged to the Green Hornet, the 1940s radio crime fighter. Now garage doors that open automatically are available to everyone. Three elements make up the garage door opening system. One is a battery-powered radio transmitter about the size of a cigarette pack that is carried in the car. The second is a radio receiver in the garage. The third is the opening mechanism that hangs from the garage ceiling. This mechanism can open doors that swing on hinges or, as shown here, doors that fold upwards on metal tracks into the garage ceiling.

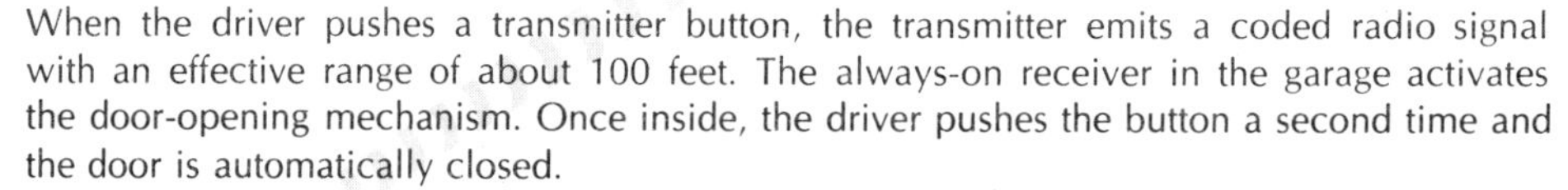

When the driver pushes a transmitter button, the transmitter emits a coded radio signal with an effective range of about 100 feet. The always-on receiver in the garage activates the door-opening mechanism. Once inside, the driver pushes the button a second time and the door is automatically closed.

Muscle for moving the garage door comes from the reversible 1/3-horse-power electric motor fastened to the garage ceiling. The motor powers an endless chain. An L-shaped arm mounted on a trolley bearing is attached to the chain. The end of the arm is bolted to the garage door. When the L-shaped arm moves toward the garage interior, it pulls the folding door open. Reversing direction, it closes the door.

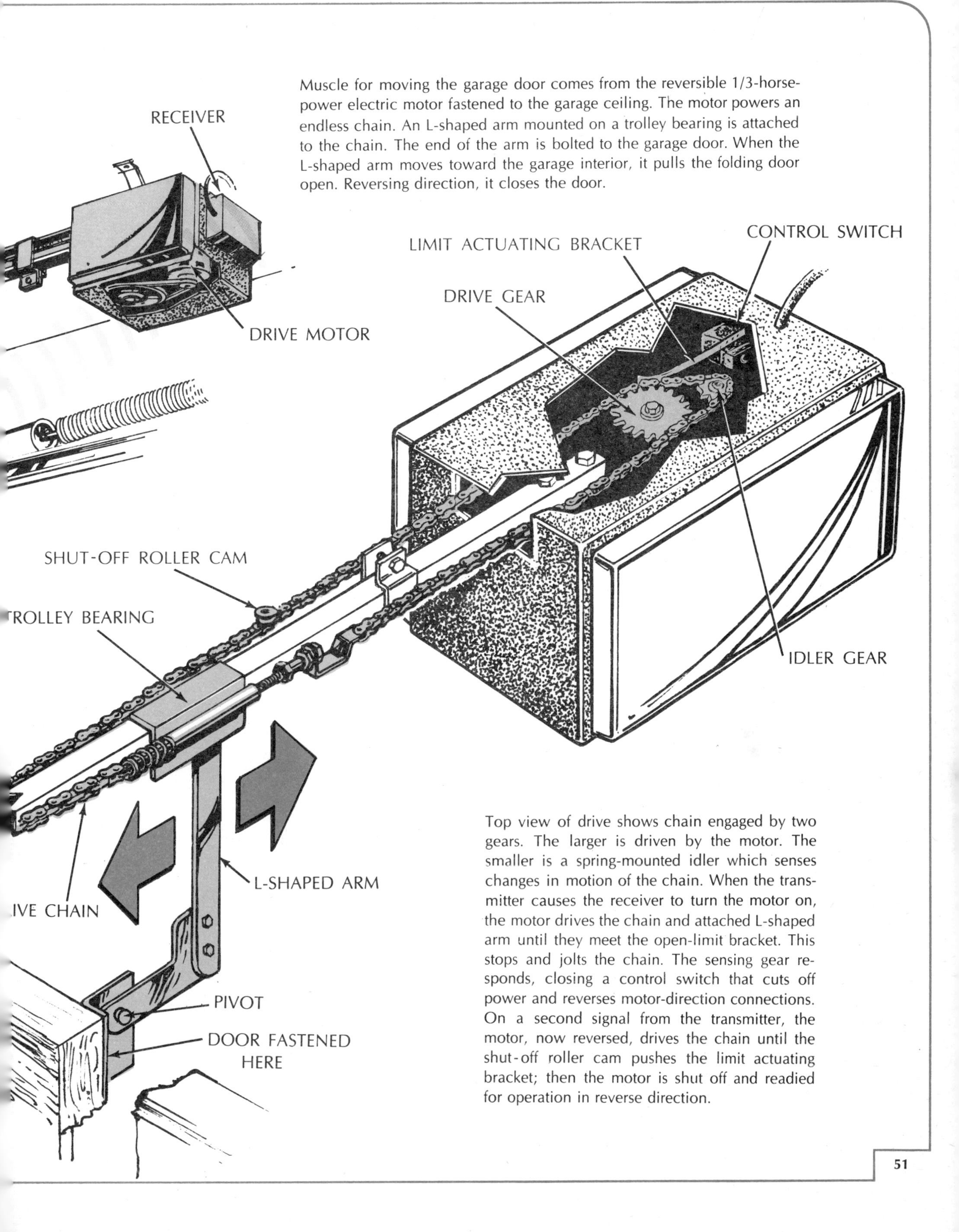

Top view of drive shows chain engaged by two gears. The larger is driven by the motor. The smaller is a spring-mounted idler which senses changes in motion of the chain. When the transmitter causes the receiver to turn the motor on, the motor drives the chain and attached L-shaped arm until they meet the open-limit bracket. This stops and jolts the chain. The sensing gear responds, closing a control switch that cuts off power and reverses motor-direction connections. On a second signal from the transmitter, the motor, now reversed, drives the chain until the shut-off roller cam pushes the limit actuating bracket; then the motor is shut off and readied for operation in reverse direction.

Aerosol Dispenser

The word *aerosol* means "a suspension of colloidal particles in a gas." The most common natural aerosols are clouds, fog, and air pollution. All hold in suspension tiny particles so small that it would take billions to fill a teaspoon. On a more practical level, aerosols can be paints, deodorants, shaving creams, and any material that can be atomized or foamed. An aerosol dispenser is usually cylindrical. It has a concave bottom, that will withstand high internal pressure, and a bell-shaped top. The inner mechanism consists of a spring-loaded push-button valve and a dip tube that extends down from the valve into a liquid or powder called the "dispensable" that is mixed with the propellant.

The propellant is always a substance that goes from liquid to gas at a low temperature. In other words, the liquid has a low boiling point. This boiling-point temperature is measured relative to atmospheric pressure at sea level (14.69 pounds per square inch) and an atmospheric temperature of about 70° F. High pressure inside the dispenser raises the propellant's boiling point, allowing it to remain a liquid at room temperatures.

One common propellant is Freon-12 (CCl_2F_2), a man-made fluorocarbon. In manufacture, a quantity of Freon-12 is mixed under pressure with a dispensable (paint, deodorant, air freshener). The pressure holds the propellant's boiling point high so that it enters the can as a liquid. With a specific amount of space left above the liquid Freon, the can is sealed. Soon, the temperature outside the dispenser causes some of the Freon to boil out of the mixture, forming a pressure pocket of gas above the mixture. At a specific pressure, which may be several times normal atmospheric pressure, the boiling point of the mixture rises to a point higher than the outside temperature. This halts gas production. The dispenser is then in a state of *thermal equilibrium* (temperature balance).

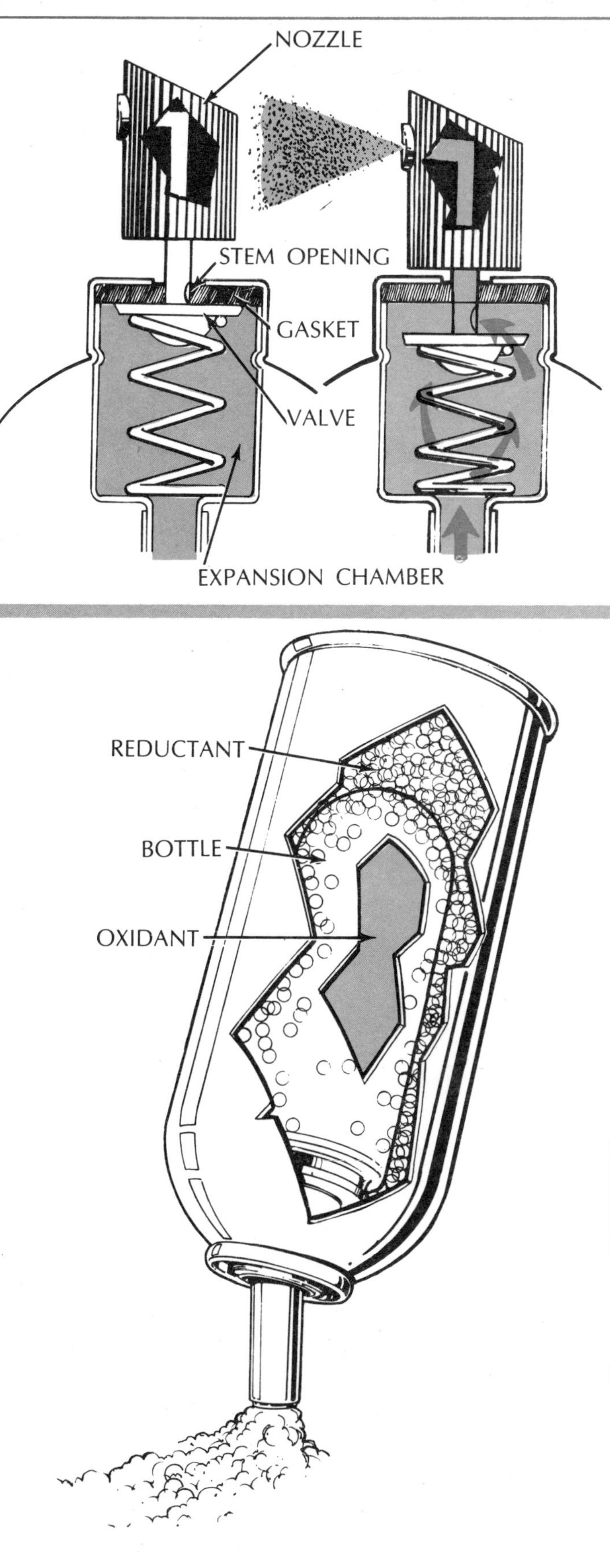

Spray Aerosol

Pressing the valve push-button breaks the air-tight seal between a rubber gasket and a tiny opening in the valve stem. The opening drops into the expansion chamber that is fed by the dip tube. Essentially, this action allows the liquid to respond to two forces: the high pressure of the gas pocket within the dispenser and the relatively low pressure outside. Internal gas pressure forces some of the dispensable up the dip tube, through the expansion chamber, into the valve, and out through atomizing nozzle where it becomes a fine-mist "spray." Escape of the dispensable drops the fluid level within the can. This creates increased space above the liquid and drops internal pressure as well as the boiling point of the Freon. Immediately, the propellant boils and re-pressurizes the gas space. This process yields a fairly consistent spraying pressure until the dispenser's contents are exhausted.

Hot-lather Foam Aerosol

Design of the hot lather dispenser is similar to that for the spray aerosol dispenser except that it contains a bottle and a different expulsion mechanism. The suspended bottle is filled with hydrogen peroxide. Soap propellant and a reductant (fuel) are forced into the dispenser, itself. A spring-loaded plug blocks the oxidizer bottle's mouth and wedges the nozzle body against a gasket, keeping the dispenser's contents pressurized. When the dispenser is turned upside down and a push is applied against the nozzle side, the spring-loaded plug rides up the bottle's opening while the nozzle body unseals the core opening. Immediately, propellant pressure forces the soap and reductant through a hole in the oxidizer bottle's supports and into a mixing chamber. Escaping hydrogen peroxide mixes with the soap/reductant. Under pressure, this mix is forced past the gasket, out and around the spiral nozzle core, and into the user's waiting hand. Though the lather is barely lukewarm as it falls into the hand, in seconds it achieves 150° F. as the oxidant and reductant complete their reaction.

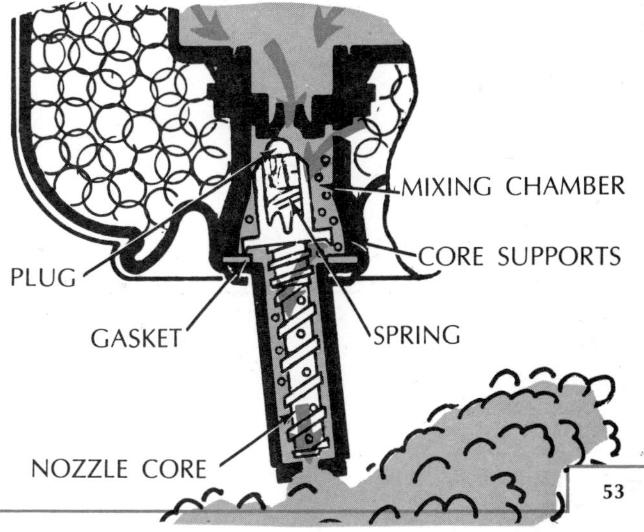

Eight-track Stereo Tape Player

Soon after Thomas Edison invented the cylinder phonograph in 1877, a Danish inventor named Valdemar Poulsen came up with the idea that voices and music could be magnetically recorded onto a steel wire. Poulsen succeeded in 1893, dubbing his invention the "Telegraphone." But Poulsen's invention flopped commercially. Not until after World War II did magnetic recorders, using tape instead of wire, begin to compete with phonographs. For many years, while magnetic recording was made difficult by reel-to-reel operation and uncooperative tape ends, recording was largely the professional's and afficionado's province. But the commercial development of stereo in the 1950s and cassettes and multi-track players in the 1960s did much to popularize tape recorders for use in the automobile and the home.

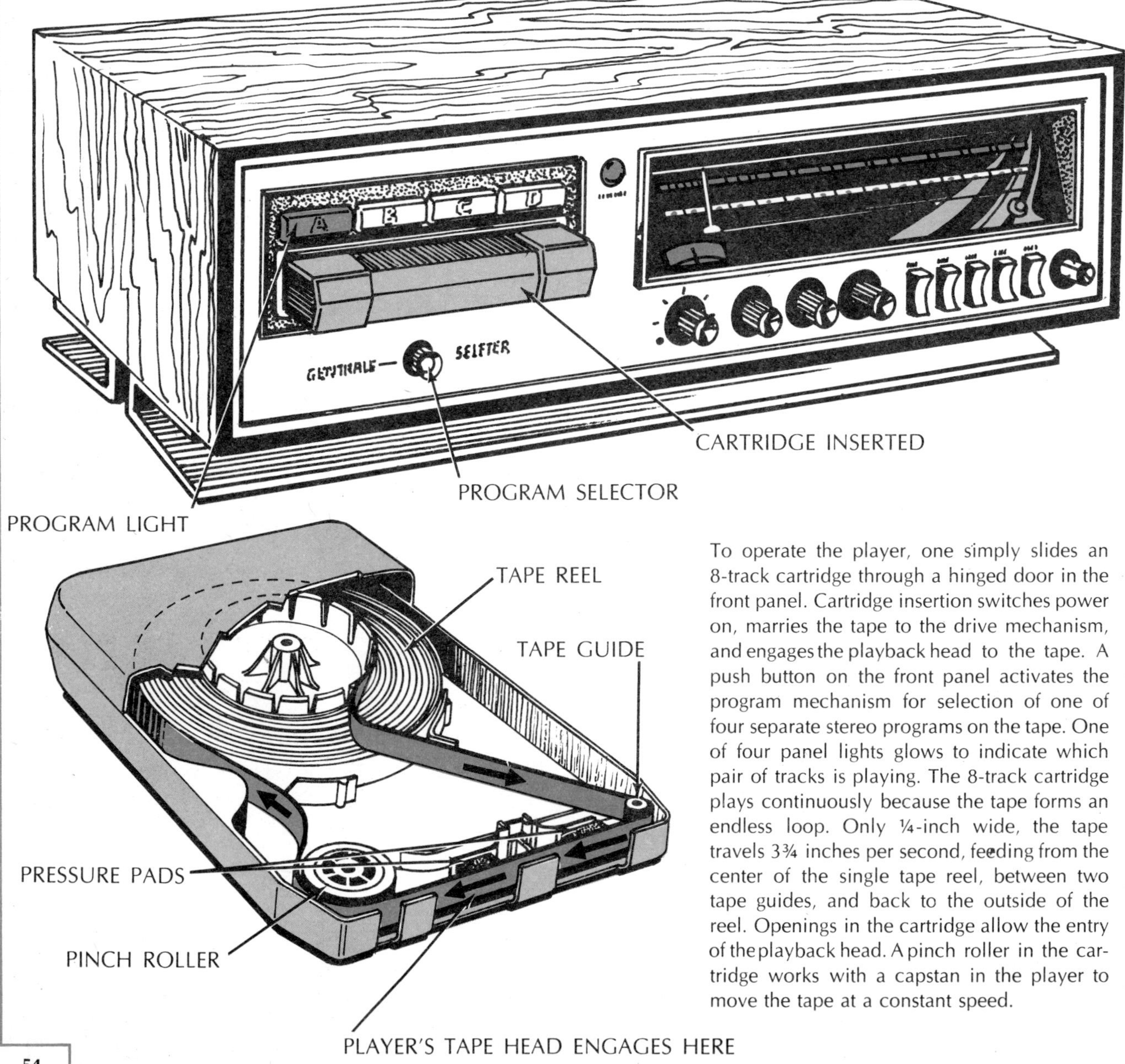

To operate the player, one simply slides an 8-track cartridge through a hinged door in the front panel. Cartridge insertion switches power on, marries the tape to the drive mechanism, and engages the playback head to the tape. A push button on the front panel activates the program mechanism for selection of one of four separate stereo programs on the tape. One of four panel lights glows to indicate which pair of tracks is playing. The 8-track cartridge plays continuously because the tape forms an endless loop. Only ¼-inch wide, the tape travels 3¾ inches per second, feeding from the center of the single tape reel, between two tape guides, and back to the outside of the reel. Openings in the cartridge allow the entry of the playback head. A pinch roller in the cartridge works with a capstan in the player to move the tape at a constant speed.

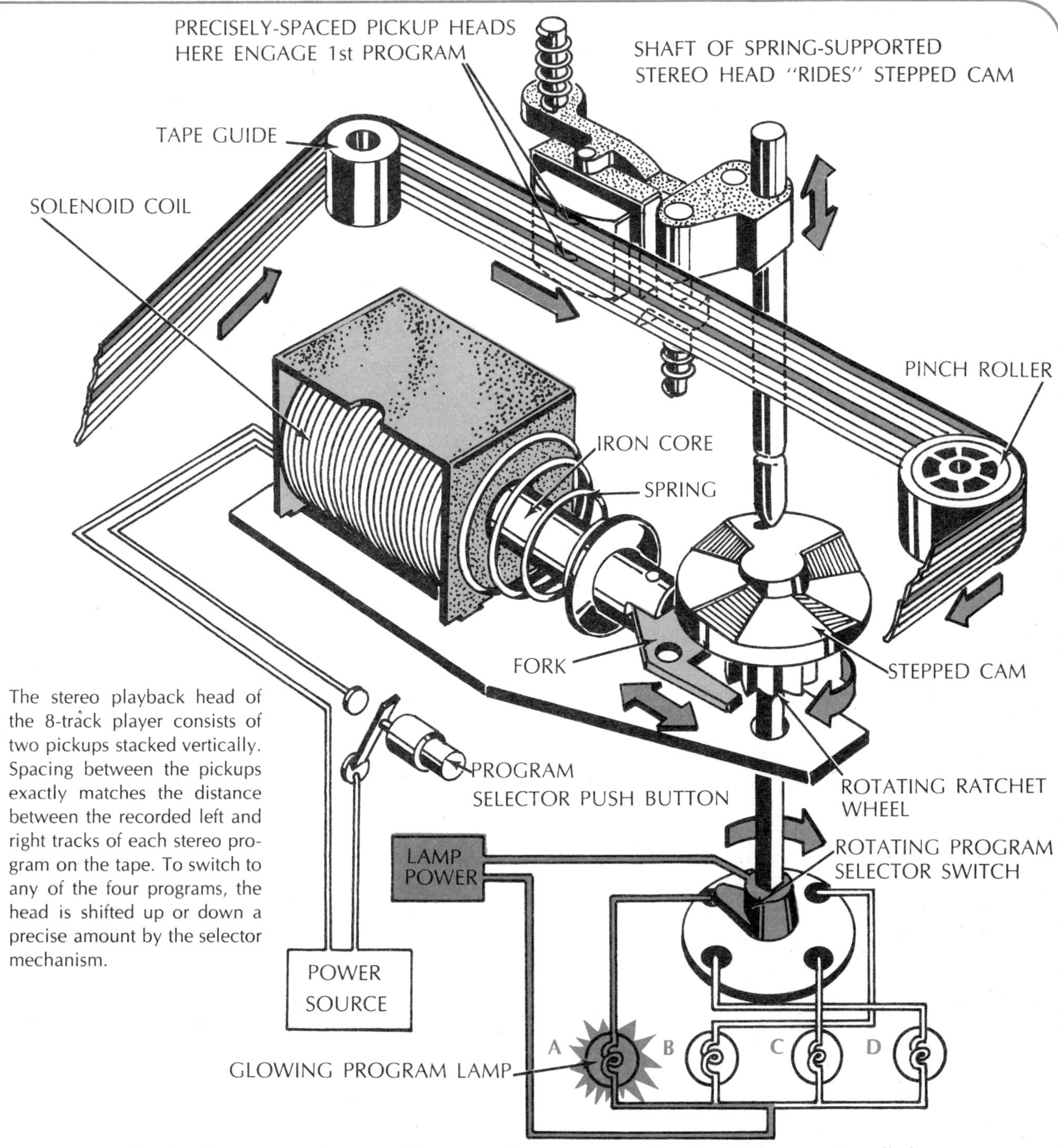

The stereo playback head of the 8-track player consists of two pickups stacked vertically. Spacing between the pickups exactly matches the distance between the recorded left and right tracks of each stereo program on the tape. To switch to any of the four programs, the head is shifted up or down a precise amount by the selector mechanism.

Pressing the program selector push button applies power to a solenoid coil. The magnetic pull of the energized coil yanks a cylindrical iron core into the coil center, compressing a spring. A tang on the fork-like extension of the solenoid core thus moves behind a ratchet-wheel casting. Atop the ratchet wheel is a cam with steps at four different levels. A shaft fixed to the spring-supported stereo head assembly rides the cam surface. Releasing the push button de-energizes the solenoid. Spring force pushes the core and fork forward so that the tang rotates the ratchet wheel, turning the stepped cam a precise distance. The stereo head shaft rides the cam. This steps the head assembly to one of the program levels matching a pair of tracks on the tape. A small rotating switch on the ratchet wheel shaft applies power to the panel lamp matching the program selected.

Parking Meter

Slot machines that sell parking time were first introduced on the streets of Oklahoma City in 1935. Today's meters gulp all manner of change and can be set to vouch for parking privileges in time increments from minutes up to days. A parking meter leads a rough life. Although the extremes of climate and nature take their toll, the meter's worst enemies are people. Maintenance men often skimp on cleaning and oiling. Motorists try to force in bogus coins or whack the meters for misbehavior. Thieves spare no tortures. Yet most meters last long enough to provide sizable revenues to cities that purchase the machines by installments, spread over a several-year period.

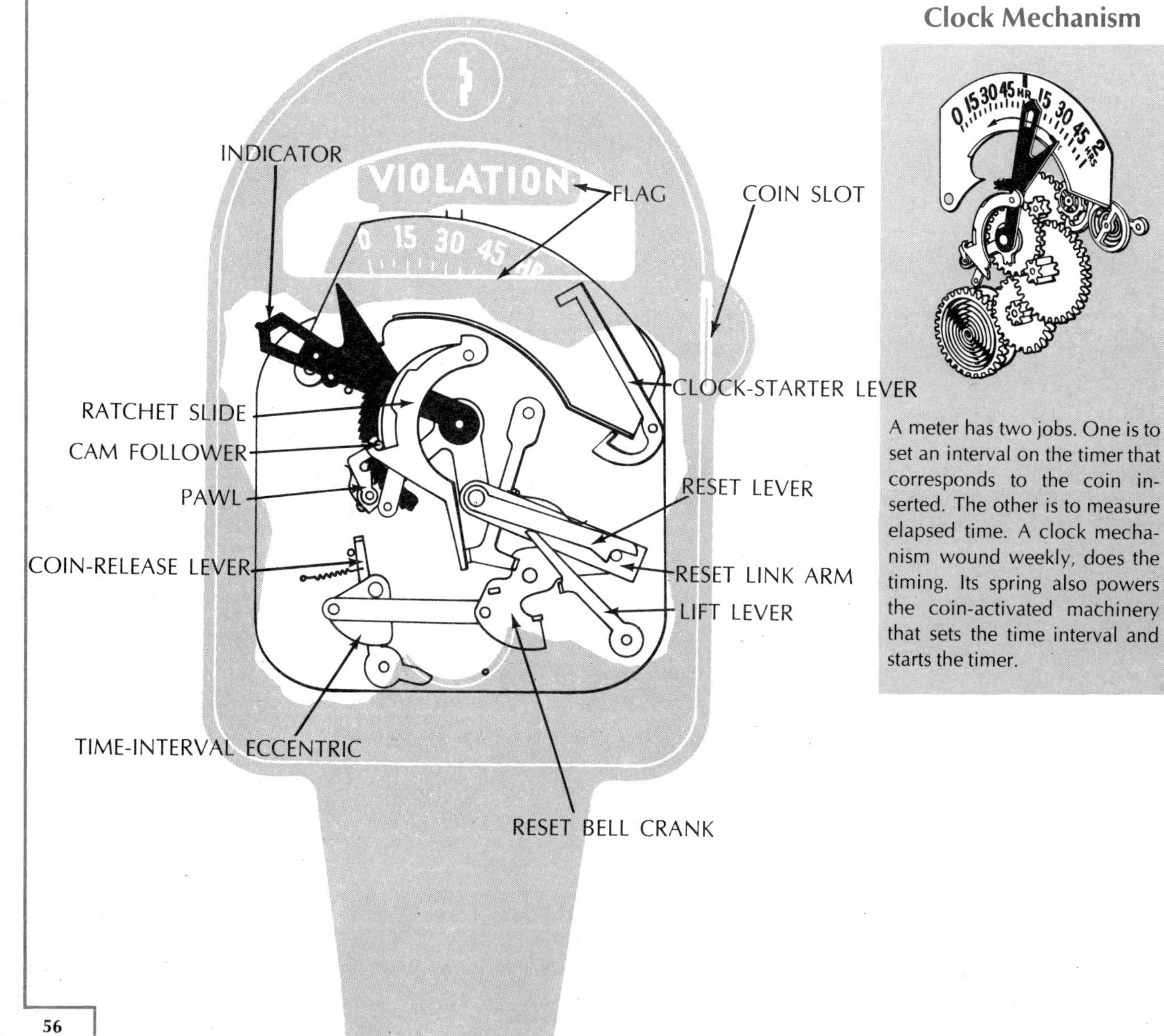

A meter has two jobs. One is to set an interval on the timer that corresponds to the coin inserted. The other is to measure elapsed time. A clock mechanism wound weekly, does the timing. Its spring also powers the coin-activated machinery that sets the time interval and starts the timer.

Dime

TRIP LEVER SHAFT
TRIP LEVER
DIME
BELL CRANK
ECCENTRIC

The smallest coin, a dime, falls neatly into a pocket formed by dogs on the reset bell crank. Nudged, a sensitive trip lever pivots slightly. Its turning shaft trips the clock mechanism, rotating the time-interval eccentric.

PAWL
INDICATOR TEETH
DIME

Driven by the eccentric, the bell crank shoves the reset link arm. The link arm swings the pawl around. And the pawl, engaged with the indicator teeth, pushes the indicator all the way over to give maximum time.

Penny

Larger than a dime, the penny shoves the lift lever aside as it falls into the coin pocket.

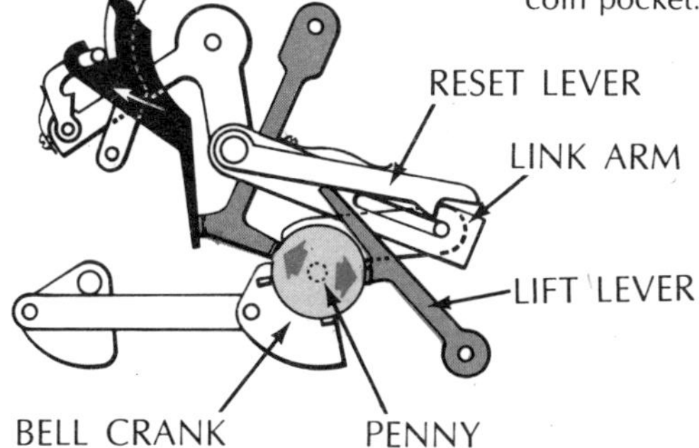

PAWL
SLOT
PENNY

The pivoting lift lever raises the reset lever on the link arm, uncovering a slot. As the eccentric drives the bell crank, the pin on the crank slides in the slot before swinging the pawl. The indicator goes to the 12-minute mark because 108 minutes' motion was lost as the pin rode in the slot.

Nickel

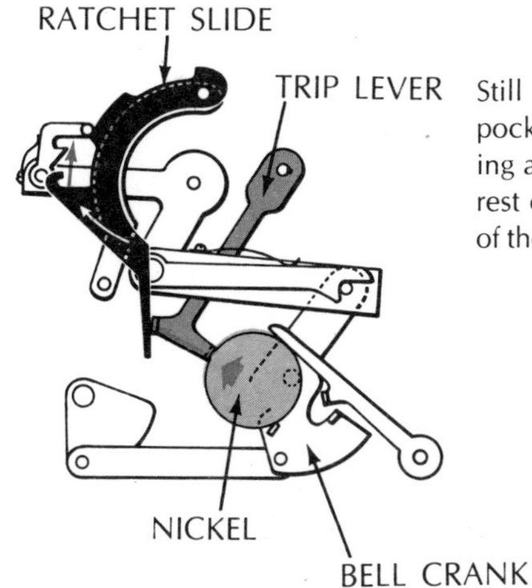

Still larger, the nickel can't fit into the pocket. It slides to the left, shouldering aside the trip lever and coming to rest on one bell-crank dog and a foot of the trip lever.

Shoved far over, the trip lever raises the outer, lobed plate of the ratchet slide. As the pawl carries the indicator teeth along, it disengages for an hour's worth of time while the cam follower is riding over the lobe on the ratchet slide.

Voting Machine

Though expensive, voting machines pay for themselves by reducing routine election costs and minimizing the number of recounts. Another advantage is that they give quick returns after the polls close. They require little maintenance, last a long time, and – according to some politicians – help draw voters to the polls. Interlocks prevent pulling the lever for too many candidates or voting both YES and NO on a question. A paper roll permits write-in votes. The first voting machine was used in Lockport, New York, in 1892. Today machines are used in nearly every state.

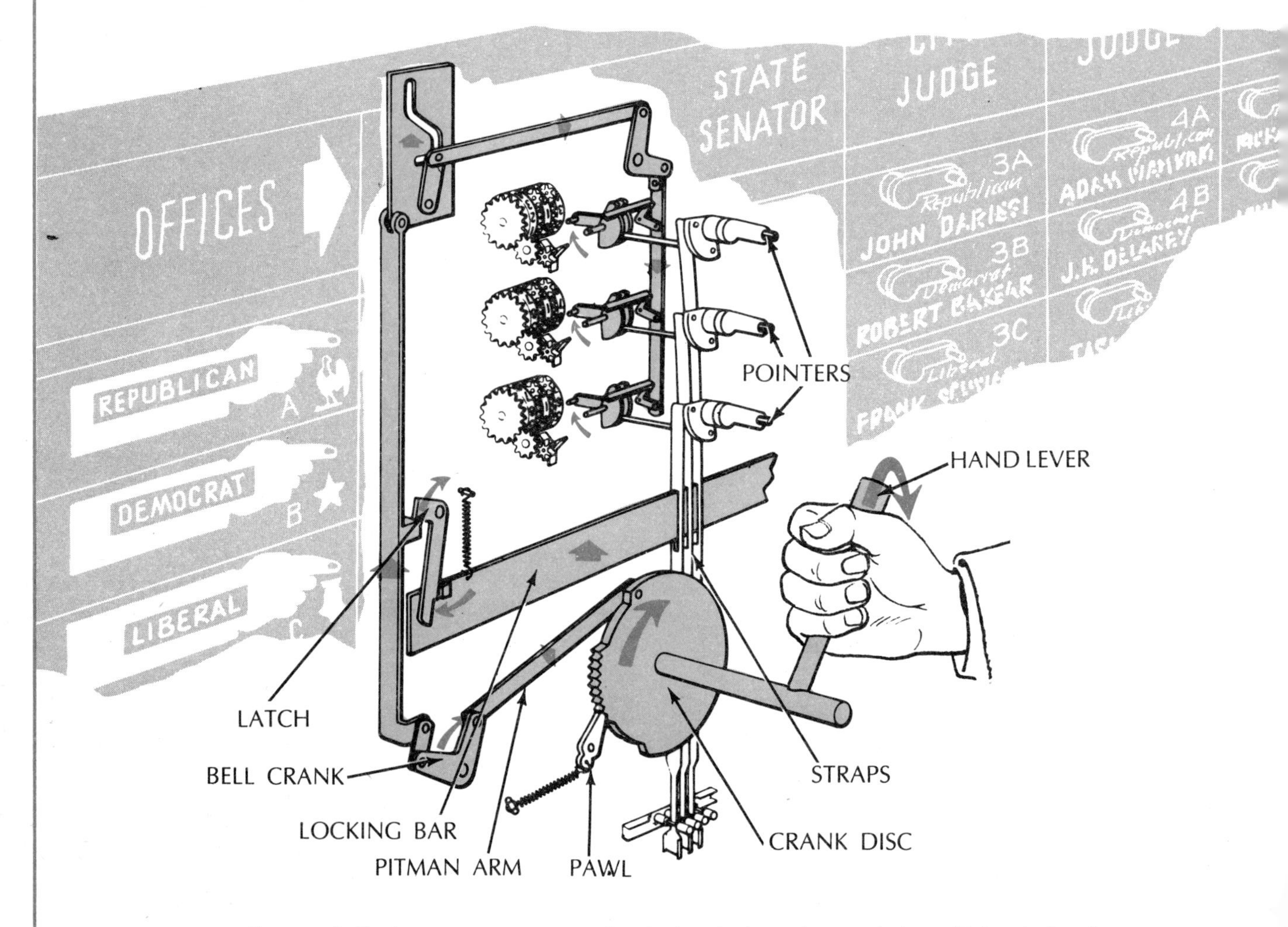

To cast a ballot in secret, a person entering the booth closes the curtain by swinging the hand lever to the right. This also turns a crank disc, working two pitman arms (only one is shown above). Through bell cranks, the arms trigger a latch on each side, releasing a spring-mounted locking bar, which snaps up. (The bar runs through a slot in each of the metal straps connected to the selection-lever pointers and locks them until it is released.) Now the pointers are free, ready for voting. Ratchet teeth on the crank disc, with a pivoted pawl, prevent reversing the lever once it has been moved. When the pawl slips past the last tooth, it is free to flip the other way for the return of the lever, as shown on the next page.

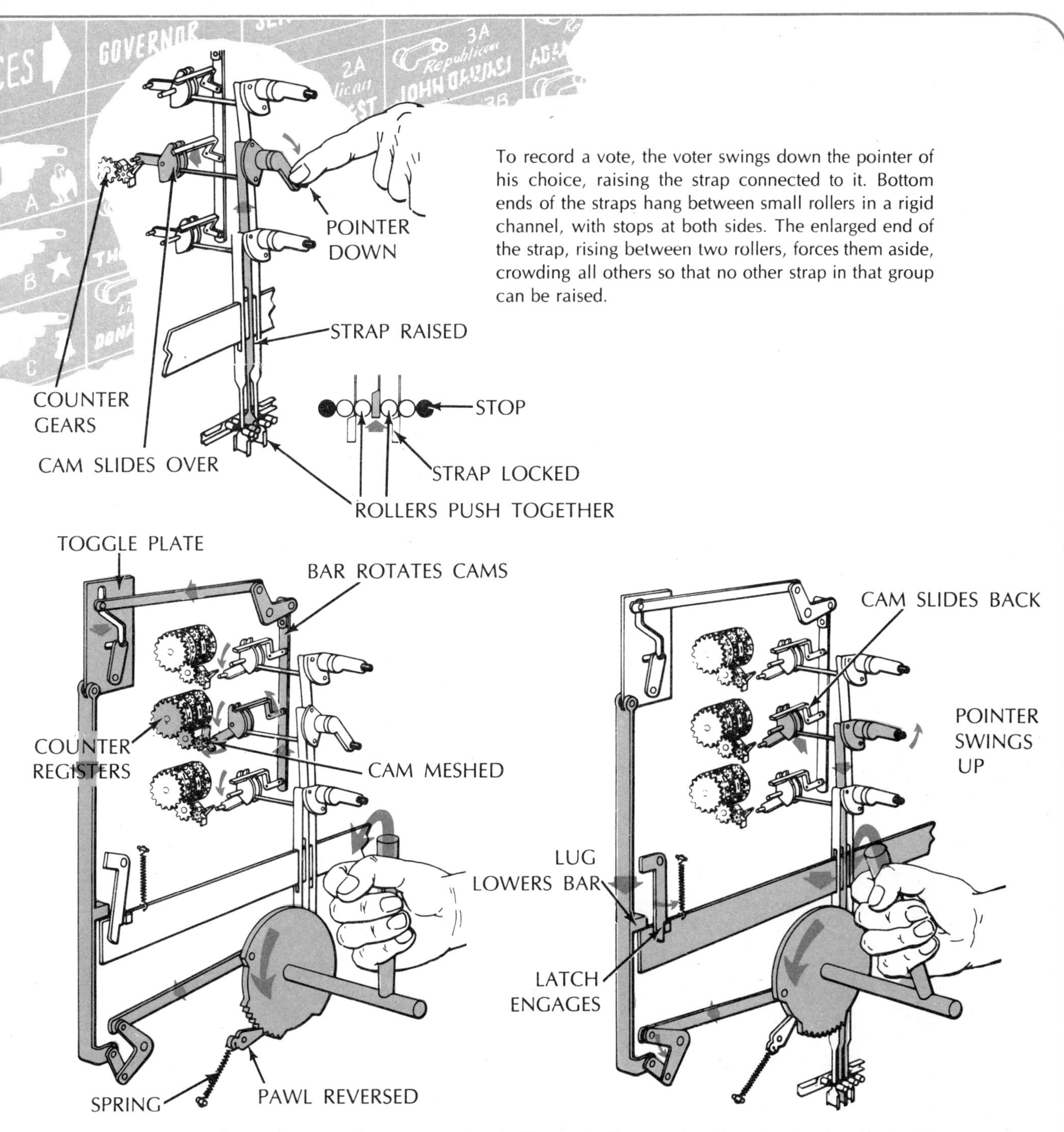

To record a vote, the voter swings down the pointer of his choice, raising the strap connected to it. Bottom ends of the straps hang between small rollers in a rigid channel, with stops at both sides. The enlarged end of the strap, rising between two rollers, forces them aside, crowding all others so that no other strap in that group can be raised.

To change his vote, the voter need only flip up the incorrectly selected pointer and swing down another. But stops and rollers, previously set up, let him vote only for the right number of candidates (one mayor, four councilmen, for example). Switching a pointer down slides the lever's registry cam into engagement with a three-digit counter that totals the votes for that candidate. When the voter has made all his selections, he swings back the hand lever, with the ratchet pawl now reversed. The crank disc and its linkage first pull down a toggle plate (top left). Through a bell crank this rotates all pointer cams, but only those in mesh with counters register votes. Next, lugs pull down the locking bar. The bar pulls down the straps, flipping up all down-swung pointers. Finally the mechanism opens the curtain.

Saber Saw

Amateur and professional carpenters alike value the saber saw for two major reasons: portability and versatility. This lightweight machine can be accurately guided with just one hand. It is rugged enough to cut through hardwood planks, yet responsive enough to perform intricate cuts. And with the proper blade, it can saw through a wide variety of materials. Additional convenience features found on many models include a tilting mechanism for making angled cuts, and trigger control for varying the cutting speed and a lock button for maintaining a steady speed.

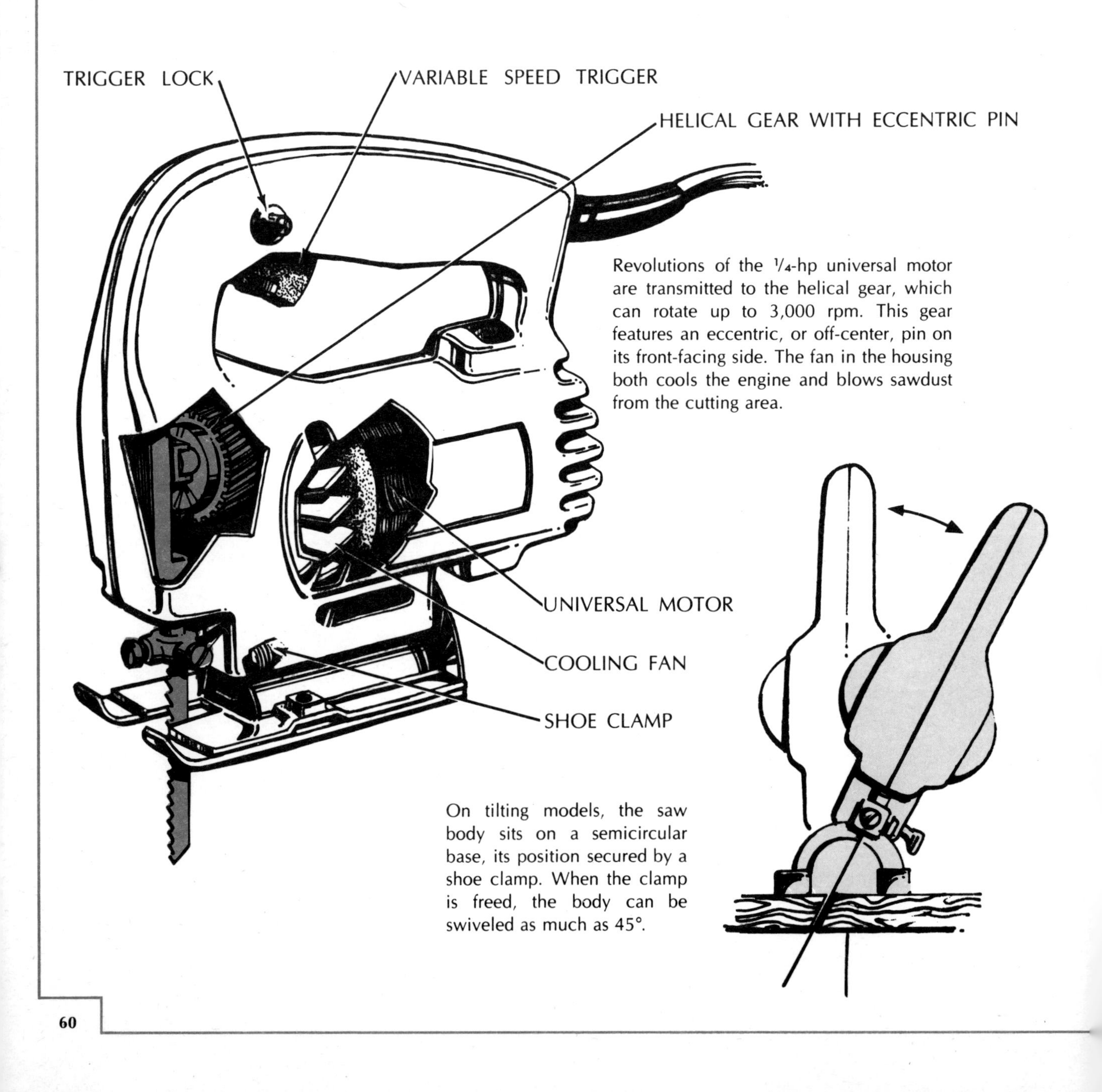

Revolutions of the 1/4-hp universal motor are transmitted to the helical gear, which can rotate up to 3,000 rpm. This gear features an eccentric, or off-center, pin on its front-facing side. The fan in the housing both cools the engine and blows sawdust from the cutting area.

On tilting models, the saw body sits on a semicircular base, its position secured by a shoe clamp. When the clamp is freed, the body can be swiveled as much as 45°.

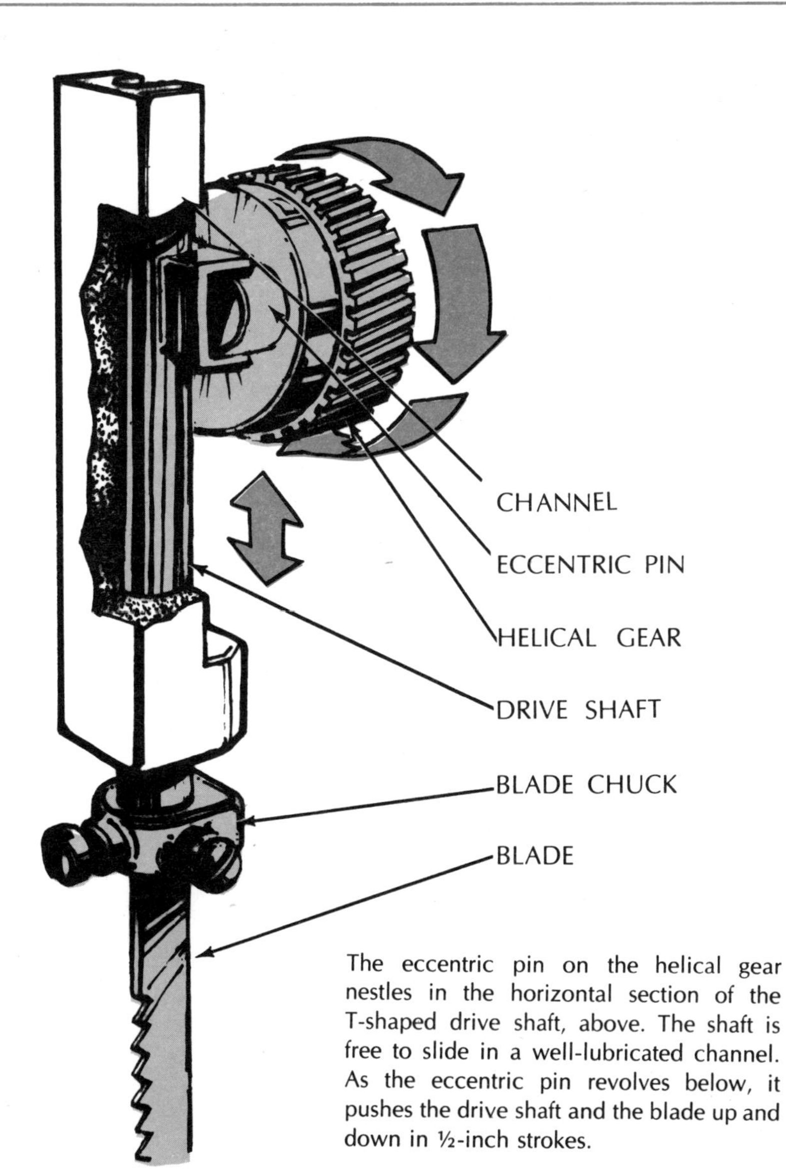

The eccentric pin on the helical gear nestles in the horizontal section of the T-shaped drive shaft, above. The shaft is free to slide in a well-lubricated channel. As the eccentric pin revolves below, it pushes the drive shaft and the blade up and down in ½-inch strokes.

Depending on its tooth configuration, a saber saw blade can cleave through materials that range from lumber to plastics to linoleum to light sheet metal. A large variety of blades is available for many cutting purposes.

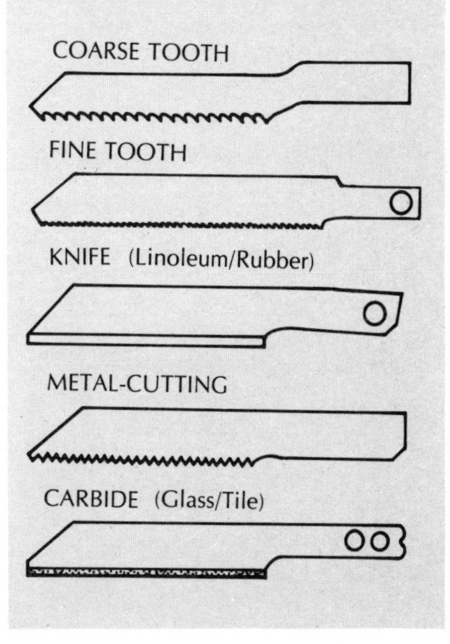

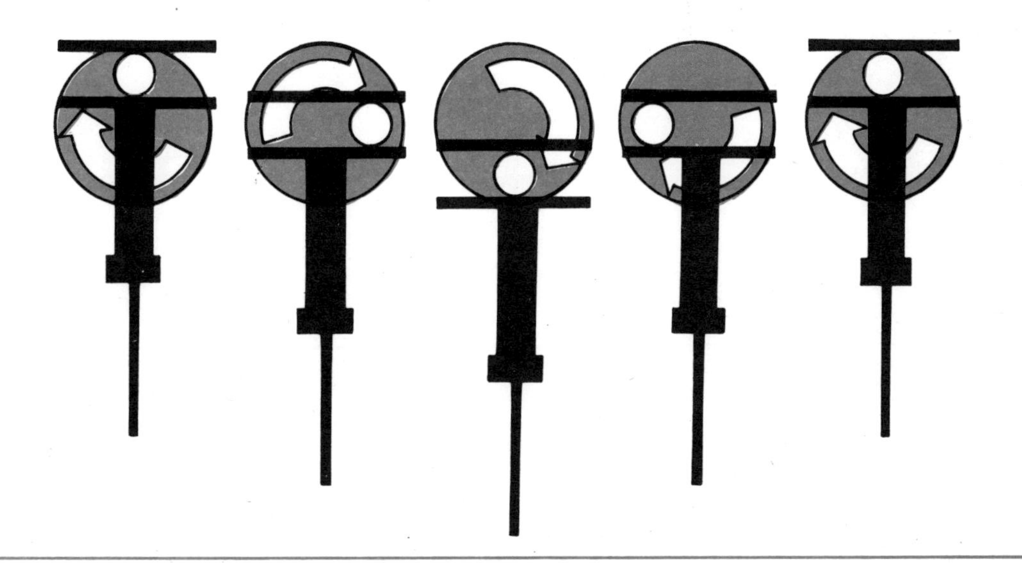

Circuit Breaker

Like fuses, circuit breakers monitor electric lines to cut off dangerously excessive current flow. Unlike fuses, breakers need not be replaced. Flipping a toggle resets them. But if the cause of excess current remains, the breaker snaps open again. Circuit breakers are permanent—they don't burn out. And they can't be doctored to carry unsafe current loads. Earlier breakers tripped open spring-loaded contacts whenever two adhered strips of dissimilar metal warped, caused by heat expanding one strip more than the other. Modern breakers combine thermal and magnetic action, or use a magnetic coil with a hydraulically delayed plunger or core. They are designed to allow current flow under temporary overloads, open the circuit if an overload continues, and break in less than $\frac{1}{100}$th of a second under the stress of a short circuit.

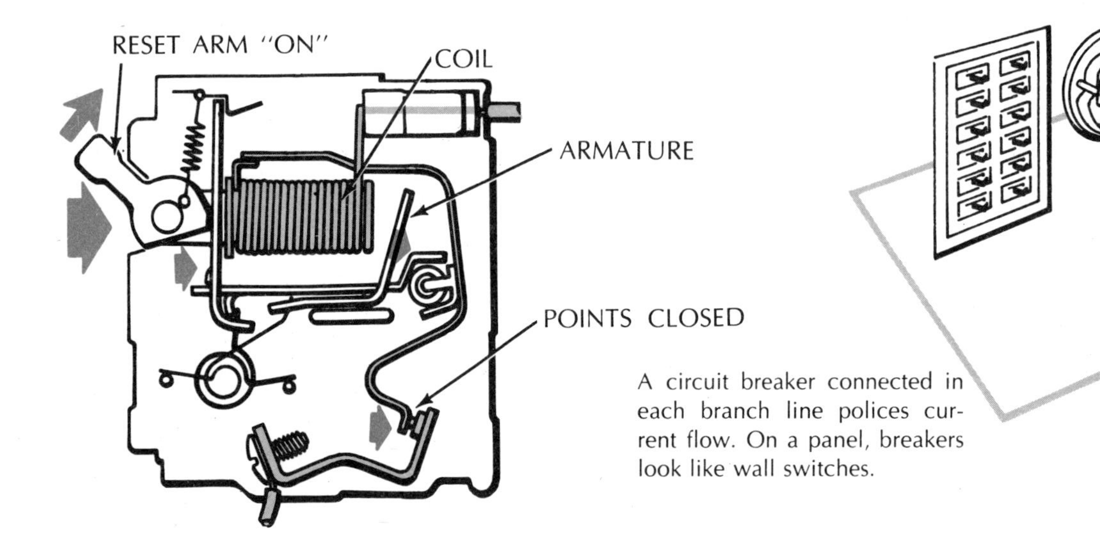

A circuit breaker connected in each branch line polices current flow. On a panel, breakers look like wall switches.

Hydraulically Delayed Magnetic Breaker

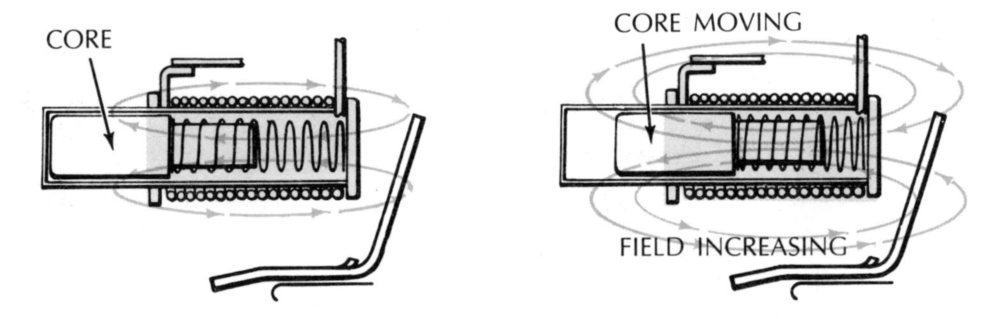

In the cutaway view, left, line current flows in a wire coil wound around a fluid-filled tube. A movable iron core is held back by a spring. Up to rated current, the magnetic field is too weak to move the core. In the cutaway view, right, excessive current, such as when the motor starts, boosts the field and draws the core in. But the spring and fluid slow it. As the motor revs up, its current draw lessens, and the spring pushes the core back.

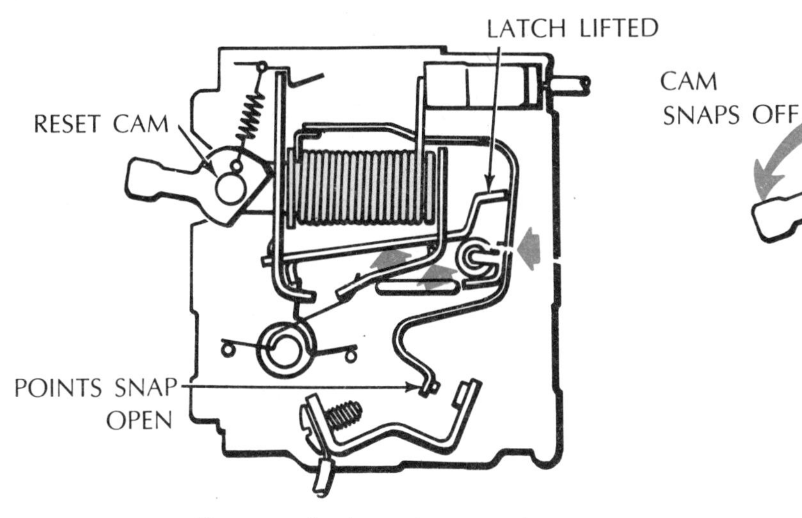

If an overload continues, such as when a motor tackles too big a job, or if the current is heávy enough to pull the core in fast despite hydraulic delay, the core's presence at the coil end intensifies the magnetic field, attracting the armature.

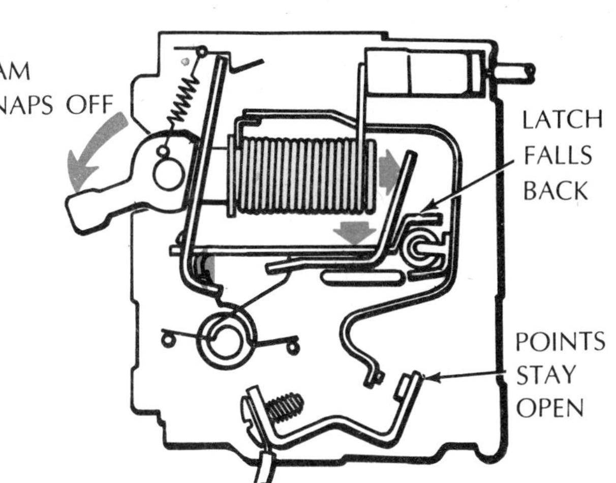

When a short circuit occurs, as when the line is faulty, high current in the coil creates a strong field that yanks the armature in before the core can budge. Armature levers on the roller unlatch, letting the spring contact snap open and the reset cam flip back.

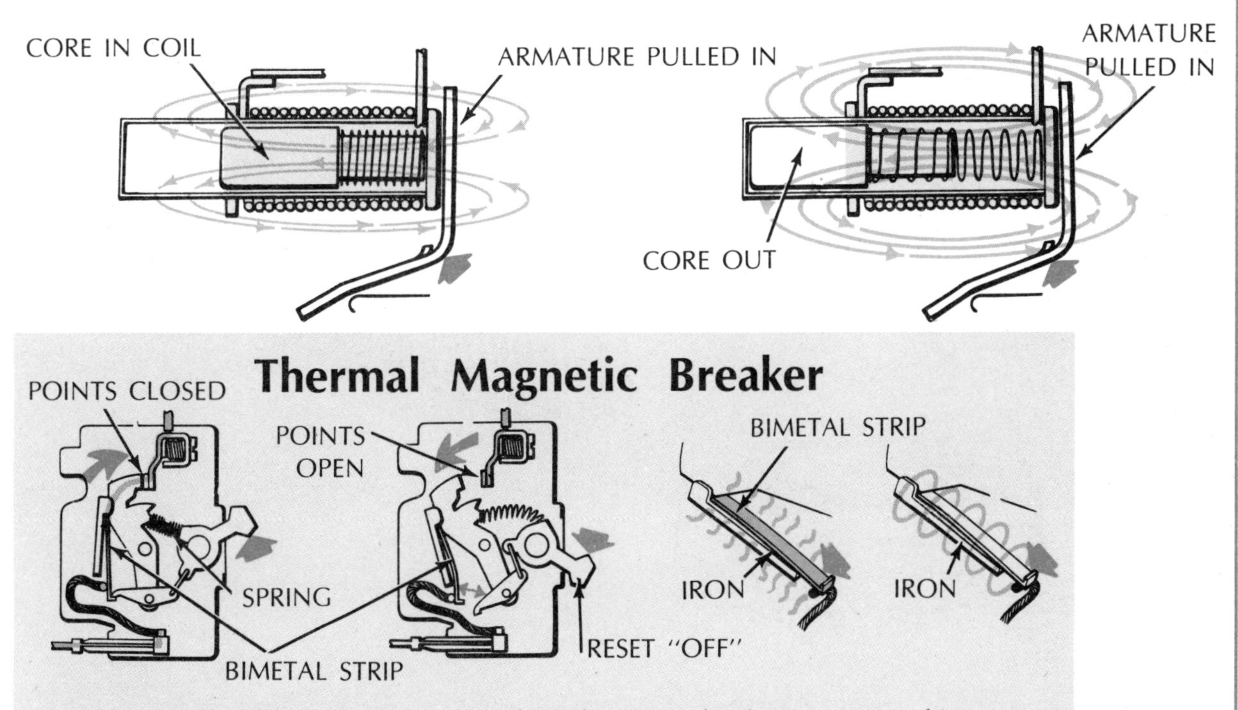

Thermal Magnetic Breaker

Fastened beside the bimetal strip in these thermal magnetic breakers is a piece of iron. Ordinary overloads, after a brief delay (enough to let a motor attain speed), heat the bimetal element. Bending, the element frees a toggle latch, causing the contacts to snap open. On big overloads or short circuits, the high current surrounds the strip with a strong magnetic field. Attracted to the iron piece, the strip bends as if heated but does so instantly. The contacts therefore snap open with no time delay at all.

Telephone Answering/Recording Machine

For many people on the go—especially the self-employed—incoming calls are vital to livelihood. Until recently, the available solutions were a full-time receptionist, a telephone answering service, or a long-suffering spouse. Each choice meant considerable monthly expense—financial or otherwise. Today, countless telephone calls are handled economically by answering/recording machines that can be connected easily to any telephone coupler. These machines answer incoming calls with the owner's pre-recorded message and then invite the caller to record a message that the owner will play back either upon his return or by phoning the machine through remote control. The machine conveys the personality of its owner while ensuring confidentiality and recording an error-free message.

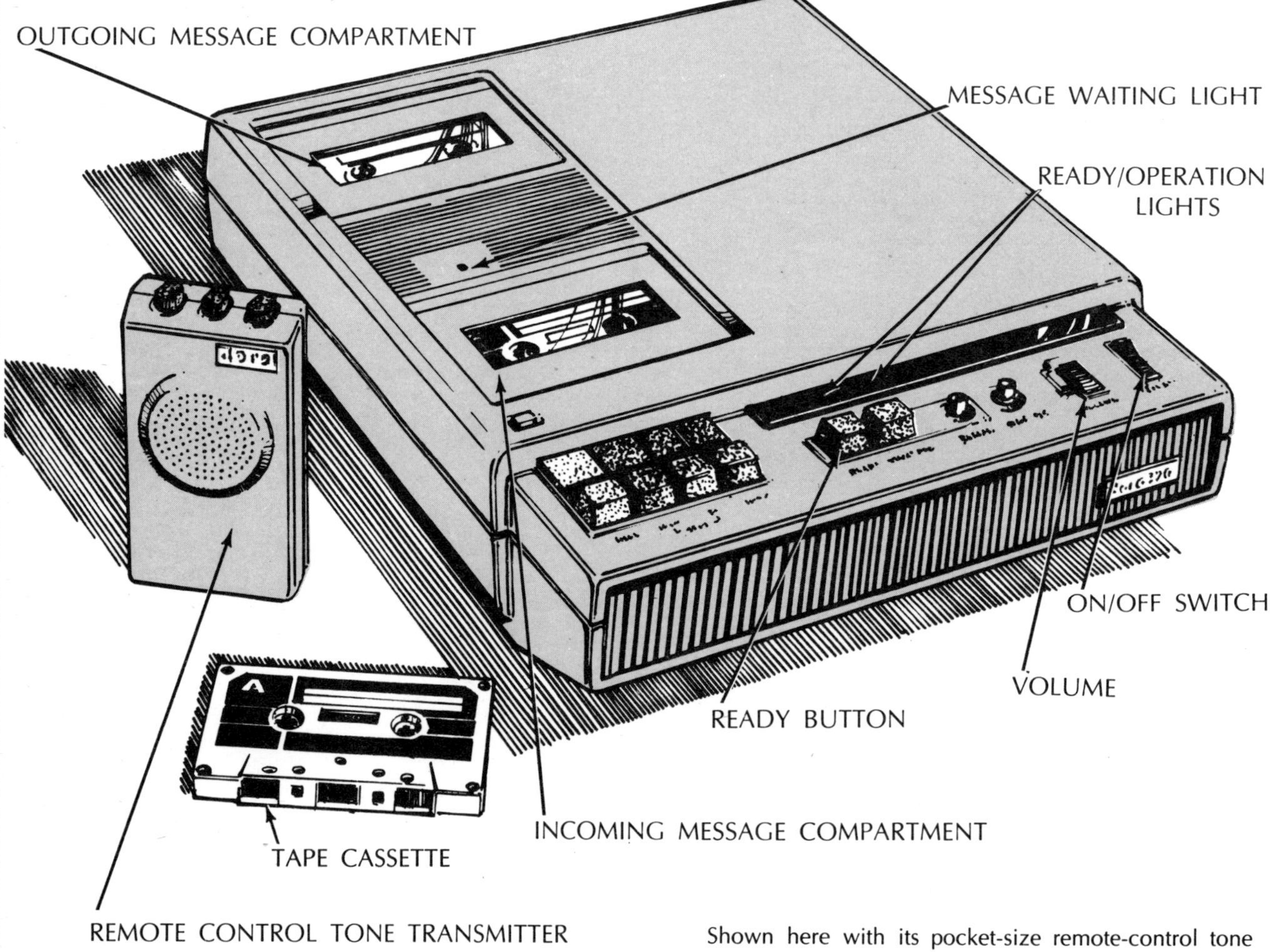

Shown here with its pocket-size remote-control tone transmitter, the telephone answering/recording machine measures just 3½-by-11½-by-13 inches. To prepare it for use, a 20-second message is recorded onto a standard, endless-loop cassette which is then inserted into the outgoing message compartment. A second cassette is then placed into the incoming message compartment. The machine thus readied, the READY button is pressed and the READY light comes on. The machine can then accept its first call.

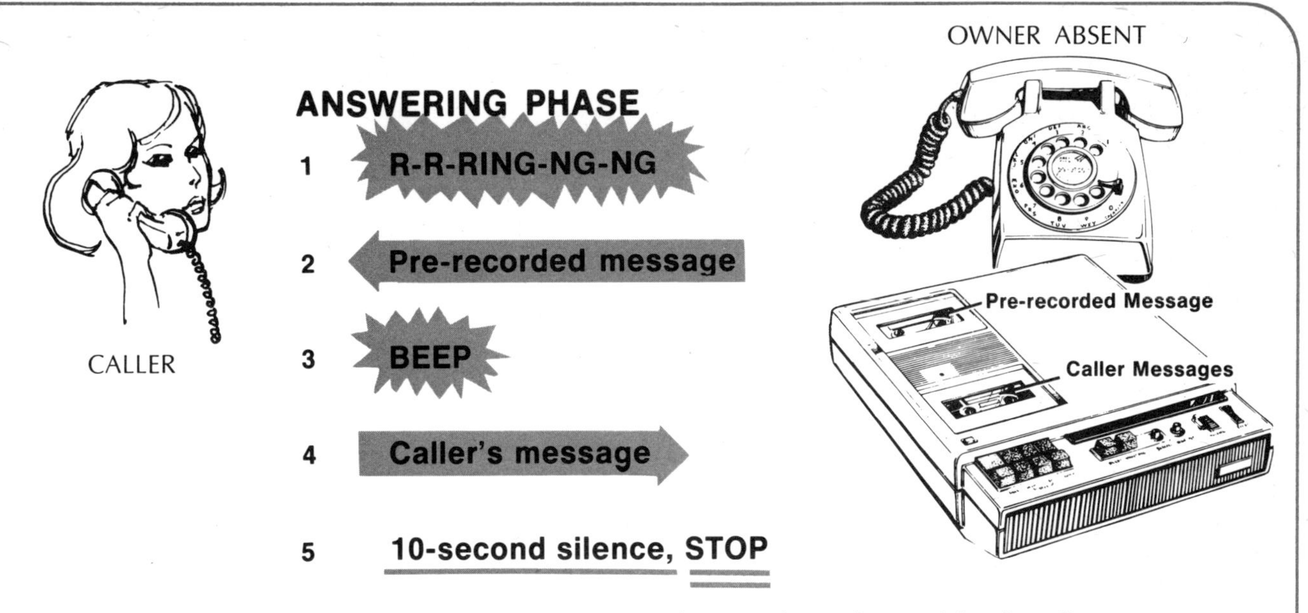

1. The very first ring sets up an electronic connection between the machine and the phone line. **2.** The MESSAGE WAITING and OPERATION lights switch on, and the outgoing-message compartment plays the pre-recorded message. **3.** After 20 seconds of the pre-recorded message, the tape stops and sends a "BEEP" to the caller, telling him that he may record his message. **4.** Now the incoming message tape starts, while an electronic voice control circuit allows the tape to continue as long as the caller keeps talking. **5.** After a 10-second silence the machine automatically disconnects. The OPERATION light goes off and the machine "hangs up" to await the next call.

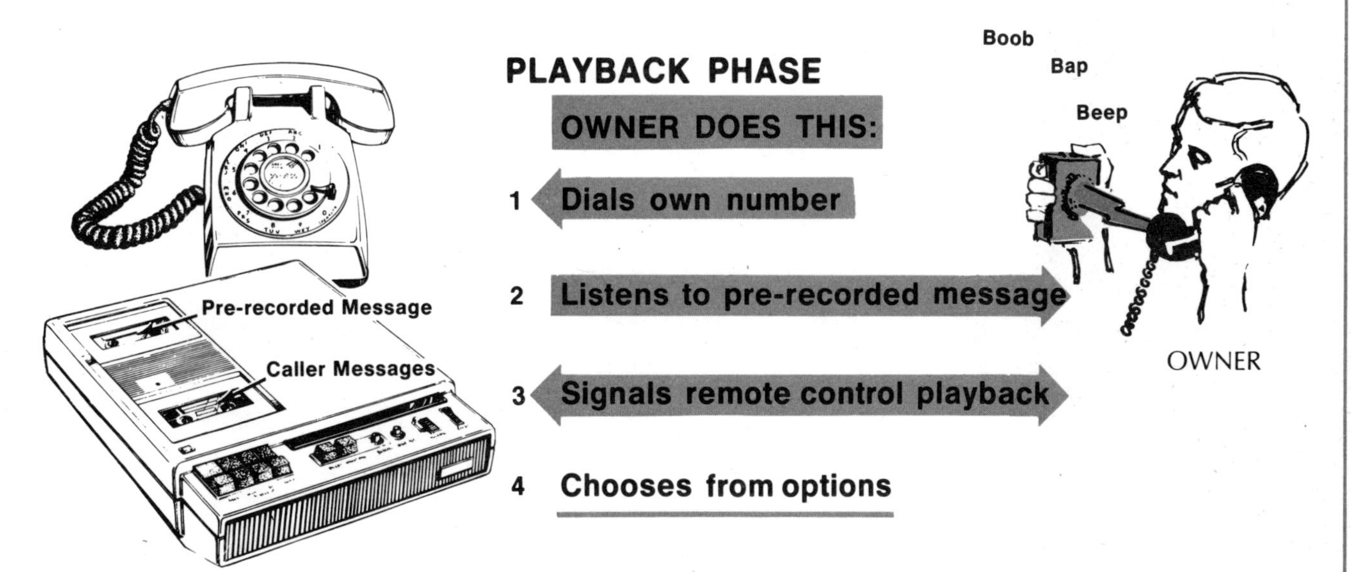

1. Owner may play back upon returning to the machine, OR he may dial his own number from an outside line and thus activate the machine. **2.** Owner listens to his own 20-second pre-recorded message. **3.** He then holds the transistorized, remote-control tone transmitter near the telephone and presses three buttons. (The transmitter sends three tones that correspond to the tone frequencies produced by standard push-button telephones. Once the machine receives the correct three-tone access code, it rewinds the incoming message tape to its beginning and, at the sound of one more tone, plays back the messages.) **4.** After hearing the messages, the owner may hang up to let the machine continue receiving messages. OR, using combinations of the access code, he may (a) replay the messages, (b) rewind so that new messages will record over the old, (c) change his pre-recorded message on the outgoing tape, (d) listen to his revised message and check it for accuracy.

Microwave Oven

Microwave ovens cook food with dazzling speed. Roasting a chicken takes only fifteen minutes—baking a potato, four. A hot dog with bun emerges piping hot in thirty seconds. The microwaves that make such lightning-fast cooking possible are very high frequency radio waves which penetrate the food and violently agitate its molecules. This creates heat. The food, in effect, cooks internally. The source of the microwaves is a magnetron, a high-frequency radar tube developed during World War II. Countertop microwave ovens operate on ordinary house current.

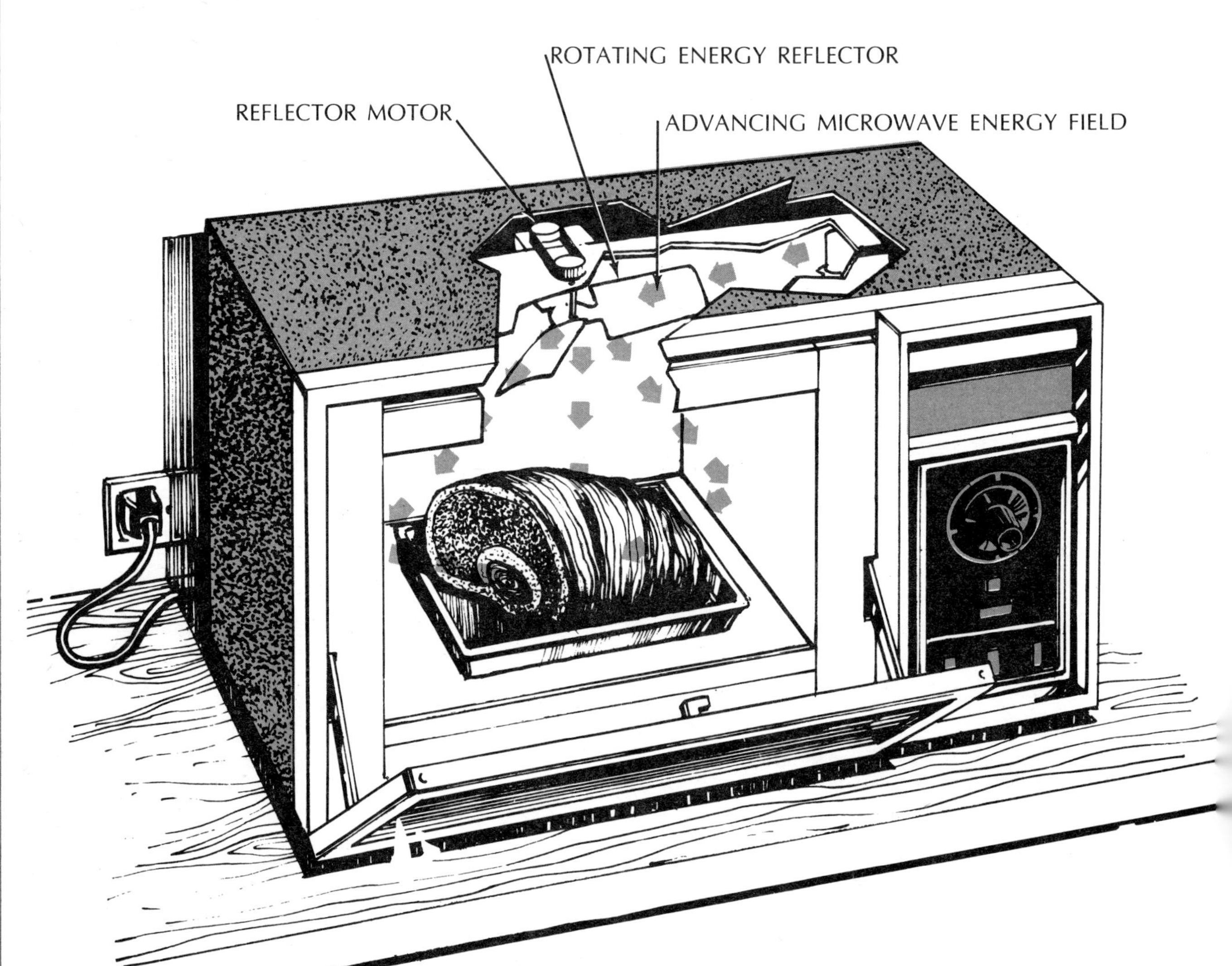

The microwaves used for radar and cooking are a potentially dangerous form of radiation. However, microwave ovens are metal lined and equipped with metal doors that must be shut securely before the oven will operate. Manufacturers are required to keep radiation leakage within Federal safety guidelines.

Microwaves generated within the magnetron travel down the wave guide (a metal duct) and are reflected into the oven by metal blades which rotate to disperse the energy evenly. The energy that does not strike the food directly is repeatedly reflected by the metal oven walls until it is absorbed. The frequencies used for cooking are in the radar class: an incredible 2,450,000,000 cycles per second. Heat is generated within food when its molecules do a flip-flop "dance" in response to the push and pull of electromagnetic energy. Molecules in a ceramic or glass dish do not respond and therefore are not heated in a microwave oven.

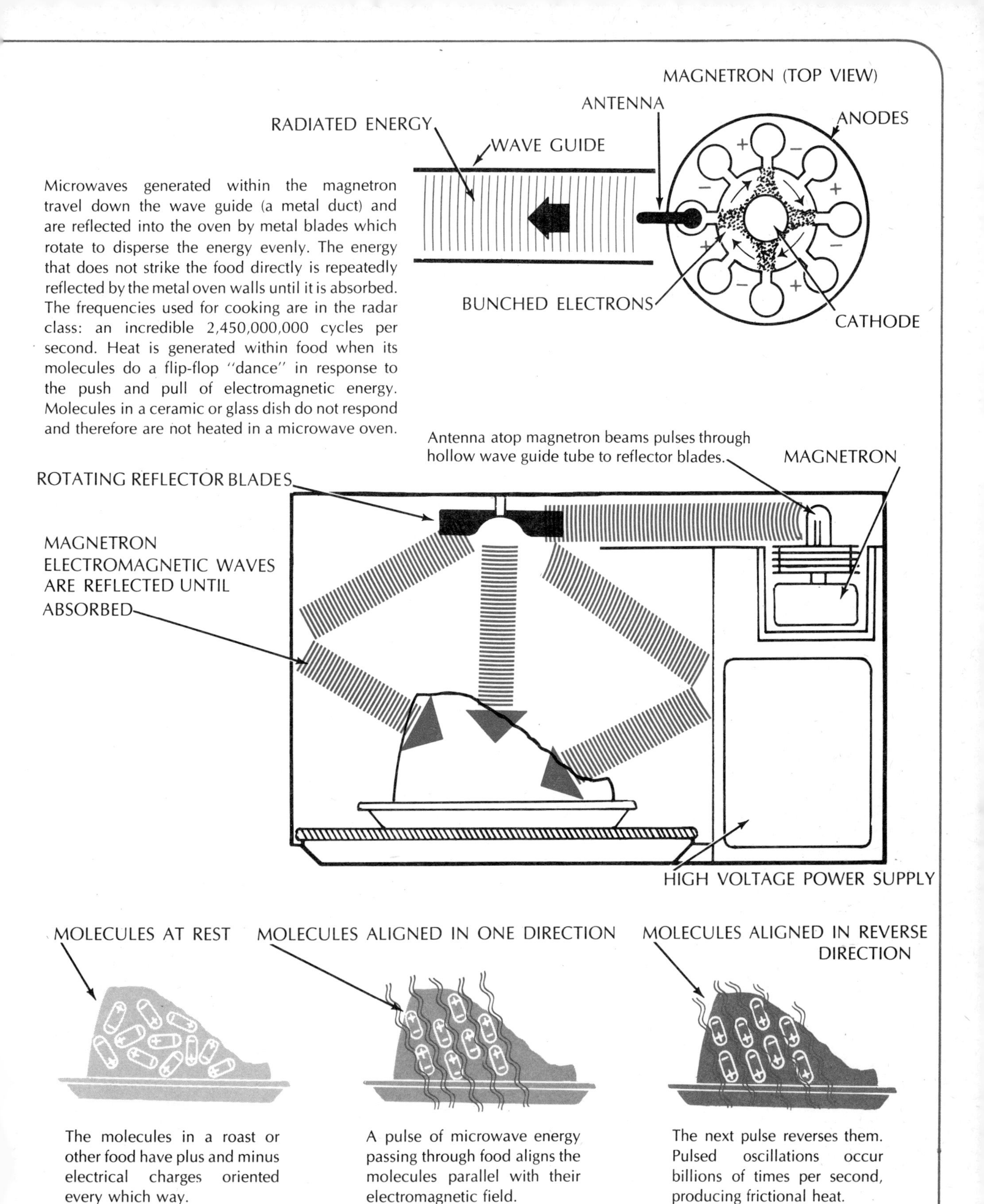

The molecules in a roast or other food have plus and minus electrical charges oriented every which way.

A pulse of microwave energy passing through food aligns the molecules parallel with their electromagnetic field.

The next pulse reverses them. Pulsed oscillations occur billions of times per second, producing frictional heat.

Room Dehumidifier

Excess household humidity is a common problem in many areas, especially during summer. The old-fashioned solution to excess humidity was a bucket of moisture-absorbing calcium chloride in the basement. The modern dehumidifier employs principles and parts similar to those of the air conditioner. But instead of cooling room air, a dehumidifier wrings out moisture. Two sensing mechanisms make the dehumidifier virtually automatic: A humidistat turns the machine on when room humidity reaches an undesirable level. A cutoff device stops the machine when its water receptacle is full.

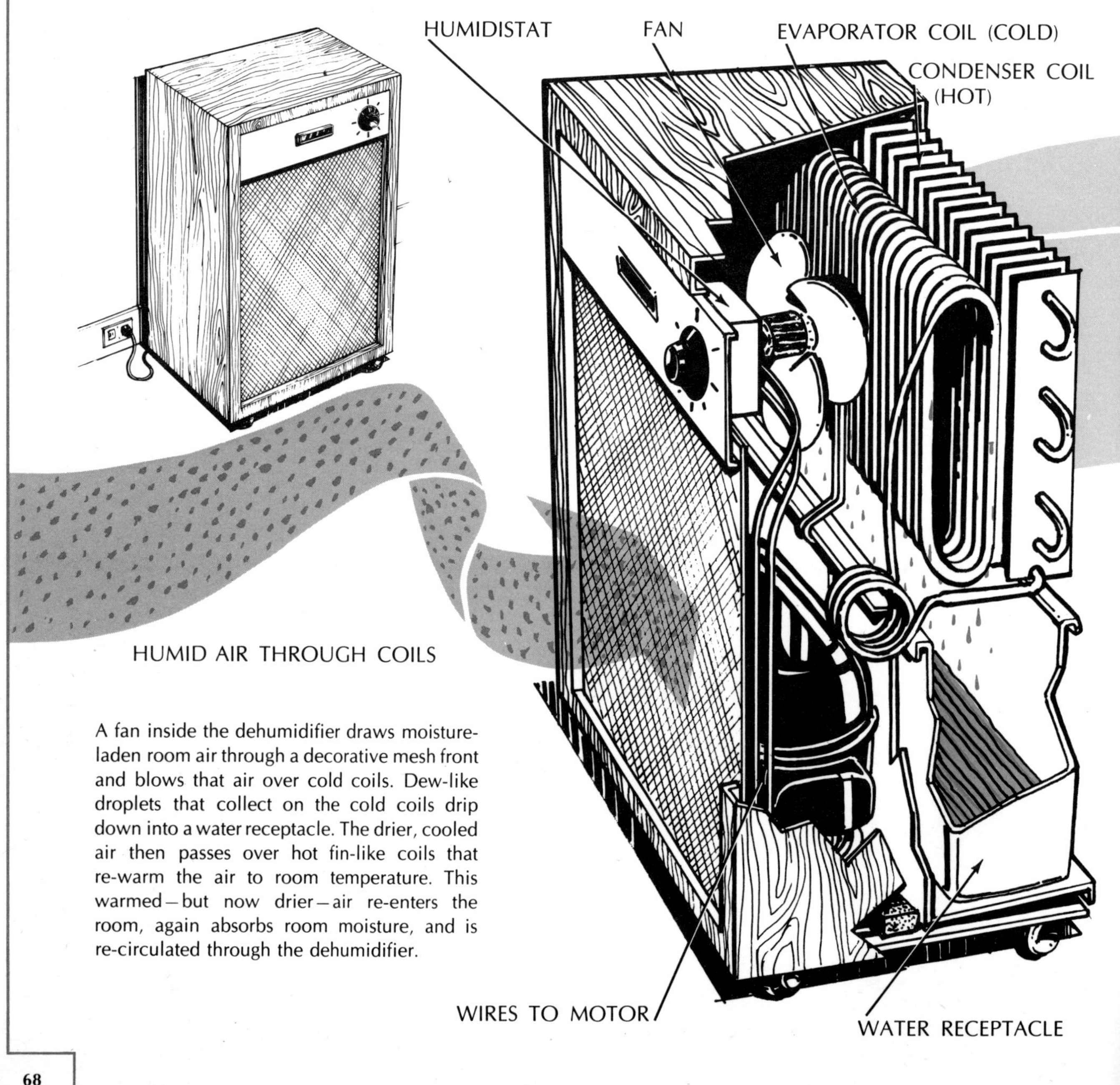

A fan inside the dehumidifier draws moisture-laden room air through a decorative mesh front and blows that air over cold coils. Dew-like droplets that collect on the cold coils drip down into a water receptacle. The drier, cooled air then passes over hot fin-like coils that re-warm the air to room temperature. This warmed—but now drier—air re-enters the room, again absorbs room moisture, and is re-circulated through the dehumidifier.

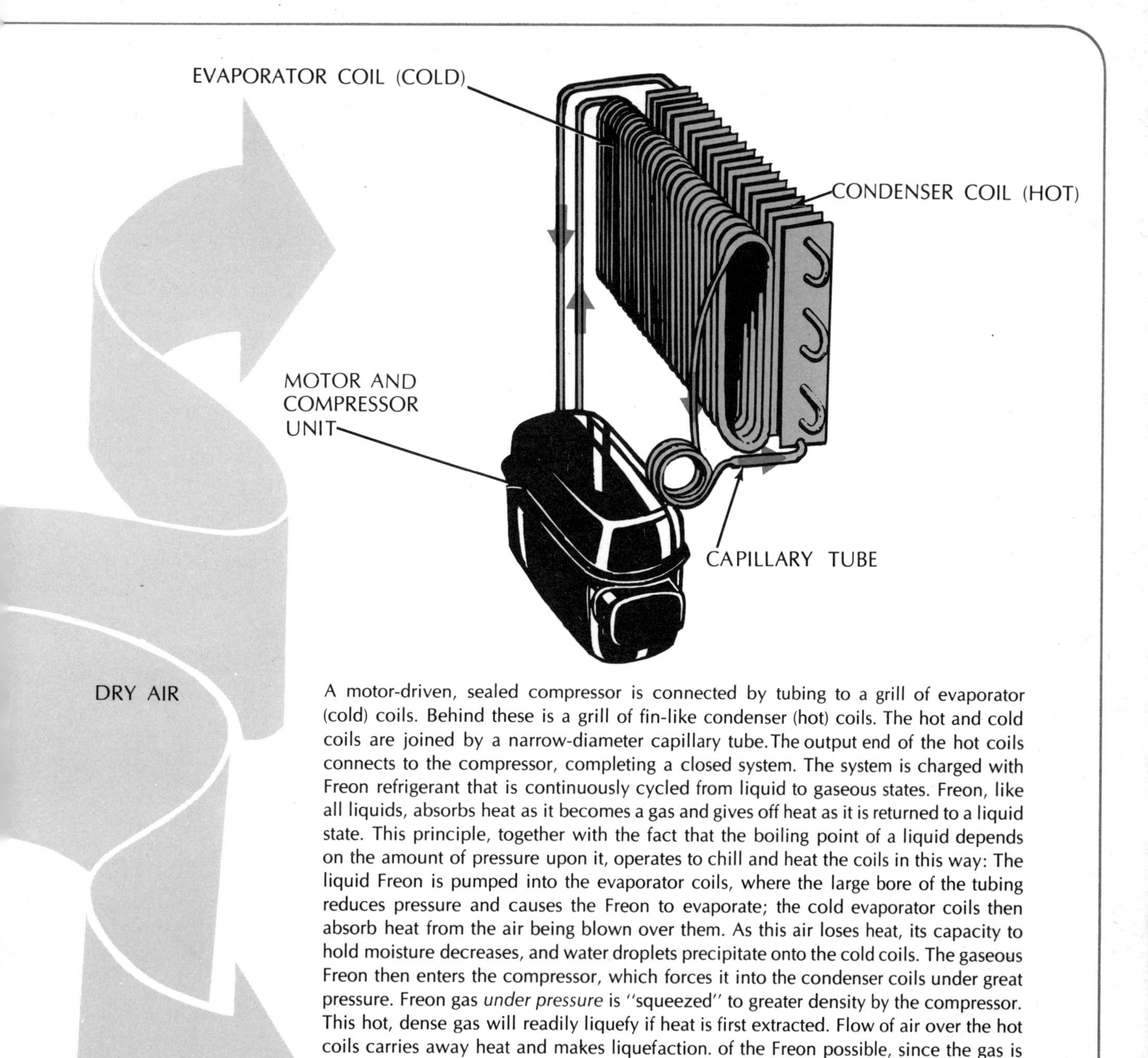

A motor-driven, sealed compressor is connected by tubing to a grill of evaporator (cold) coils. Behind these is a grill of fin-like condenser (hot) coils. The hot and cold coils are joined by a narrow-diameter capillary tube. The output end of the hot coils connects to the compressor, completing a closed system. The system is charged with Freon refrigerant that is continuously cycled from liquid to gaseous states. Freon, like all liquids, absorbs heat as it becomes a gas and gives off heat as it is returned to a liquid state. This principle, together with the fact that the boiling point of a liquid depends on the amount of pressure upon it, operates to chill and heat the coils in this way: The liquid Freon is pumped into the evaporator coils, where the large bore of the tubing reduces pressure and causes the Freon to evaporate; the cold evaporator coils then absorb heat from the air being blown over them. As this air loses heat, its capacity to hold moisture decreases, and water droplets precipitate onto the cold coils. The gaseous Freon then enters the compressor, which forces it into the condenser coils under great pressure. Freon gas *under pressure* is "squeezed" to greater density by the compressor. This hot, dense gas will readily liquefy if heat is first extracted. Flow of air over the hot coils carries away heat and makes liquefaction. of the Freon possible, since the gas is still under pressure.

FILLED-RECEPTACLE SHUT-OFF SWITCH

As the receptacle fills with water, its increasing weight gradually pushes down on the hinged supporting platform. When the receptacle is $\frac{3}{4}$ full, the platform triggers an electrical switch which shuts off the dehumidifier. Some receptacles have a drain connection and a hose that lead to the sewer.

Air Conditioner

A closed room fitted with a window-mounted air conditioner is equivalent to a refrigerator. Like a refrigerator, the air conditioner extracts heat from the air in a closed space and pumps it elsewhere. The heat extraction process is made possible by a refrigerant (usually Freon), which is endlessly cycled from a liquid to a gas and then back to a liquid as it is forced through the air conditioner system. Freon, like all liquids, absorbs heat as it becomes a gas and gives off heat as it returns to a liquid state. This principle, together with the fact that the boiling point of a liquid depends on the amount of pressure upon it, operates to chill and heat the air conditioner's Freon-filled coils—and thus the air circulated over them.

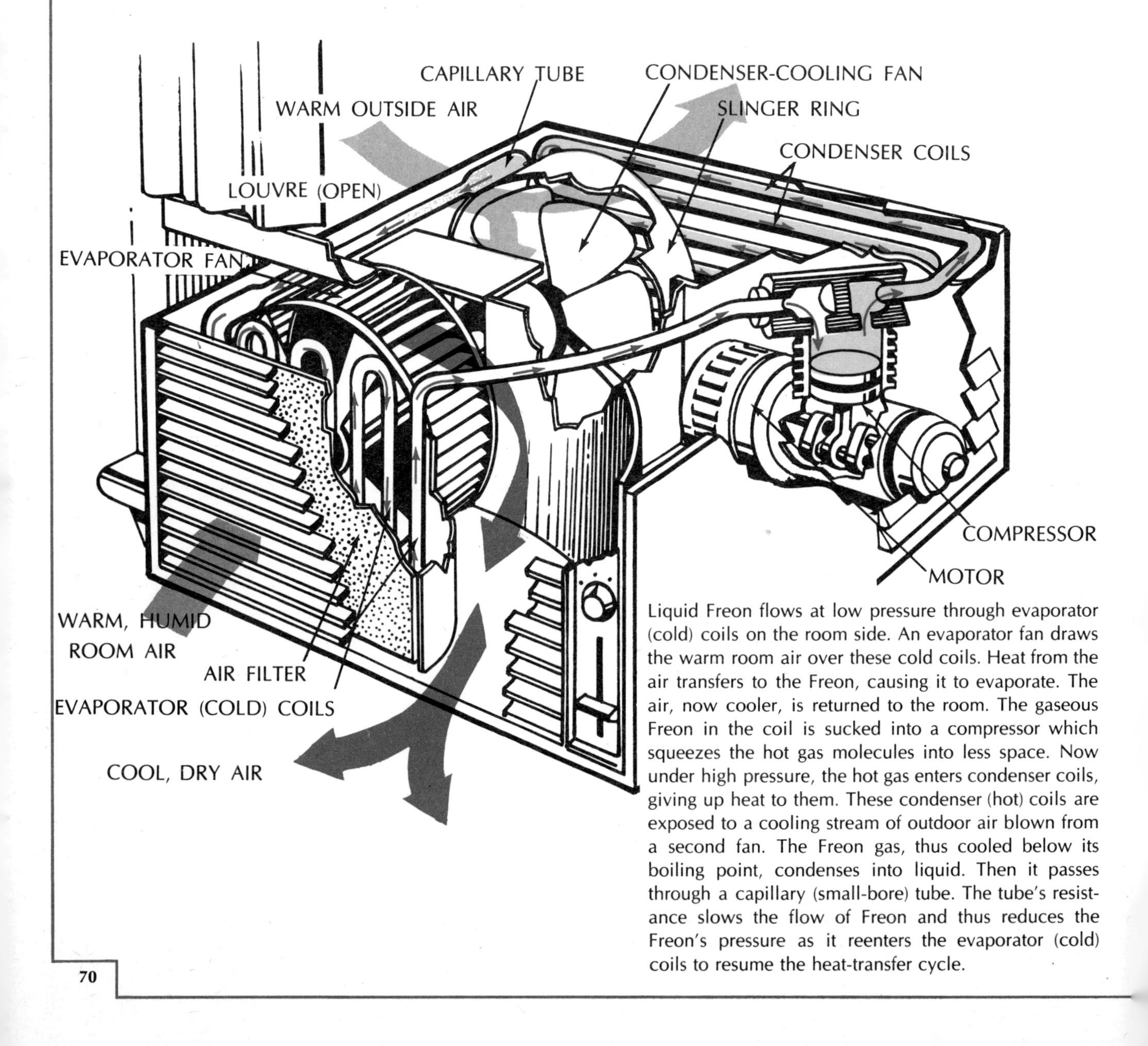

Liquid Freon flows at low pressure through evaporator (cold) coils on the room side. An evaporator fan draws the warm room air over these cold coils. Heat from the air transfers to the Freon, causing it to evaporate. The air, now cooler, is returned to the room. The gaseous Freon in the coil is sucked into a compressor which squeezes the hot gas molecules into less space. Now under high pressure, the hot gas enters condenser coils, giving up heat to them. These condenser (hot) coils are exposed to a cooling stream of outdoor air blown from a second fan. The Freon gas, thus cooled below its boiling point, condenses into liquid. Then it passes through a capillary (small-bore) tube. The tube's resistance slows the flow of Freon and thus reduces the Freon's pressure as it reenters the evaporator (cold) coils to resume the heat-transfer cycle.

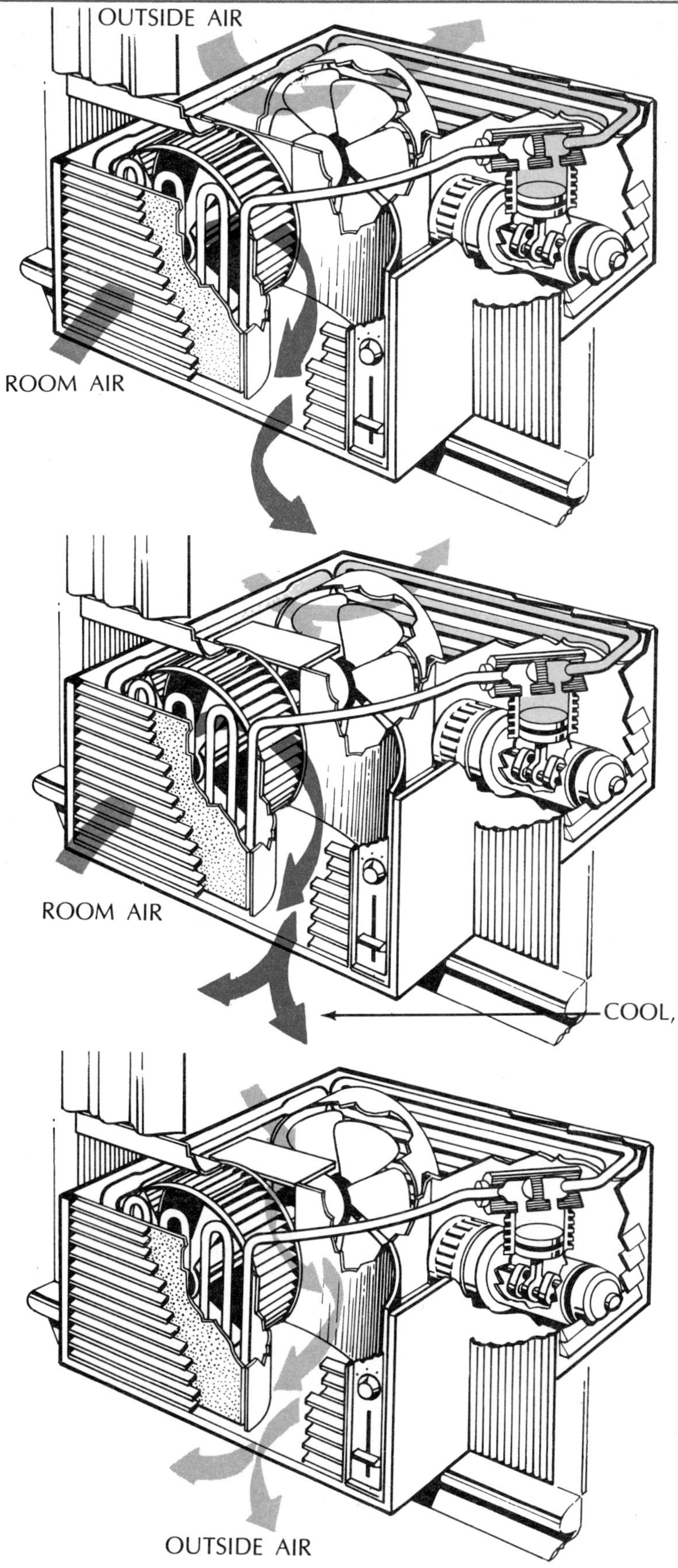

The controls of modern air conditioners allow several modes of operation. As shown at left, the conditioner can be set to cool, clean, and dehumidify room air. In this mode, the evaporator fan draws room air through a filter and over evaporator (cold) coils. As the air cools, it gives off moisture in the form of droplets that condense on the coils and drip into a pan below. The fan then returns this cooler, drier air into the room.

In the second mode, the evaporator fan draws fresh outside air through the outside louvre and mixes the outside air with cool and dry (conditioned) room air.

In the third mode, the compressor is turned off. The evaporator fan draws fresh air directly into the room without cooling it. In some designs, incoming air passes through an air filter that removes most dust particles.

Garbage Compactor

Communities across America are fast running out of land on which they can dump wastes. The garbage compactor cannot reduce the amount of trash produced—but it can reduce the amount of space trash occupies. The average American disposes of some five-and-a-half pounds of boxes, wrappings, cans, bottles and paper each day, and over ninety percent of the space taken up by these discards is air. By compacting the material with a one-ton force and thereby squeezing out the air pockets, the compactor can cram a week's trash—enough to fill three 20-gallon cans—into a bag that is only 16-by-16-by-9 inches in size.

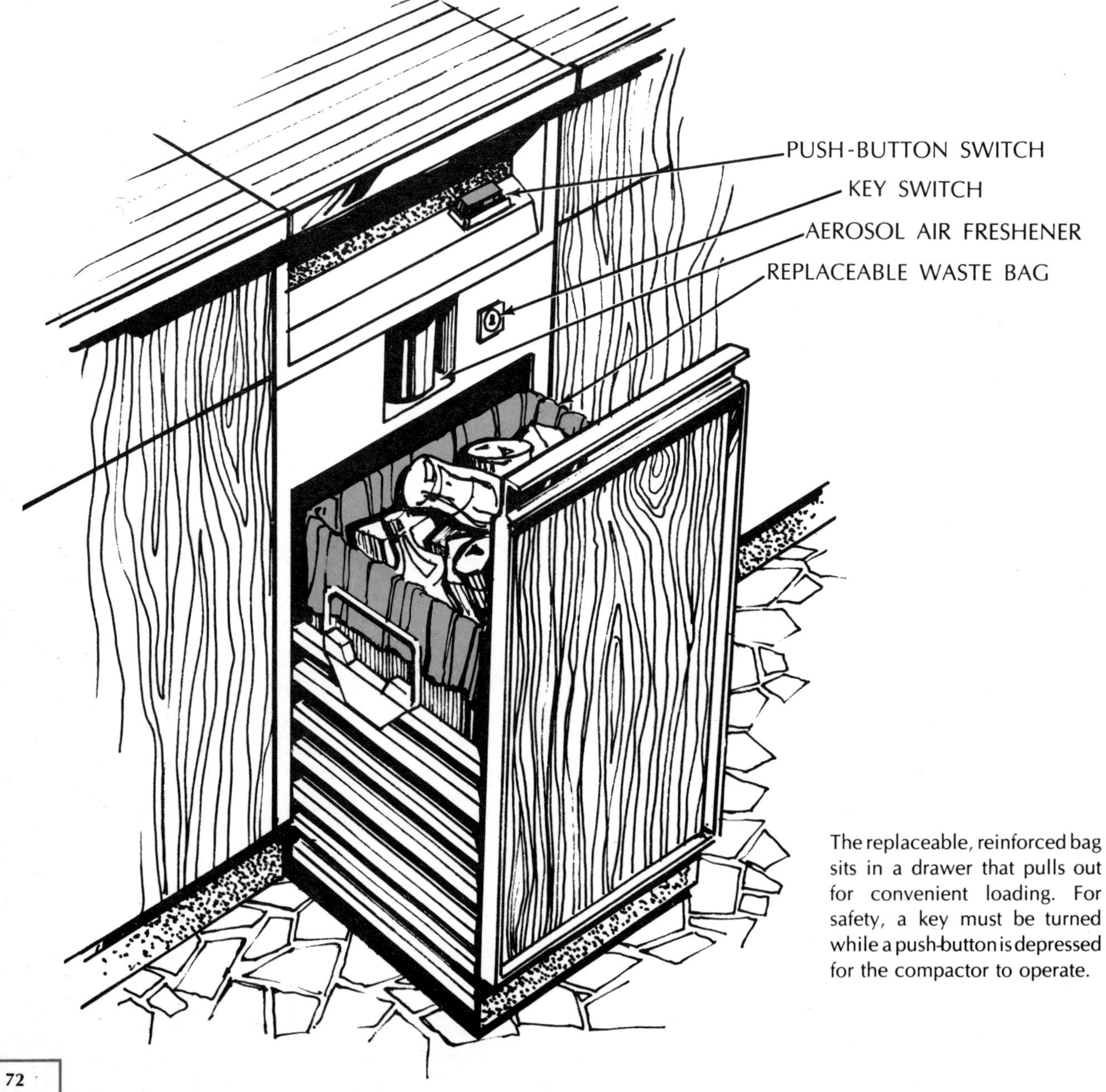

The replaceable, reinforced bag sits in a drawer that pulls out for convenient loading. For safety, a key must be turned while a push-button is depressed for the compactor to operate.

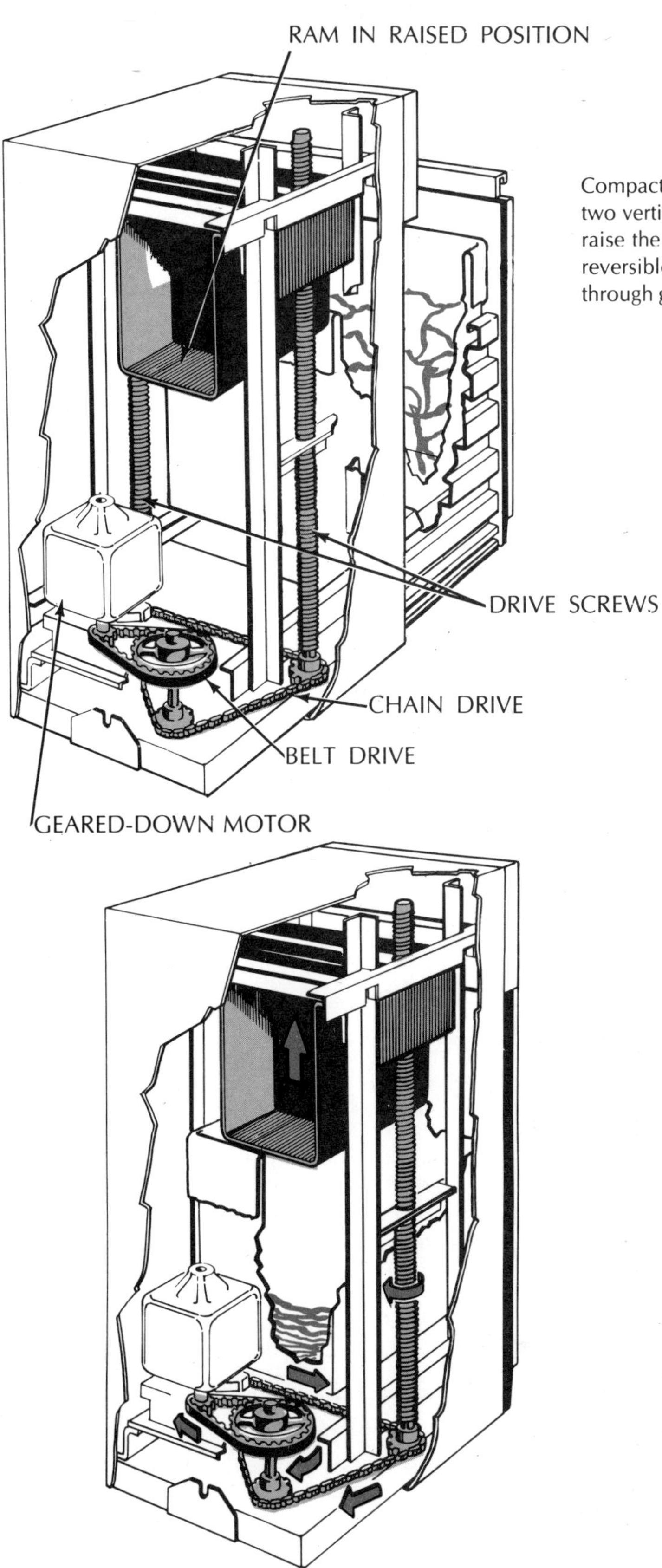

Compaction is done by a steel ram mounted on two vertical drive screws which turn to lower and raise the ram. Power comes from a geared-down, reversible motor and is transmitted to the screws through gears, a belt, and a chain.

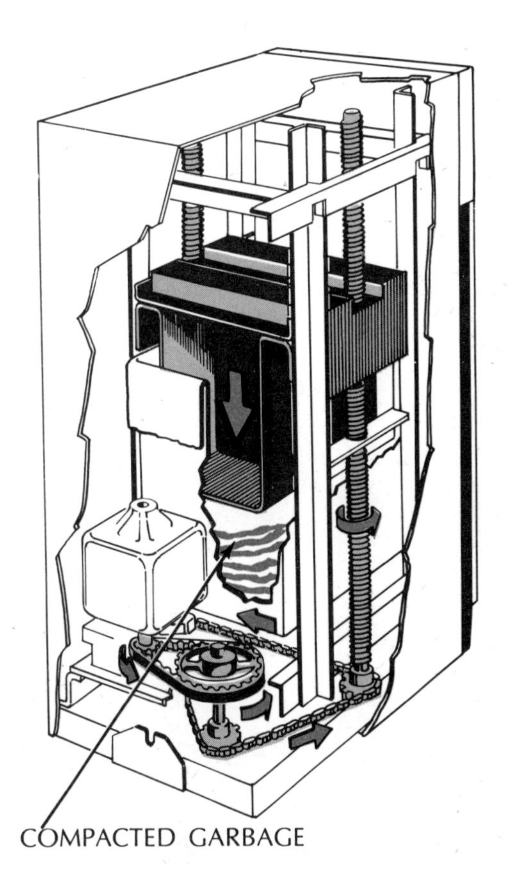

When the compactor is turned on (above), the motor turns the drive screws in a counterclockwise direction. This lowers the ram. When the ram has descended far enough to crush the trash, it trips a switch that reverses the motor. The screws now turn clockwise (left), raising the ram. When the ram reaches the top of the unit, it trips another switch that turns the compactor off. Time of cycle: less than 30 seconds.

Garbage Disposer

A garbage disposer is a handy household device for getting rid of degradable food waste. But too often, it is recklessly fouled with trash, bones, celery, paper, string, cigarette filters, and food wrappings—all recipes for a jam-up. The disposer can save time and labor in the kitchen, but it provides another important advantage: Its finely ground garbage doesn't present a health hazard once community sewer systems have provided complete treatment. The alternative, collecting organic waste at a garbage dump, can cause problems, for dumps are breeding grounds for vermin and all that goes with them.

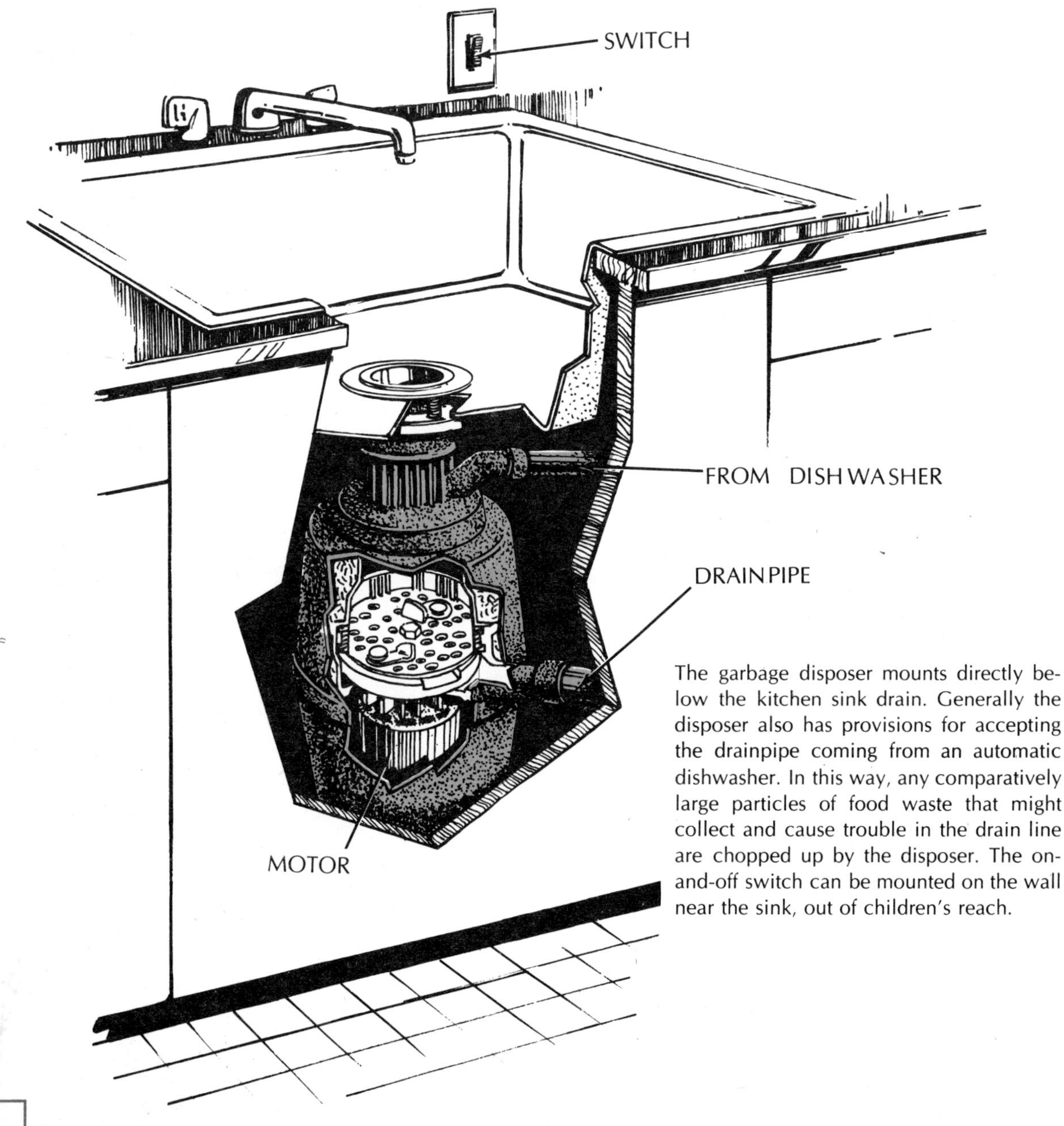

The garbage disposer mounts directly below the kitchen sink drain. Generally the disposer also has provisions for accepting the drainpipe coming from an automatic dishwasher. In this way, any comparatively large particles of food waste that might collect and cause trouble in the drain line are chopped up by the disposer. The on-and-off switch can be mounted on the wall near the sink, out of children's reach.

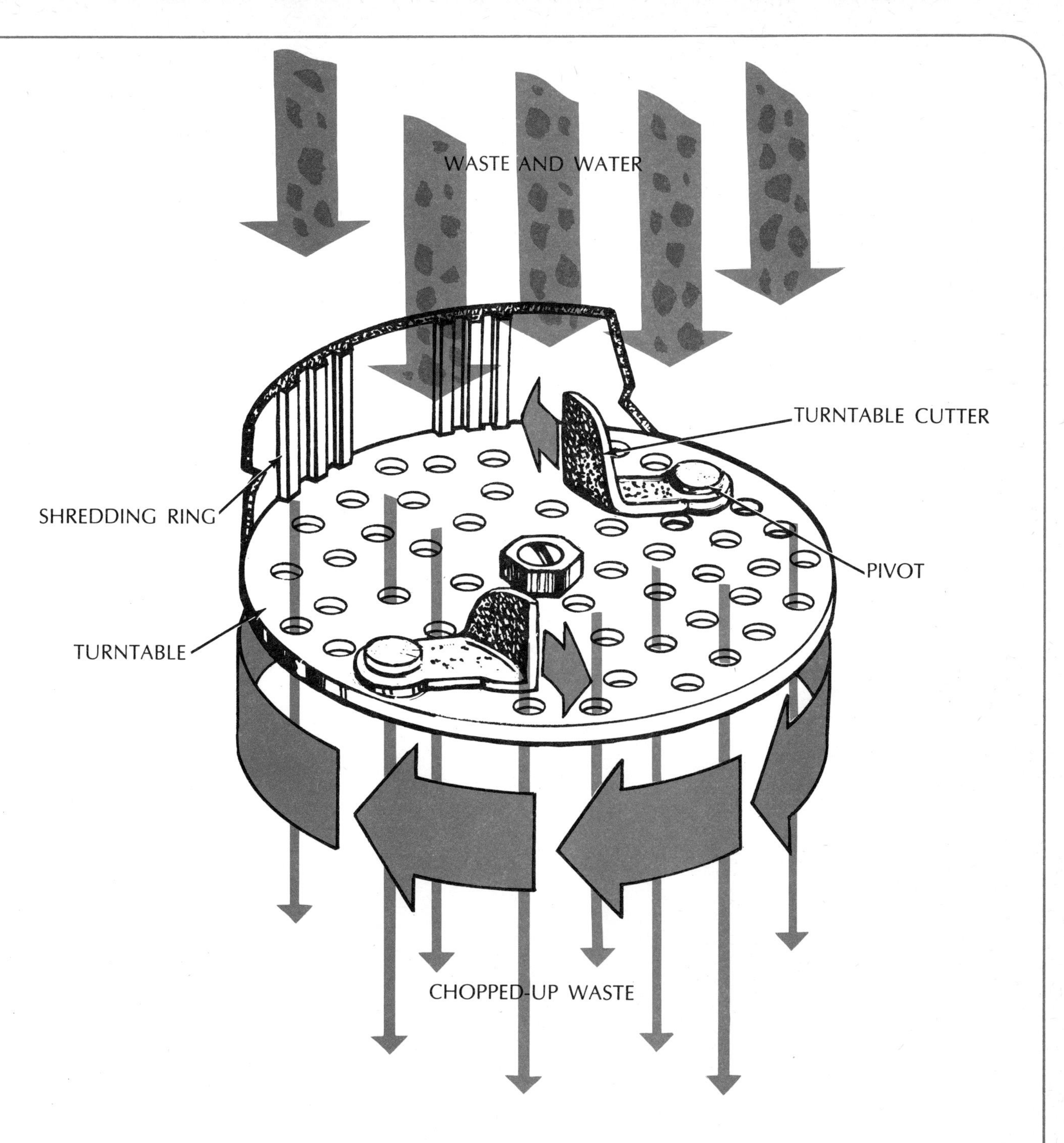

Leftover food dropped into the drain falls down onto a turntable fitted with two cutters pivoted to prevent jam-ups. Until the turntable comes up to full speed, the cutters move inward when they encounter an obstruction. As the $\frac{1}{3}$- to $\frac{1}{2}$-horsepower motor gains speed, centrifugal force hurls the cutters outward, and the food waste is caught and cut up between the cutters and the stationary shredding ring. While the disposer operates, the sink tap is kept open so that a stream of water flows through the disposer, washing finely chopped food waste down the drainpipe.

Sump Pump

A basement drainage system or sump pump is vital insurance against a flooded basement. Such a pump, set into a pit some two feet below the level of the basement floor, starts automatically when the water level in the sump rises. Some pumps have the small electric motor that powers the pumping apparatus positioned above floor level. The type shown here is less bulky, as the entire mechanism is hidden within the sump and is out of the way under a wooden cover. The motor, which runs on house current, is completely submersible, being encased in a waterproof cast-iron or bronze housing. Although small in size, such pumps can move some 2,700 gallons per hour.

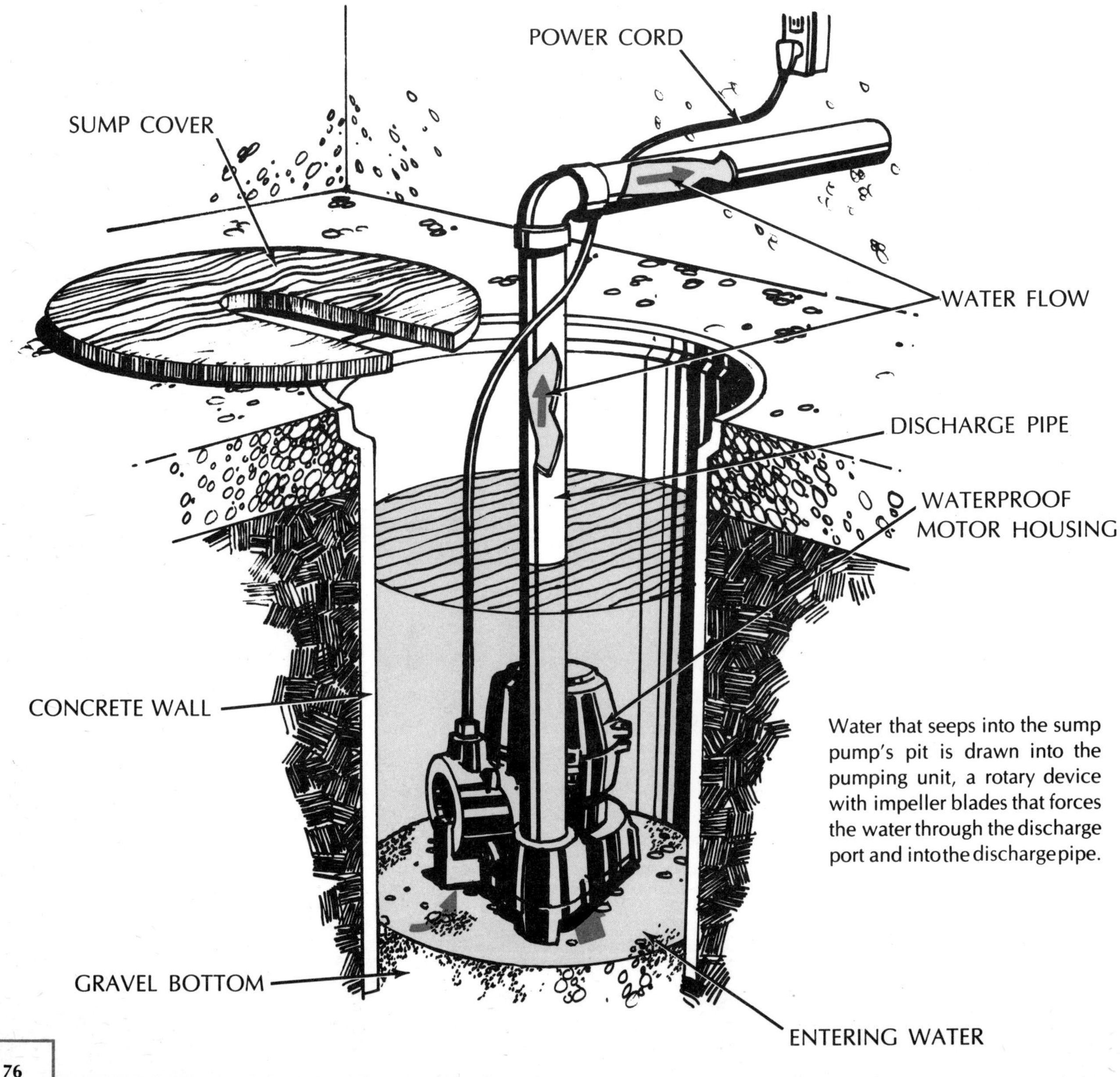

Water that seeps into the sump pump's pit is drawn into the pumping unit, a rotary device with impeller blades that forces the water through the discharge port and into the discharge pipe.

Sump pumps are switched on automatically when the water level rises above a given point. In the type illustrated the rising water enters a rectangular opening in the bottom of the switch housing. This increases the pressure on the air trapped within the housing. The air presses on the diaphragm (black line), bends it inward and pushes closed the snap-action switch, which turns on the pump's electric motor.

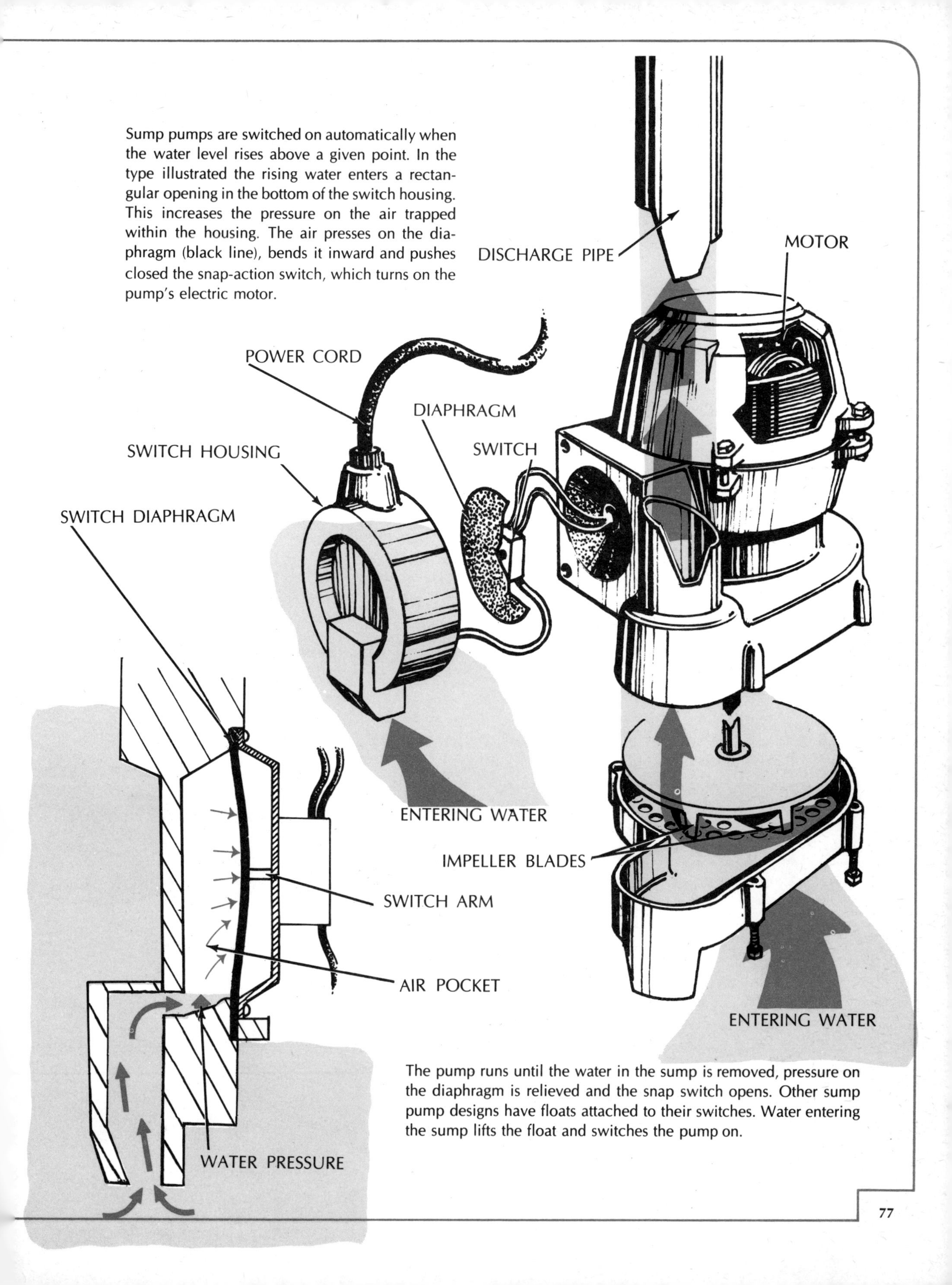

The pump runs until the water in the sump is removed, pressure on the diaphragm is relieved and the snap switch opens. Other sump pump designs have floats attached to their switches. Water entering the sump lifts the float and switches the pump on.

Truck Scale

The platform scales that weigh trucks work on the same principle as the personal scales in a medical examining room. Both have levers that divide the weight, balancing many pounds on the platform with one pound at the indicator mechanism. A 50-ton truck scale, like the one illustrated here, has four pairs of main levers, so linked that 7,000 pounds on the platform exerts a one-pound force at the indicator. The scale used by a medical doctor, with one pair of levers, has a 40 to 1 "tip ratio."

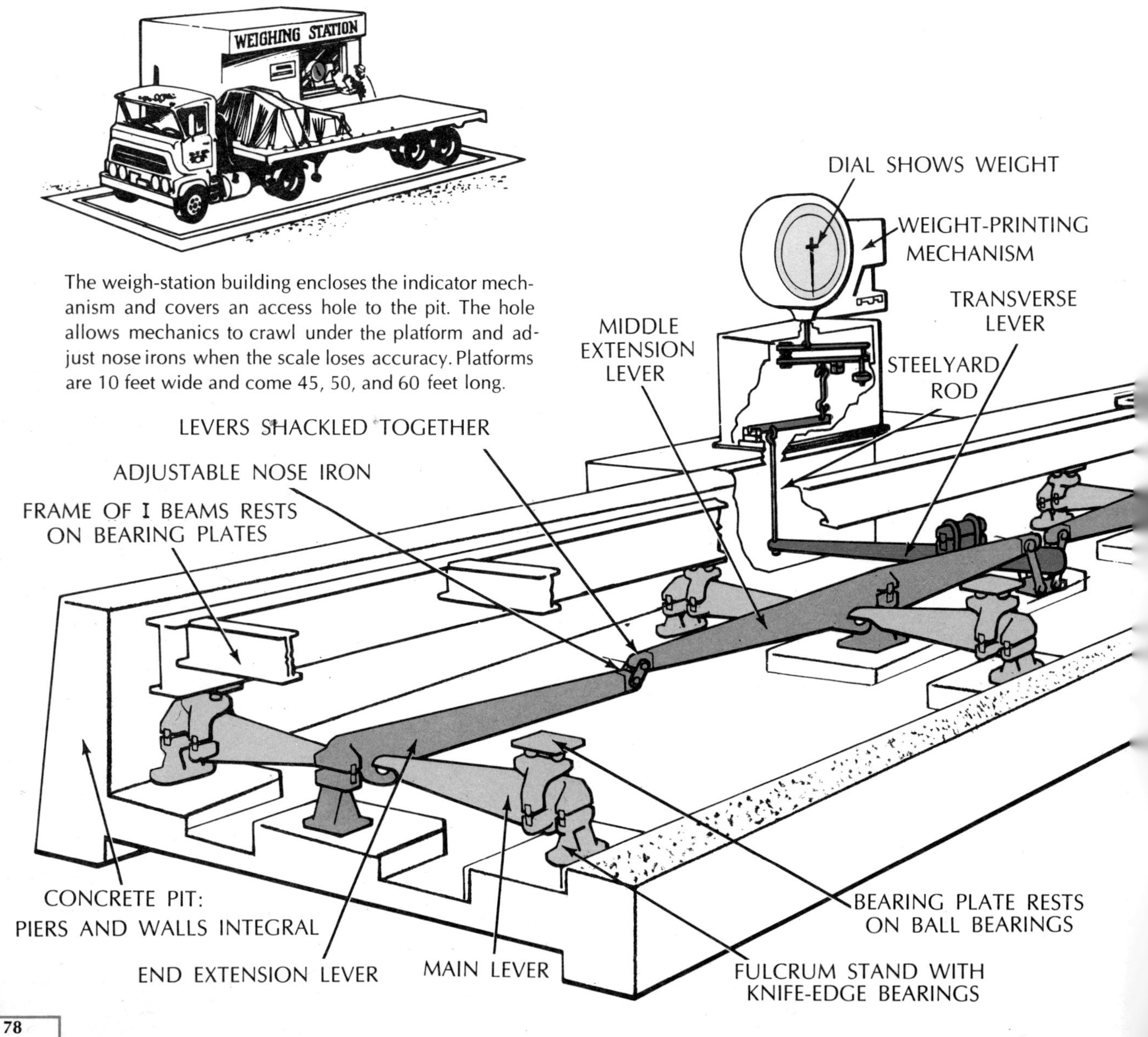

The weigh-station building encloses the indicator mechanism and covers an access hole to the pit. The hole allows mechanics to crawl under the platform and adjust nose irons when the scale loses accuracy. Platforms are 10 feet wide and come 45, 50, and 60 feet long.

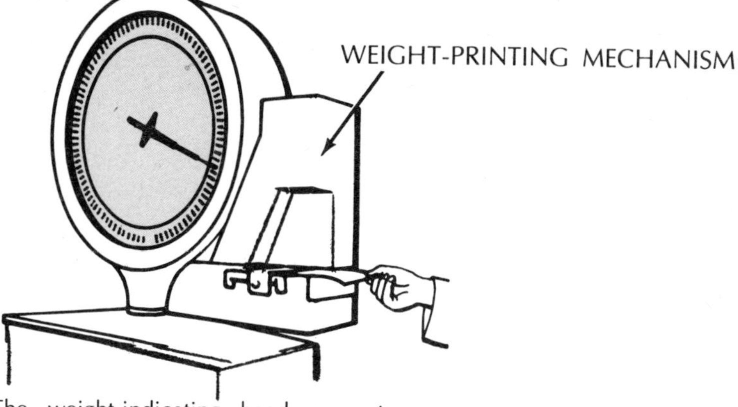

The weight-indicating head converts a downward pull on the steelyard rod into a movement of the dial needle. Weight is read from a dial numbered at 10-pound intervals. The model shown here has an attachment that also prints the weight on a ticket. The printer, usually motor-driven, works off the dial through a system of cams and levers. The printer can be equipped to put the time, date, and an identifying number on the weight ticket, too.

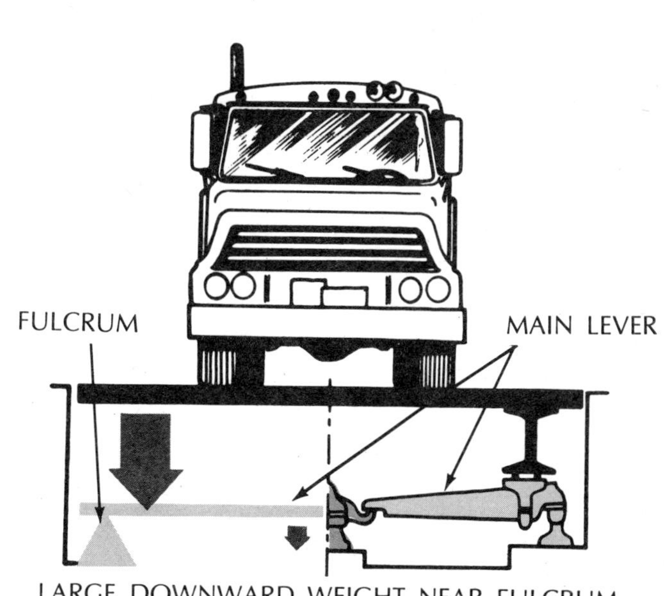

LARGE DOWNWARD WEIGHT NEAR FULCRUM CAUSES SMALL DOWNWARD PUSH AT FAR END

Weight on the platform is passed through the frame to eight main levers. One end of each lever pivots on a fulcrum stand; the other bears down on an extension lever. Applied close to the fulcrum, most of the weight goes to the stand. Only a little weight goes on to the extension lever.

REINFORCED-CONCRETE DECK

PUSH NEAR FULCRUM HAS DIMINISHED EFFECT AT ENDS

The four extension levers, tipped by main levers resting on them, apply a combined upward pull on the transverse lever. This lever pivots, pulling down the steelyard rod. The rod then actuates the indicator head. In the course of being shifted from platform to steelyard rod, the weight is divided at each step by the levers. Most of the weight is absorbed by the lever stands, and only a small part is passed on. At the indicator head, only one pound for every 7,000 pounds on the platform is left to move the dial's needle to provide the reading.

FULCRUM

MIDDLE EXTENSION LEVER

Automatic Ice-maker

In bygone days, ice-making was a tough winter chore. Skilled sawyers cut blocks of ice from frozen-over ponds and lakes. Sledged to an ice house, the blocks were insulated with layers of sawdust and burlap that would preserve the ice until the next winter. In the cities, "ice men" lugged ice blocks to their customers. Finally, refrigerator-freezers cost the sawyers and the ice men their jobs. But even with modern freezers, ice-making often means grappling with frozen trays, mopping up spattered ice melt, and balancing trays full of surging water on the way to the freezer. Today, the automatic ice-maker provides the ultimate in convenience. It installs inside the freezer and occupies about the same space as two regular ice trays.

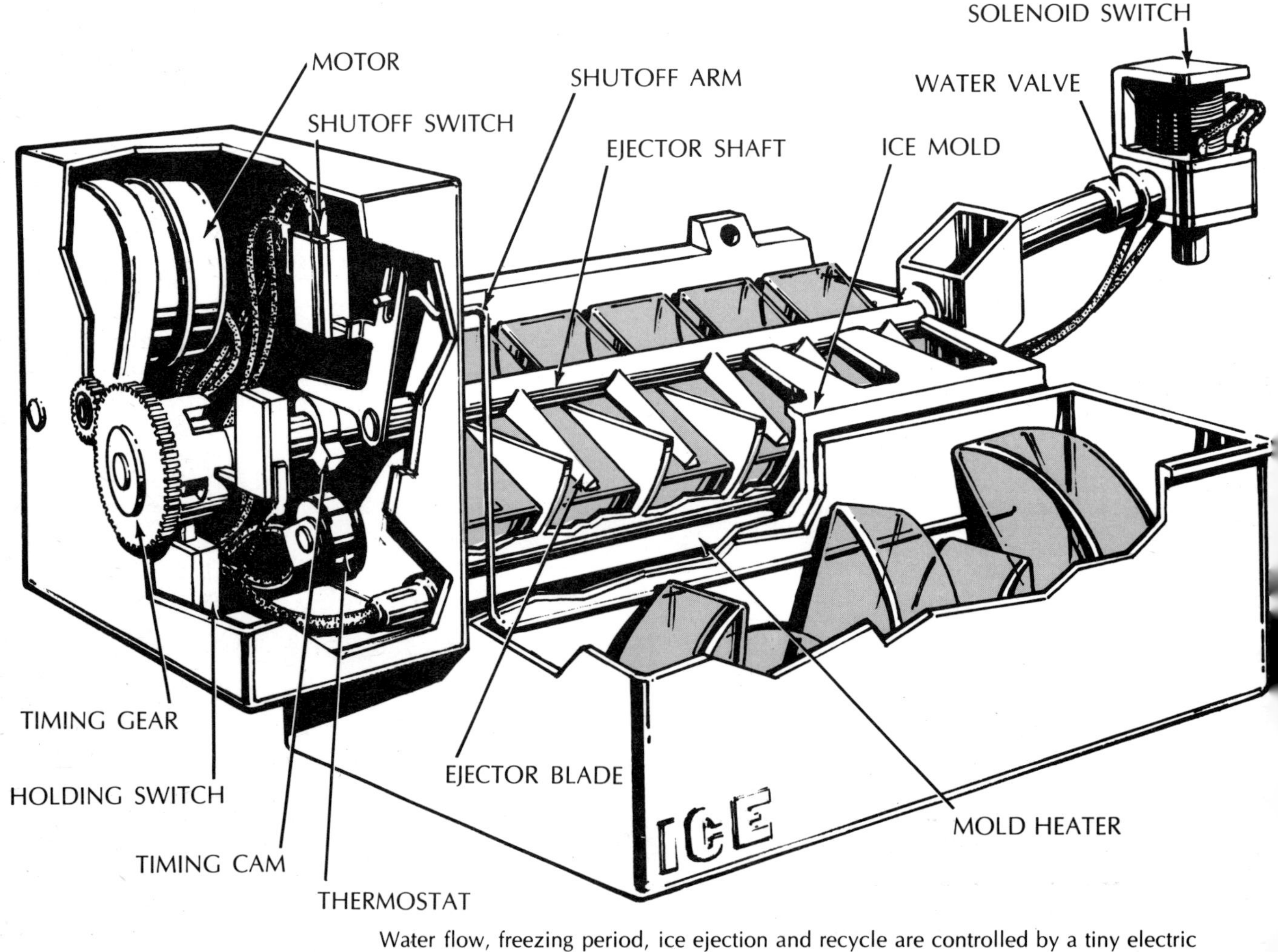

Water flow, freezing period, ice ejection and recycle are controlled by a tiny electric motor that slowly turns a timing gear installed on an ejector shaft that runs through the ice mold. The solenoid switch controls the water-flow cycle. The holding switch holds power applied to the motor during ice ejection. The shutoff switch removes power when the ice bin is full.

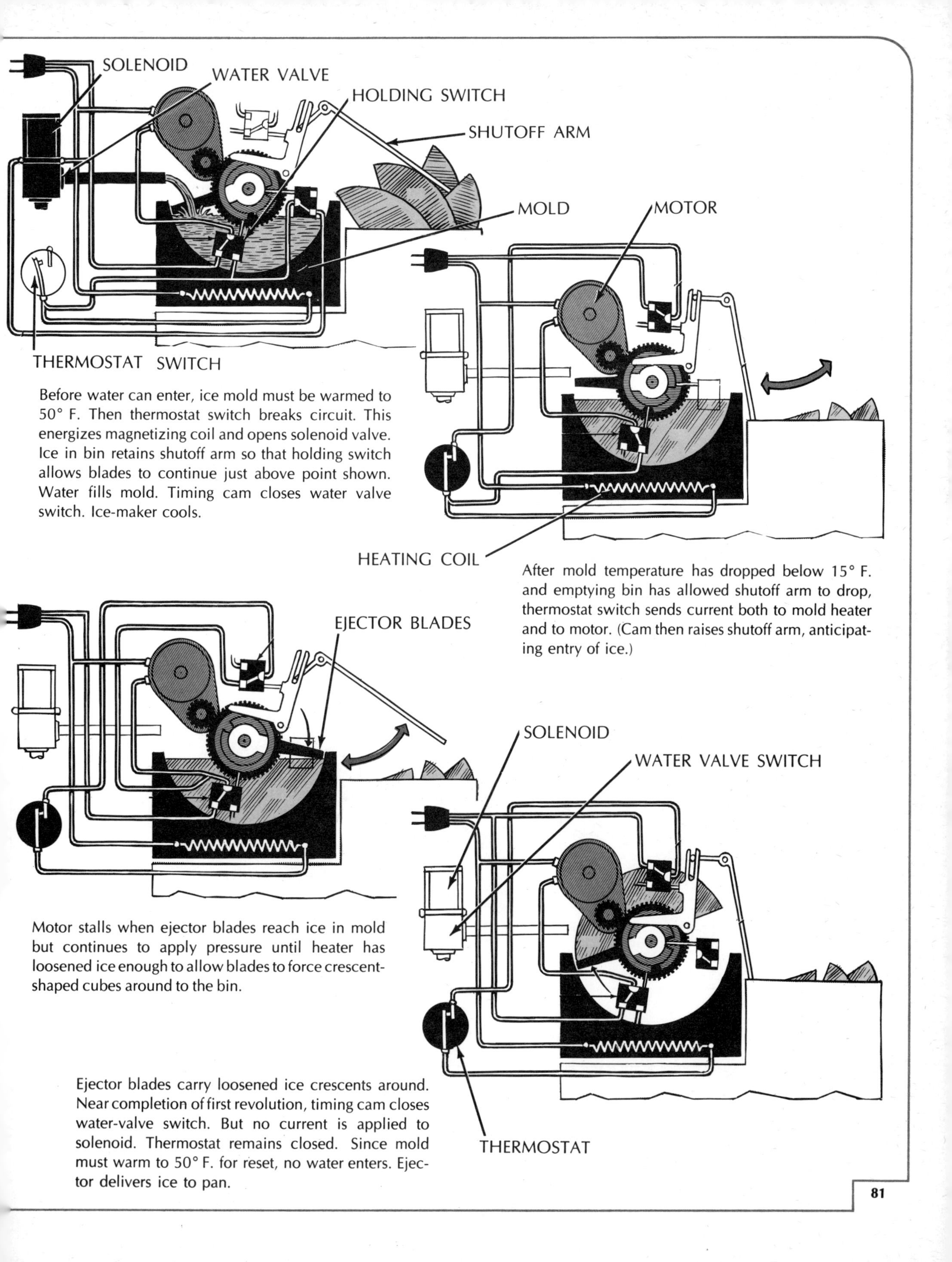

Before water can enter, ice mold must be warmed to 50° F. Then thermostat switch breaks circuit. This energizes magnetizing coil and opens solenoid valve. Ice in bin retains shutoff arm so that holding switch allows blades to continue just above point shown. Water fills mold. Timing cam closes water valve switch. Ice-maker cools.

After mold temperature has dropped below 15° F. and emptying bin has allowed shutoff arm to drop, thermostat switch sends current both to mold heater and to motor. (Cam then raises shutoff arm, anticipating entry of ice.)

Motor stalls when ejector blades reach ice in mold but continues to apply pressure until heater has loosened ice enough to allow blades to force crescent-shaped cubes around to the bin.

Ejector blades carry loosened ice crescents around. Near completion of first revolution, timing cam closes water-valve switch. But no current is applied to solenoid. Thermostat remains closed. Since mold must warm to 50° F. for reset, no water enters. Ejector delivers ice to pan.

Room Humidifier

Governed by a humidistat that senses the air's moisture content, the room humidifier maintains desired levels of room humidity. In principle, the humidifier merely supplies an airstream for evaporation of water and for expulsion of the vaporized water into the room air. The humidifier's cabinet is a single piece of moulded plastic that has a simulated wood exterior and a plain interior that serves mainly as a water reservoir. A large fan fitted to the rear wall of the cabinet draws room air through a filter screen and forces the air through a water-logged rotating belt made of flexible plastic foam. The airstream absorbs moisture from the belt and emerges through a lid vent into the room. Warm air currents then circulate the newly humidified air throughout the home.

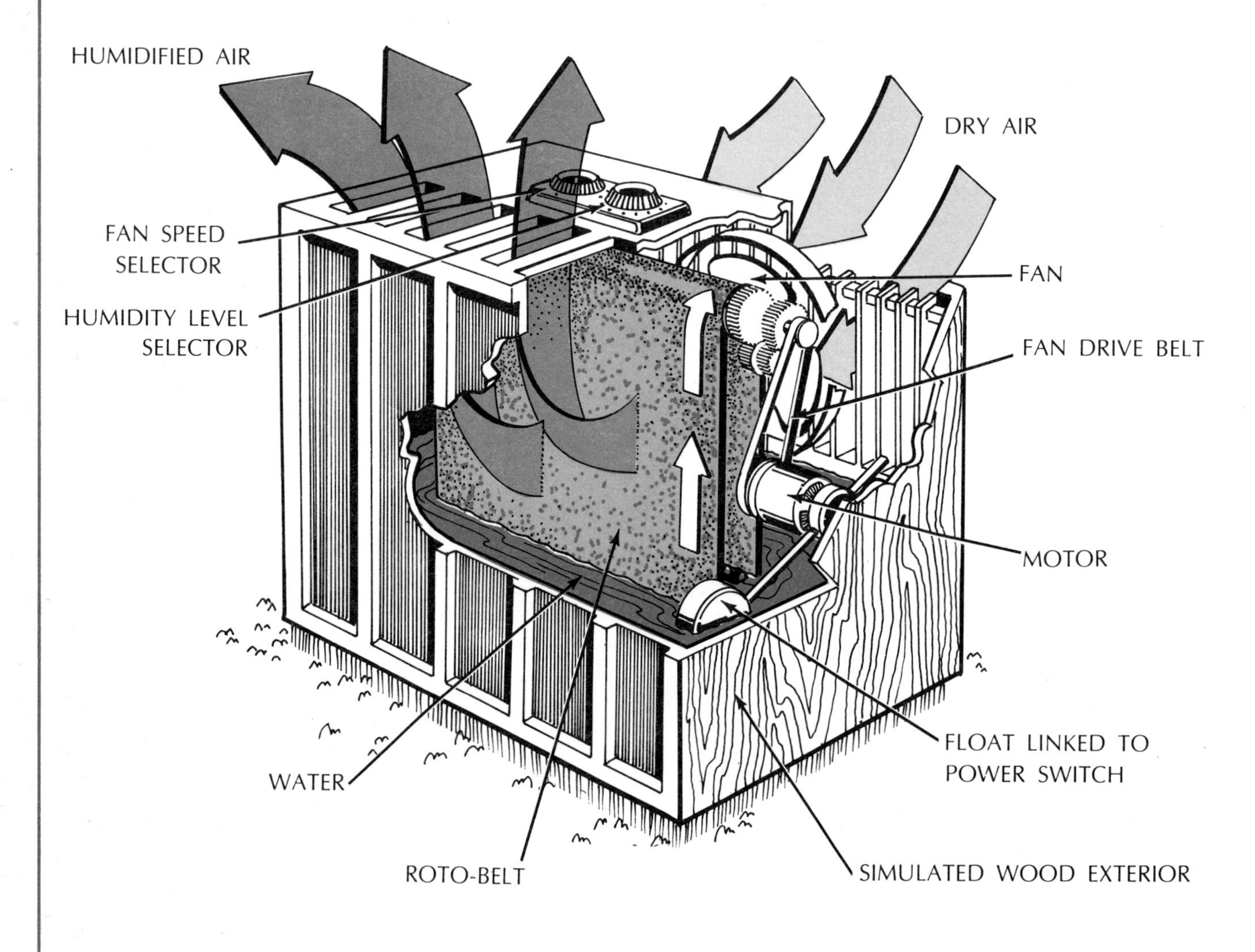

Deluxe models like the one shown employ an electronic speed control for the fan. A solid-state switch, connected in series with the ac line and the fan motor, is opened and closed by alternating current in the power line. The speed control setting determines the current flow to the motor and, thus, the speed of fan.

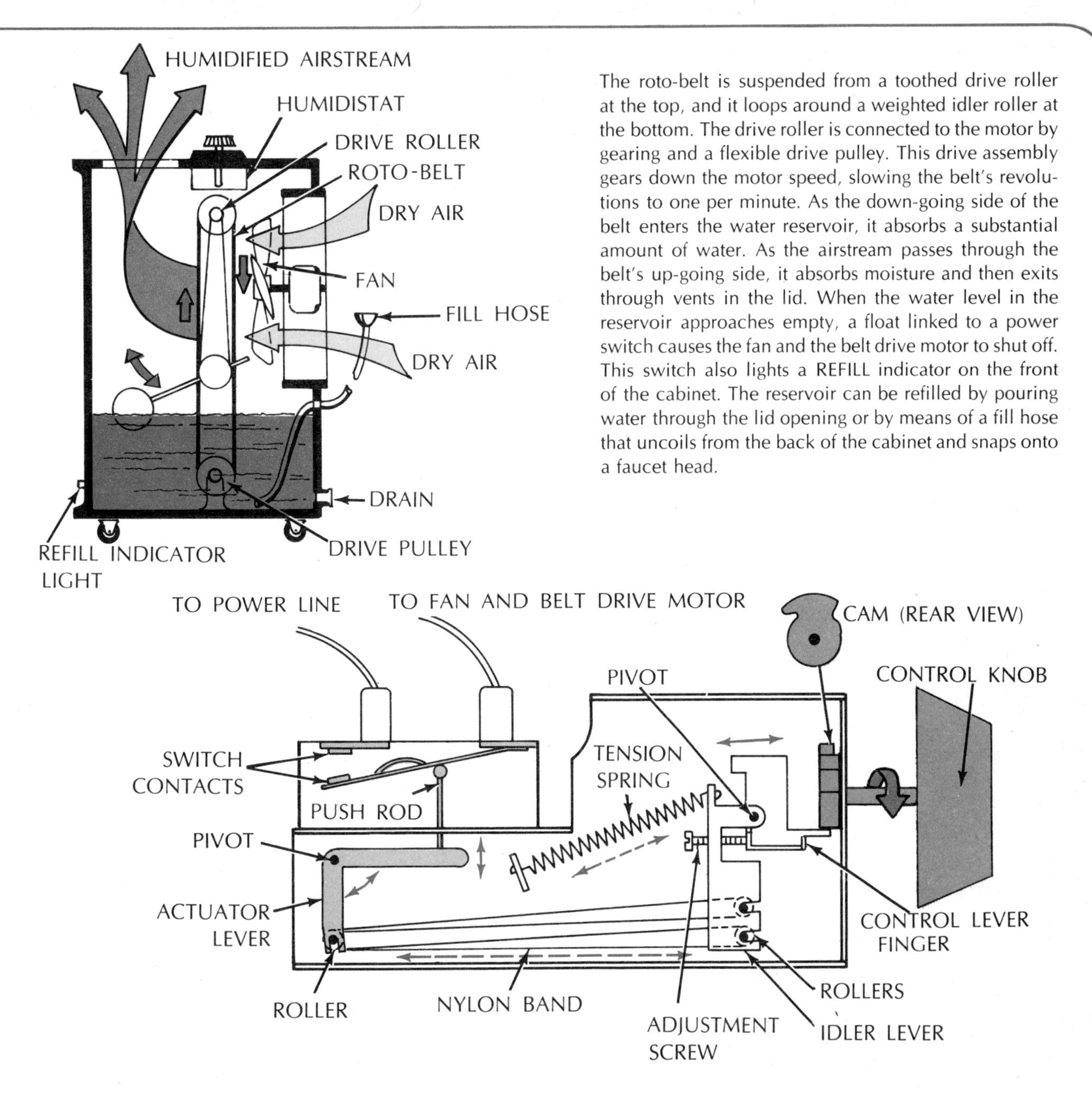

The roto-belt is suspended from a toothed drive roller at the top, and it loops around a weighted idler roller at the bottom. The drive roller is connected to the motor by gearing and a flexible drive pulley. This drive assembly gears down the motor speed, slowing the belt's revolutions to one per minute. As the down-going side of the belt enters the water reservoir, it absorbs a substantial amount of water. As the airstream passes through the belt's up-going side, it absorbs moisture and then exits through vents in the lid. When the water level in the reservoir approaches empty, a float linked to a power switch causes the fan and the belt drive motor to shut off. This switch also lights a REFILL indicator on the front of the cabinet. The reservoir can be refilled by pouring water through the lid opening or by means of a fill hose that uncoils from the back of the cabinet and snaps onto a faucet head.

The humidistat controls the functioning of the fan and the belt drive motor by opening and closing electrical contacts in response to humidity changes. The humidistat's sensing device is an inch-wide nylon band which lengthens when damp and contracts when dry. The band fits around three rollers. Its midpoint passes over a roller attached to an actuator lever. Band ends pass over two rollers attached to an idler lever. Selecting a desired humidity level with the control knob (shown right) determines the amount of tension on the band. When room humidity is above the preset minimum, the band remains slack. But as room humidity drops, the band gives up moisture to the arid air and contracts. Since the idler lever submits to balancing tension from a spring and from the contracting band, the contracting band now draws in the lower arm of the actuator lever. This swings the free arm of the actuator upward, forcing a pushrod to close switch contacts. Power then flows to the fan and the belt drive motor. When room humidity reaches the desired minimum level, the band absorbs moisture and slackens, allowing a switch contact arm to break the contact and cut off power.

Furnace Humidifier

With the onset of winter, windows and doors are closed tight and the thermostat is turned up. Since air heated to 72°F. will hold four times as much moisture as it will at 32°F., heating the air increases its capacity to hold moisture; in effect, the air becomes drier. Thus, as the air is heated its "relative" humidity decreases. This relatively drier air absorbs moisture like a sponge and causes withered plants, loosened furniture joints, static electric shocks, and dry-feeling noses, throats and skin. A furnace humidifier can correct unpleasant dry air conditions by supplying moisture for comfortable humidity levels.

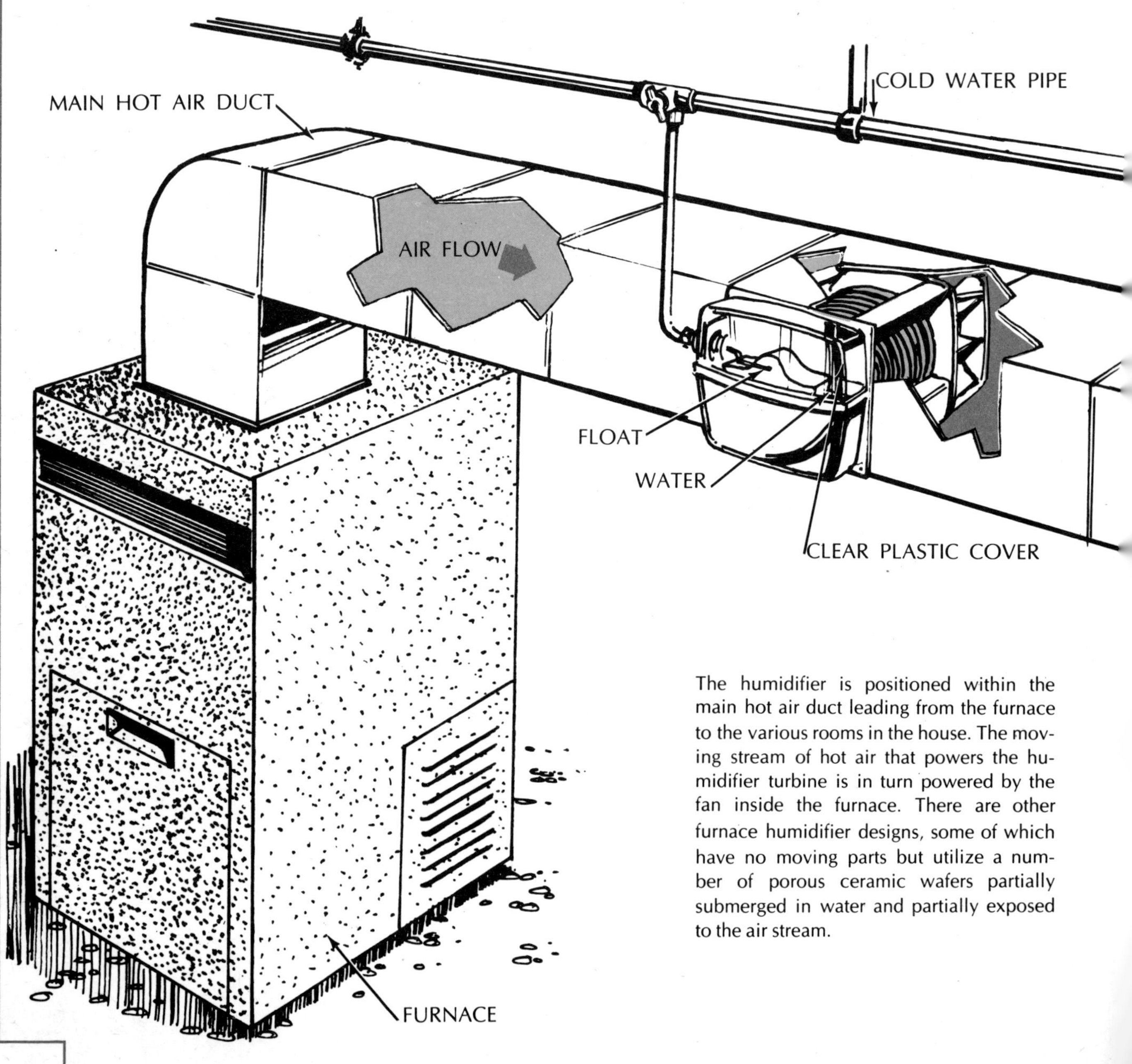

The humidifier is positioned within the main hot air duct leading from the furnace to the various rooms in the house. The moving stream of hot air that powers the humidifier turbine is in turn powered by the fan inside the furnace. There are other furnace humidifier designs, some of which have no moving parts but utilize a number of porous ceramic wafers partially submerged in water and partially exposed to the air stream.

By far the most popular kind of humidifier consists of a simple air turbine mounted on a shaft along with a number of metal mesh discs, which are partially immersed in a pan filled with water. Moving air from the furnace strikes the turbine causing it and the attached discs to rotate. In this way the hot, dry air coming out of the furnace is continuously exposed to a large water surface. When the furnace is at rest, no air flows up through the duct, and the turbine does not rotate. The level of the water within the pan is automatically maintained by a float and valve system, which in the design illustrated is visible through a clear plastic cover. Water is supplied by a pipe permanently connected to the cold water line. The unit needs no attention, other than periodic removal and cleaning of the metal discs.

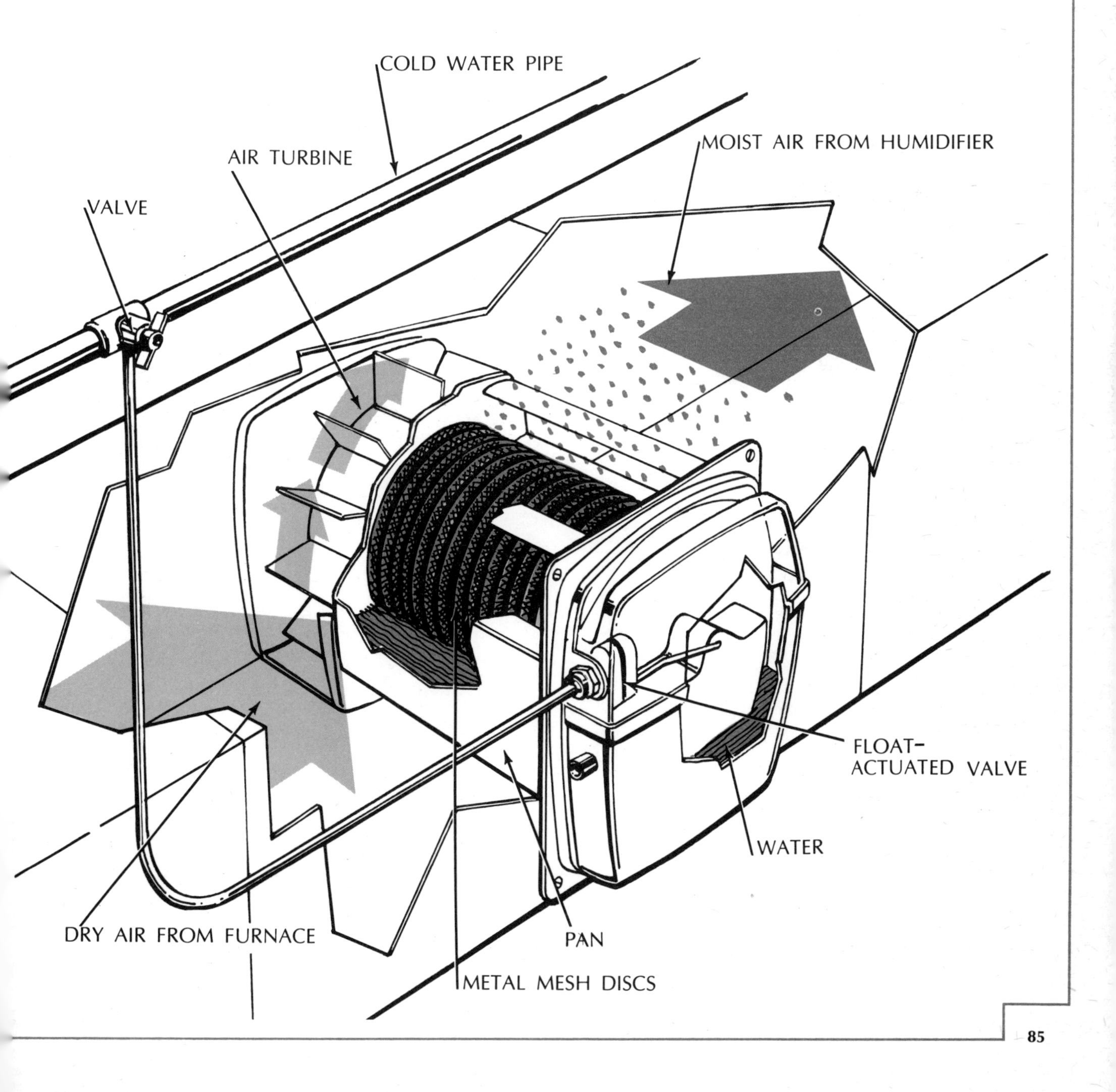

Electronic Air Cleaner

The average person every day passes about 45 pounds of air through his lungs. This air is often filled with soot, dust, pollen, and odors. With an electronic air cleaner, the quality of indoor air can be dramatically upgraded. Such a unit traps about 95 percent of all contaminants, including particles so small that 25,000 of them would fit into the eye of a needle. Since ordinary mesh filters in furnaces and air conditioners remove only about eight percent of airborne debris, the electronic cleaner is a boon, especially for persons who suffer from respiratory problems.

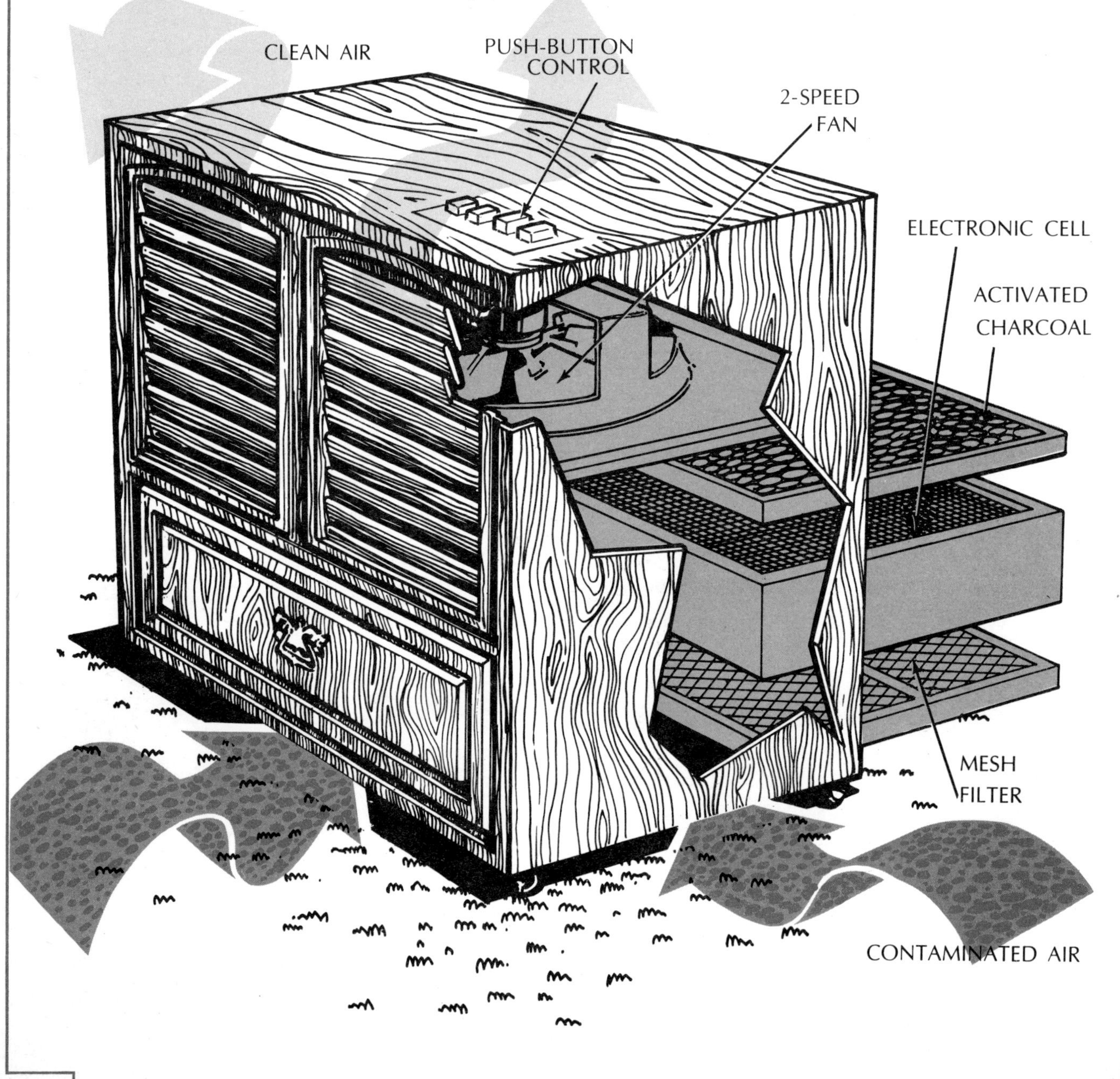

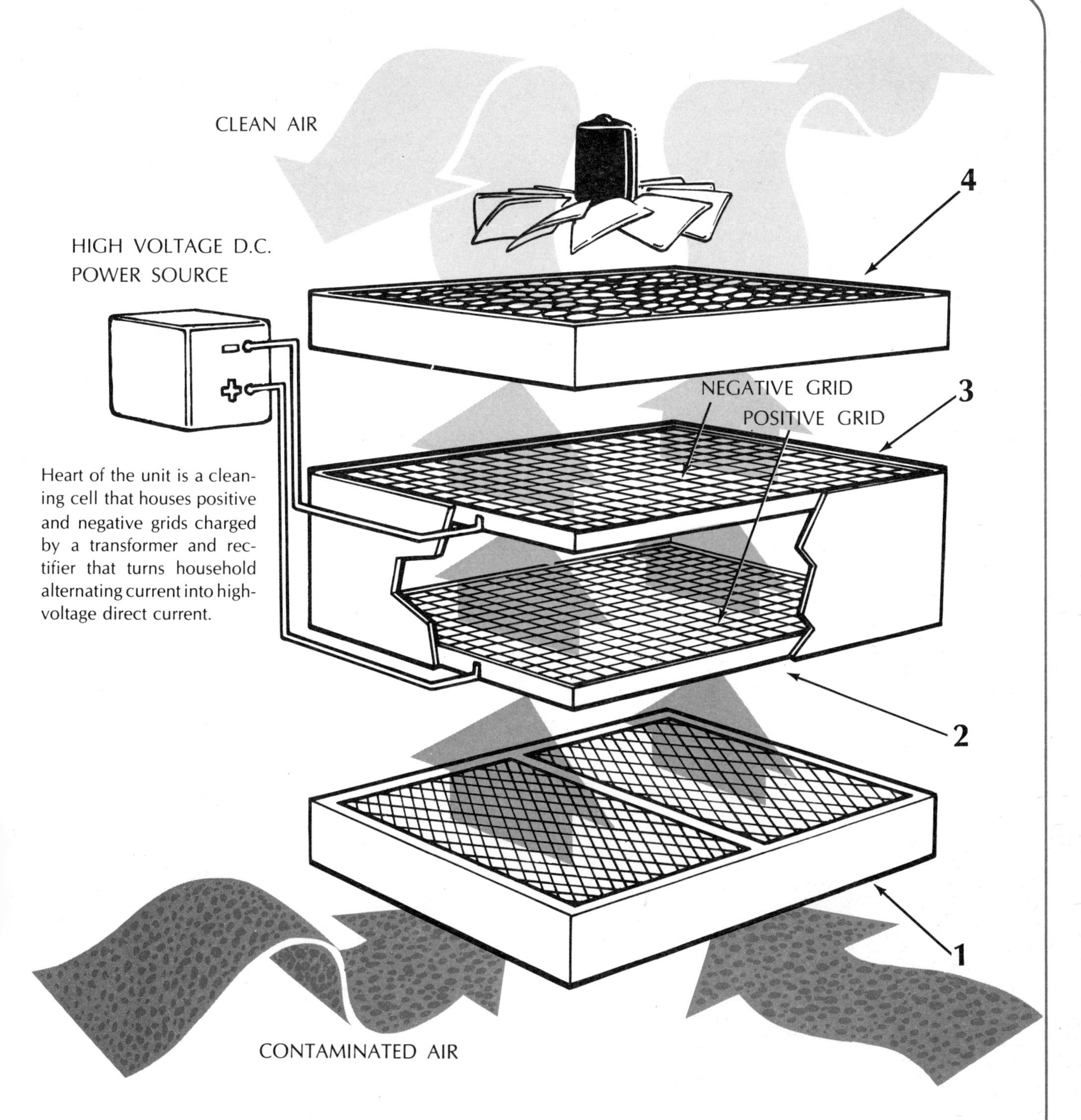

1. Drawn through the machine by a fan, the air first passes through a mesh filter similar to that found in an air conditioner. This filter removes soot and other large particles. **2.** Then, the air sifts through the positively-charged grid. As it does, the smaller particles that still remain are given a positive charge through a process called ionization. **3.** The ionized particles continue upwards through the negatively-charged grid where the electrostatically charged particles are attracted and trapped by the grid. **4.** Now 95 percent cleansed of particles, the air filters through a layer of activated charcoal, one of nature's most effective odor absorbers, before finally being returned to the room.

Optical Scanner

At incredible speed, optical scanners can read characters, whether typed, printed, or written in longhand and convert them into electrical signals that a computer can process. Many businesses use optical scanners 24 hours a day to whiz with great accuracy through tons of paper work. The first reading machines appeared in 1952 after the government found itself short of key punch operators to translate information onto punched cards for processing by old-style computers. These early readers were called OMR's (Optical Mark Readers). They recognized only the presence or absence of black pencil marks on pre-selected areas on a card or sheet of paper. But soon came the OCR's (Optical Character Readers) that can now recognize *several thousand* printed or hand written characters *per second.*

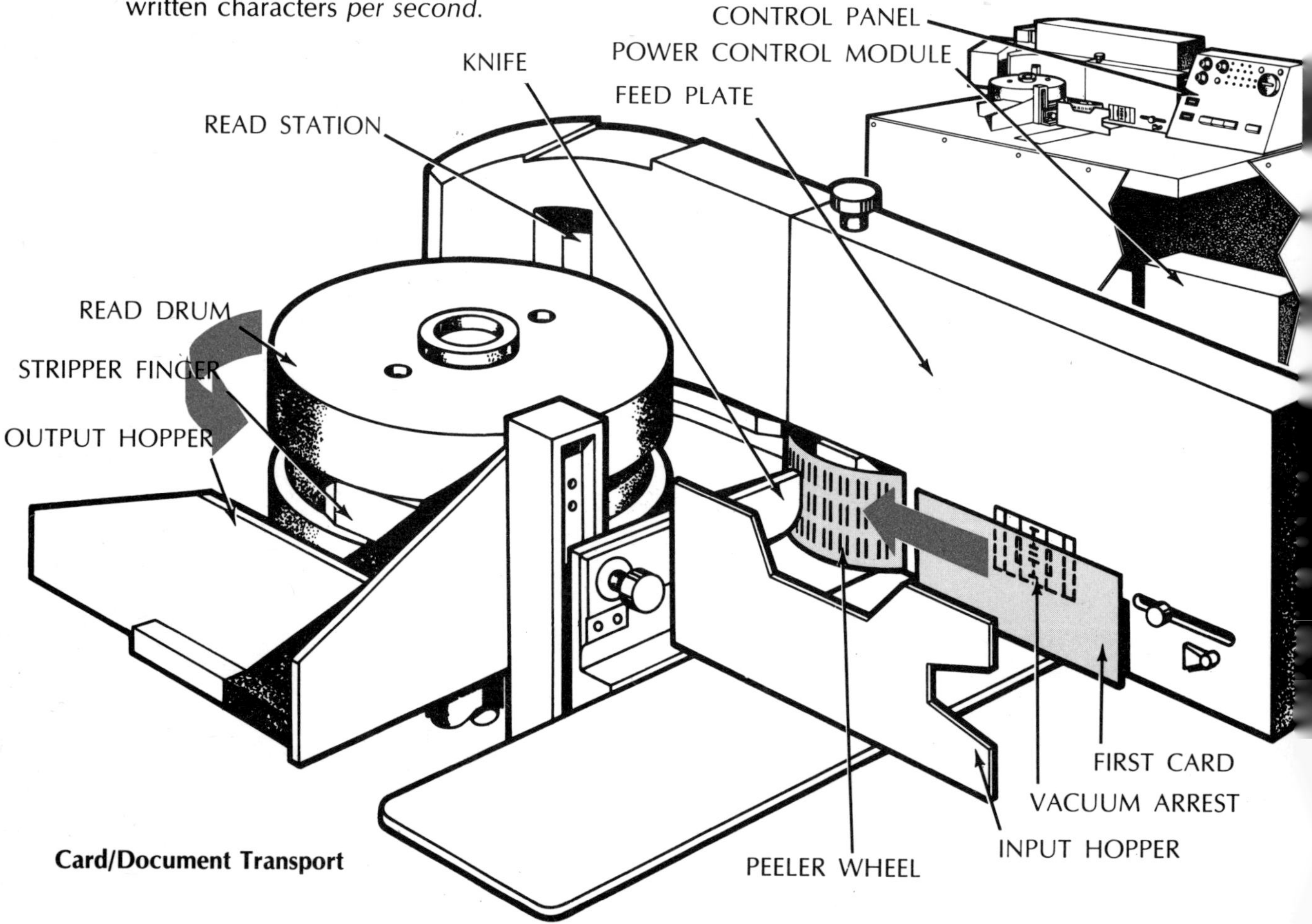

Card/Document Transport

A stack of cards is loaded by hand into the input hopper. The feed cycle is initiated when the vacuum arrest momentarily shuts off to release the first card. As the peeler wheel advances the first card, a knife edge separates the card from the rest of the stack while curving the card so that it remains stiff as it continues its travel to the read drum. When the first card has cleared the vacuum arrest (about 1½ inches of travel), the vacuum arrest switches on again, holding the second card in place until its path is clear. Each card remains on the revolving read drum for over half a rotation (195 degrees). During this time, an entire line of information is read by the scanner. Then the card is stripped from the drum and stacked in the output hopper. This feed/read cycle occurs at an amazing 20 times per second.

"Laser" is short for **l**ight **a**mplifications by **s**timulated **e**mission of **r**adition. Thus, a laser produces and amplifies light. Laser beams are produced when electrons are driven or "pumped" into higher-than-usual energy levels or states by exposing them to electrical force. Light from a laser has special properties. It is all of a single frequency (monochromatic). Its waves are all in step (or coherent). And it is extremely intense and brilliant. Laser light can be focused by lenses the way ordinary light can be focused. The helium-neon continuous-wave laser used here consists of a gas-filled tube that has high voltage applied across two electrodes near the ends of the tube. This causes an electrical discharge to take place. The gas glows, giving off a continuous stream of monochromatic light.

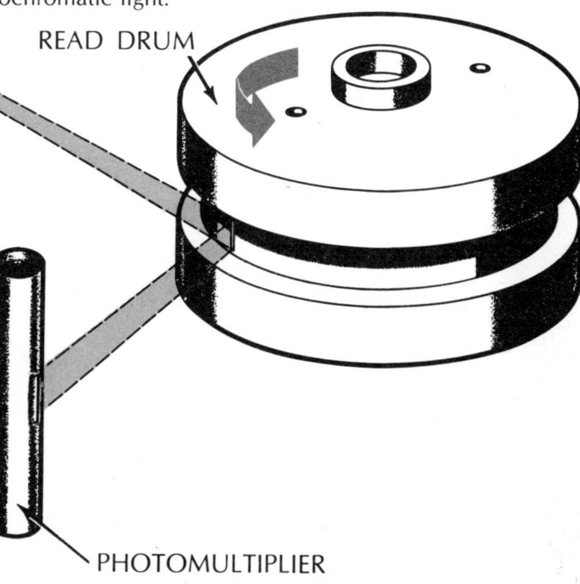

Reading is accomplished by means of a moving beam of light generated by a helium-neon continuous-wave laser. The light beam passes through a series of lenses and mirrors that shape it into a small spot. When the spot strikes a high-speed oscillating mirror, it is deflected and sweeps a vertical line at the drum surface, illuminating the characters in the read zone. A photomultiplier detects the light reflected from the document and converts dark and light reflections into electrical signals which are then sent to the OCR's memory bank.

How a Character Looks to the OCR

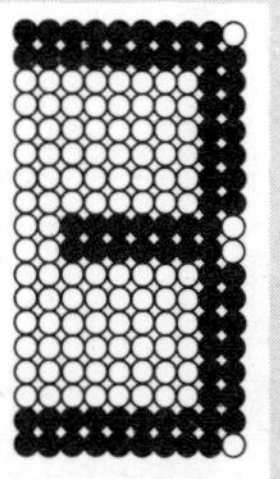

The memory bank stores properly grouped electrical signals that correspond to those produced by the photomultiplier when a character is projected upon it. The OCR's computer can then recall the proper matching signal for processing by another computer.

Automatic Coffee Percolator

A percolator brews coffee by repeatedly filtering heated water through coffee grounds. Essentially, the heated water is forced up a central tube that supports a perforated basket containing the grounds. When the heated water spurts out of the tube, it strikes the pot's lid and falls onto a perforated basket cover that distributes the water evenly over the grounds. The brewing cycle of most percolators is thermostatically controlled by a bimetal switch that breaks a circuit contact at the end of the brewing cycle. This same bimetal switch also keeps the coffee hot by occasionally closing the contact to reactivate the heating element.

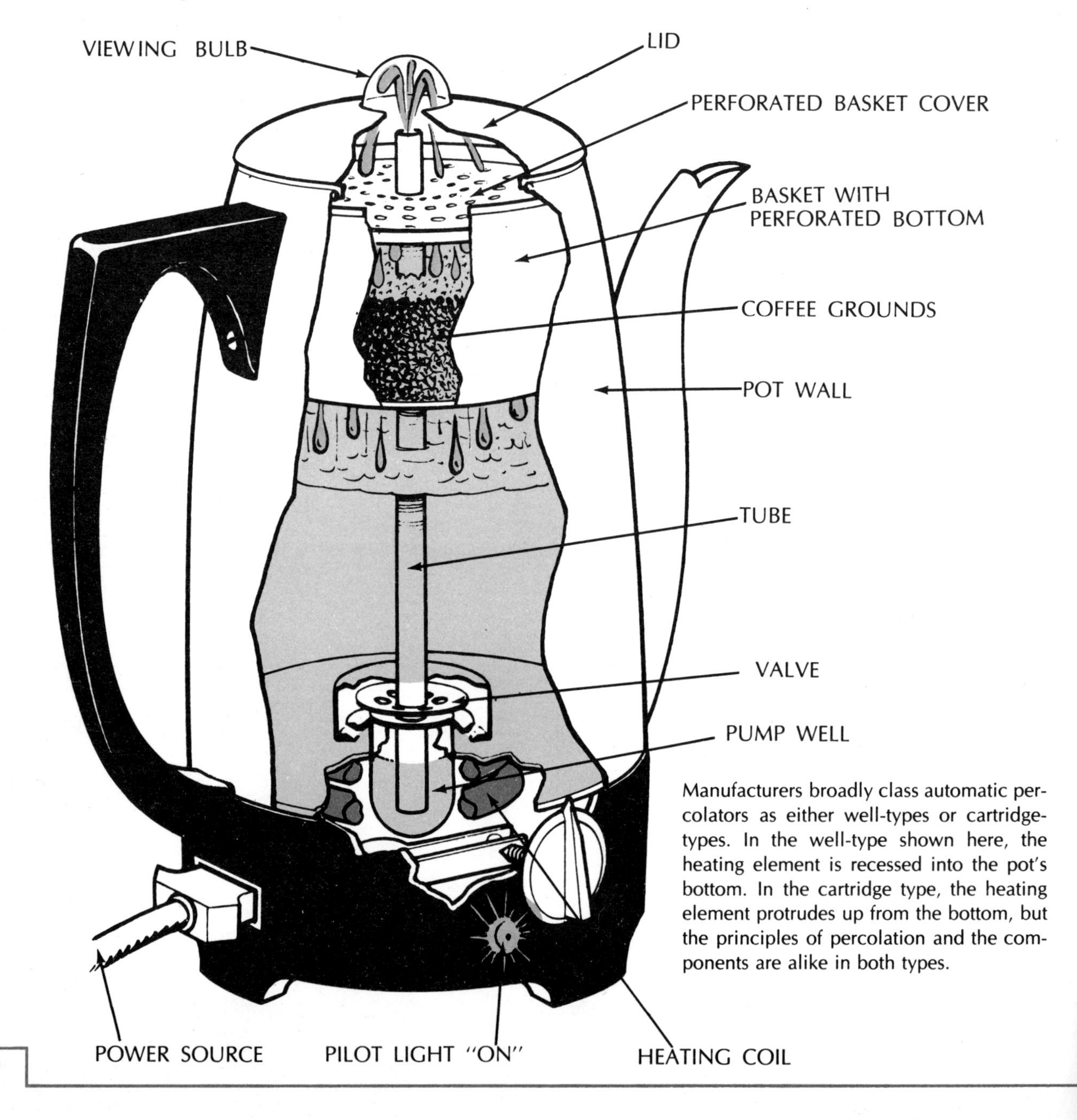

Manufacturers broadly class automatic percolators as either well-types or cartridge-types. In the well-type shown here, the heating element is recessed into the pot's bottom. In the cartridge type, the heating element protrudes up from the bottom, but the principles of percolation and the components are alike in both types.

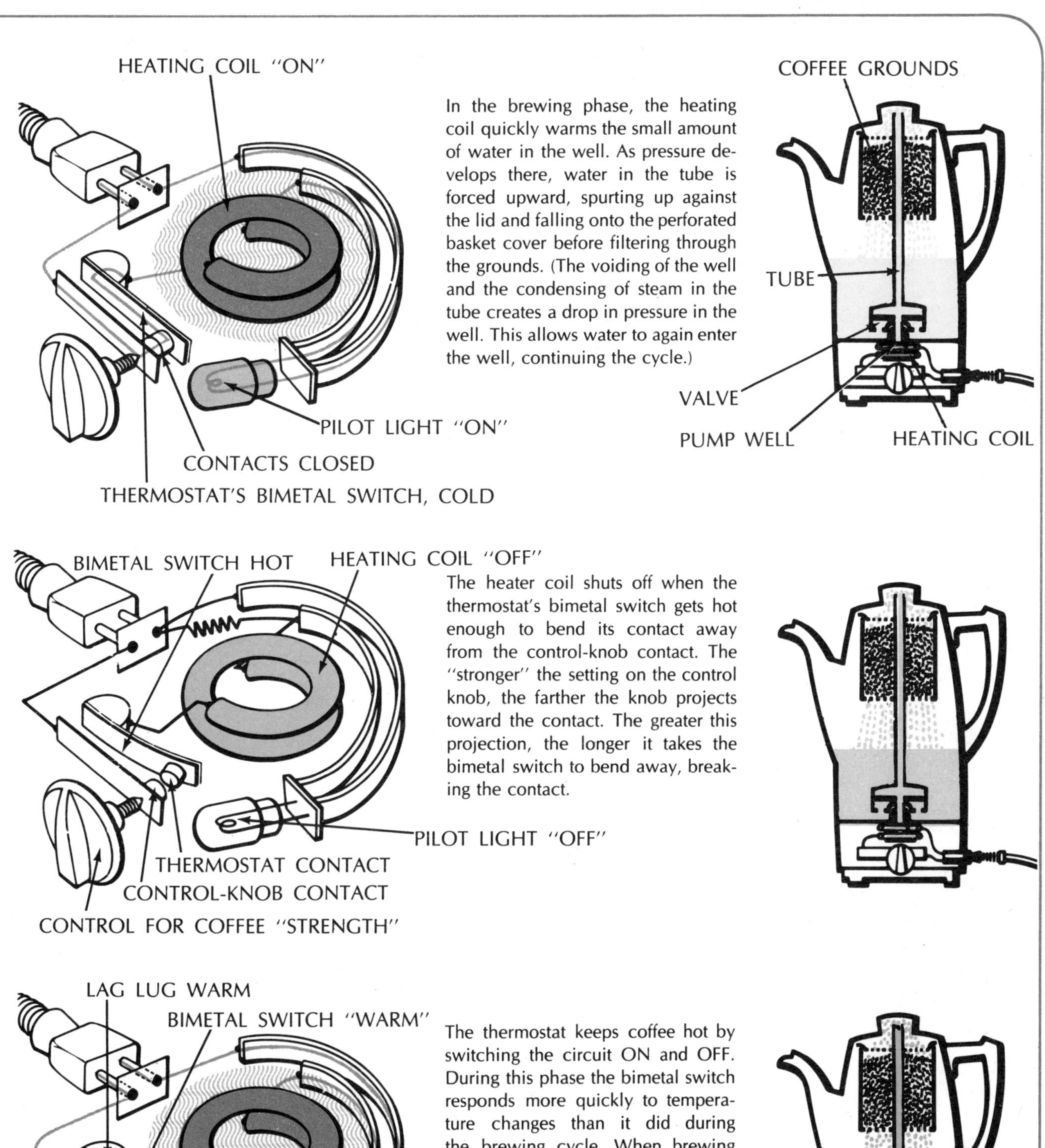

In the brewing phase, the heating coil quickly warms the small amount of water in the well. As pressure develops there, water in the tube is forced upward, spurting up against the lid and falling onto the perforated basket cover before filtering through the grounds. (The voiding of the well and the condensing of steam in the tube creates a drop in pressure in the well. This allows water to again enter the well, continuing the cycle.)

The heater coil shuts off when the thermostat's bimetal switch gets hot enough to bend its contact away from the control-knob contact. The "stronger" the setting on the control knob, the farther the knob projects toward the contact. The greater this projection, the longer it takes the bimetal switch to bend away, breaking the contact.

The thermostat keeps coffee hot by switching the circuit ON and OFF. During this phase the bimetal switch responds more quickly to temperature changes than it did during the brewing cycle. When brewing started, the switch and its attached lag lug (a chunk of metal) were cold. But since the lag lug is already hot during the "warming" phase, slight temperature changes cause the switch to make and break.

Fire Sprinkler

Of the large number of lives lost to fires in buildings, deaths have occurred only rarely in buildings equipped with automatic sprinkler systems. And these rare deaths have usually been attributed to suffocation resulting from smoldering material that didn't produce enough heat to set off a sprinkler. A completely automatic system does three things simultaneously: It sprays water onto the fire. It rings a bell in the building. And it alerts a central office or fire house, or both. In the "wet" system shown here, the most common, all feeder pipes contain water under pressure. A "dry" system (used where water might freeze) holds compressed air. When a head fuses in the dry system, air escapes, opening a valve that admits water into the pipes.

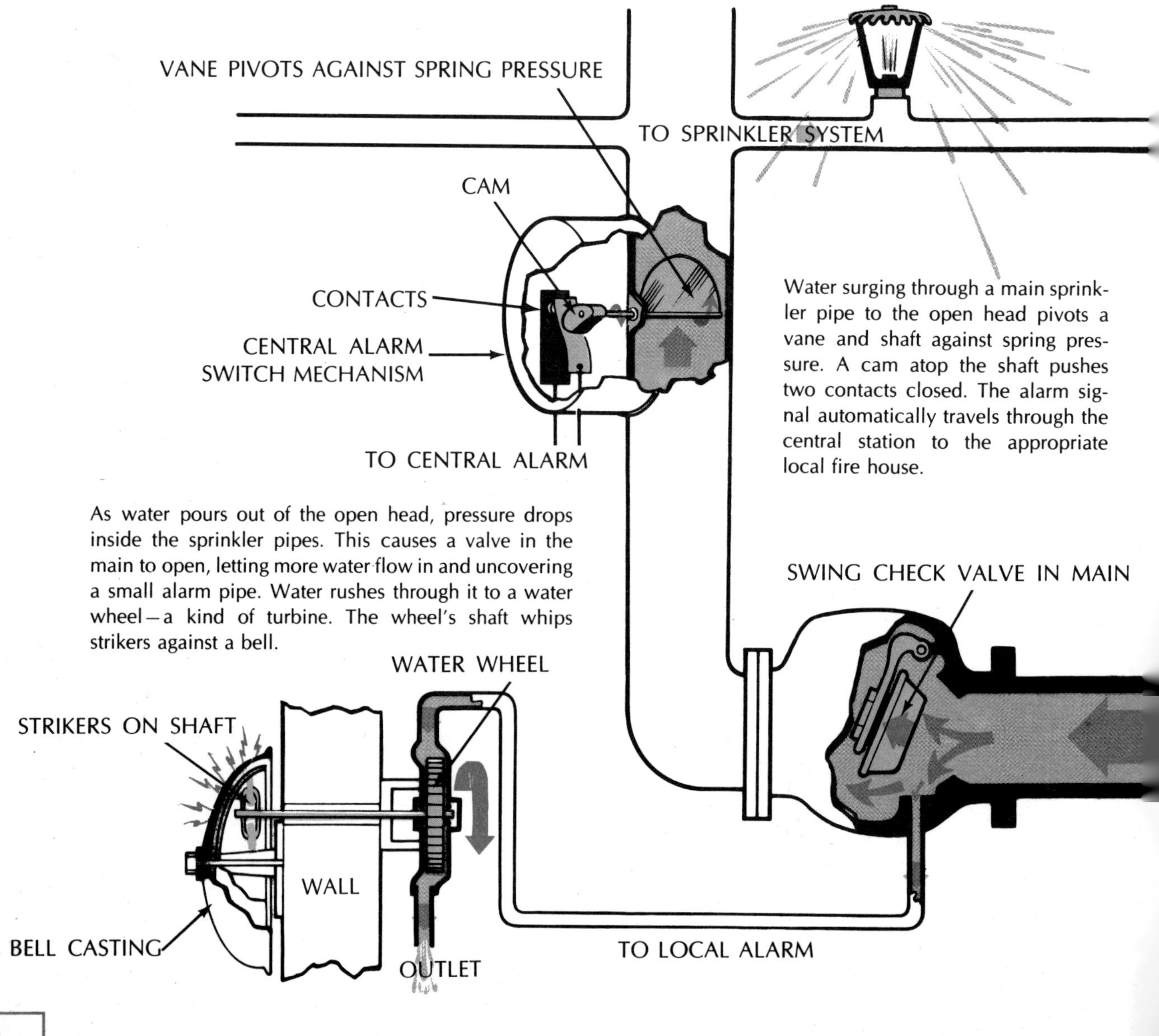

Water surging through a main sprinkler pipe to the open head pivots a vane and shaft against spring pressure. A cam atop the shaft pushes two contacts closed. The alarm signal automatically travels through the central station to the appropriate local fire house.

As water pours out of the open head, pressure drops inside the sprinkler pipes. This causes a valve in the main to open, letting more water flow in and uncovering a small alarm pipe. Water rushes through it to a water wheel—a kind of turbine. The wheel's shaft whips strikers against a bell.

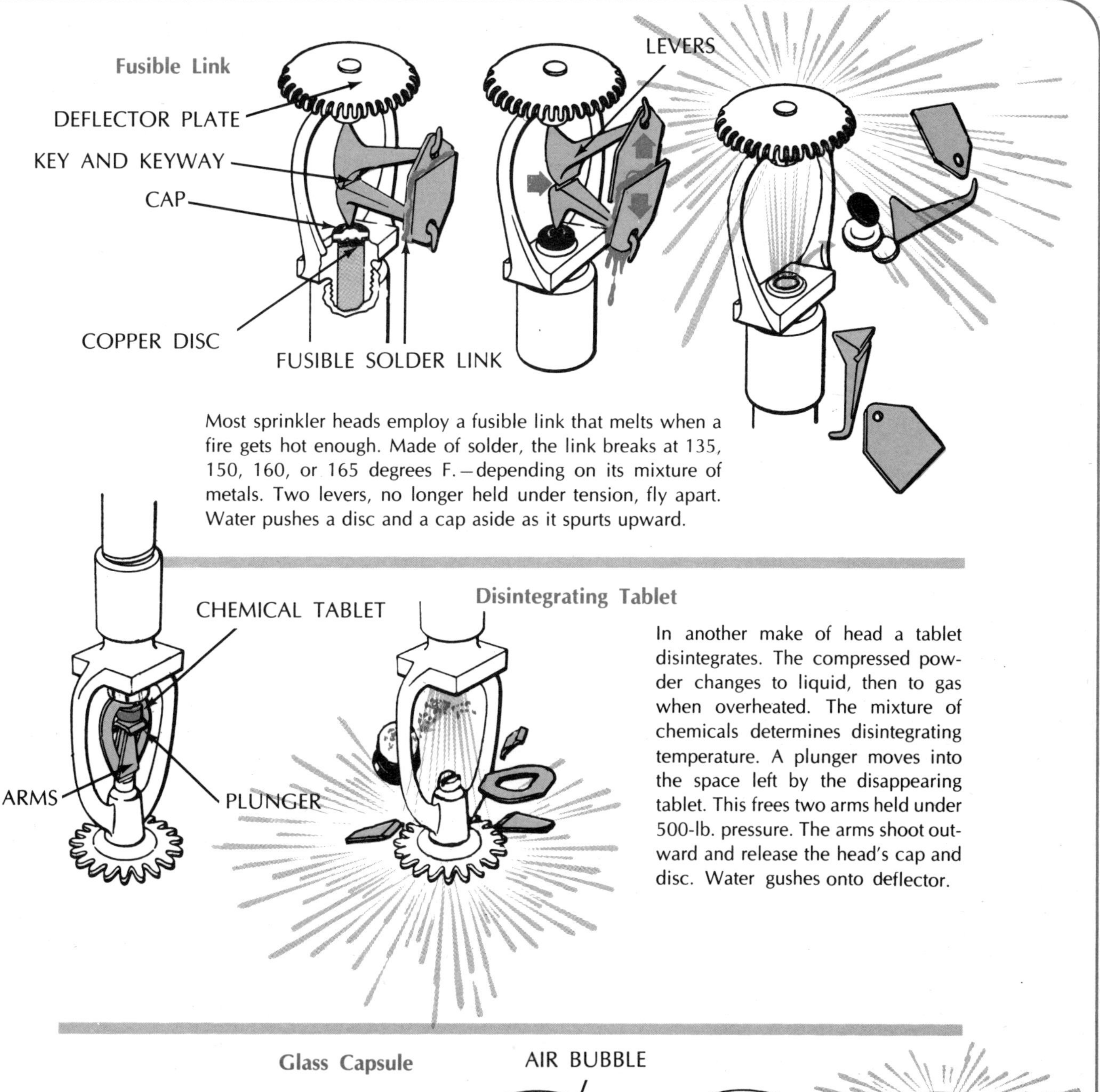

Most sprinkler heads employ a fusible link that melts when a fire gets hot enough. Made of solder, the link breaks at 135, 150, 160, or 165 degrees F.—depending on its mixture of metals. Two levers, no longer held under tension, fly apart. Water pushes a disc and a cap aside as it spurts upward.

Disintegrating Tablet

In another make of head a tablet disintegrates. The compressed powder changes to liquid, then to gas when overheated. The mixture of chemicals determines disintegrating temperature. A plunger moves into the space left by the disappearing tablet. This frees two arms held under 500-lb. pressure. The arms shoot outward and release the head's cap and disc. Water gushes onto deflector.

Glass Capsule

A glass capsule bursts in still another design. Heat expands a special liquid until there is no air space left and the glass shatters. Bursting temperature is set by filling a capsule with liquid at the bursting temperature. As the liquid cools (and contracts), a bubble appears. The higher the bursting temperature, the larger the bubble. When the capsule bursts, cap and disc pop off and water rushes out.

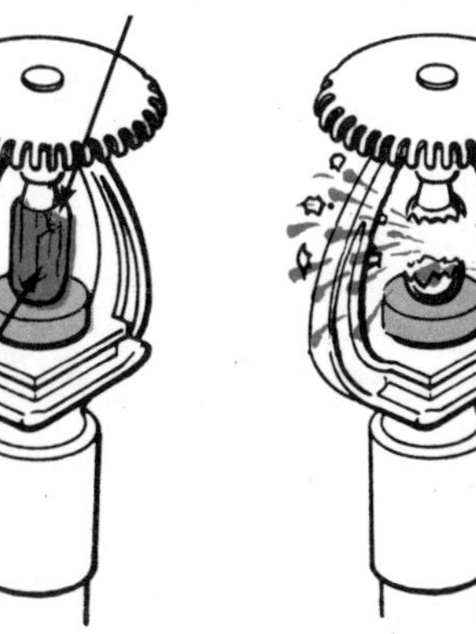

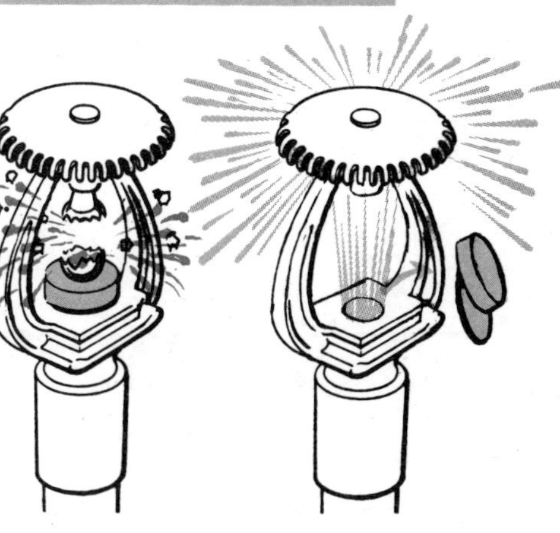

Player Piano

After years of obsolescence, the player piano has made its comeback in the form of a spinet. Music rolls of both new and oldtime songs, as well as classical pieces, are now pouring from revitalized piano roll factories. The earliest mechanical players built around 1896 functioned almost like automatons, since they sat in front of the piano like a human and touched the ivories with mechanical fingers. But consumer preferences soon dictated that the mechanism be housed inside the piano. Though electrically driven models are now available, many people opt for the foot pedal models, partly for nostalgia's sake but also because the "musician" can contribute some expression by varying foot pressure.

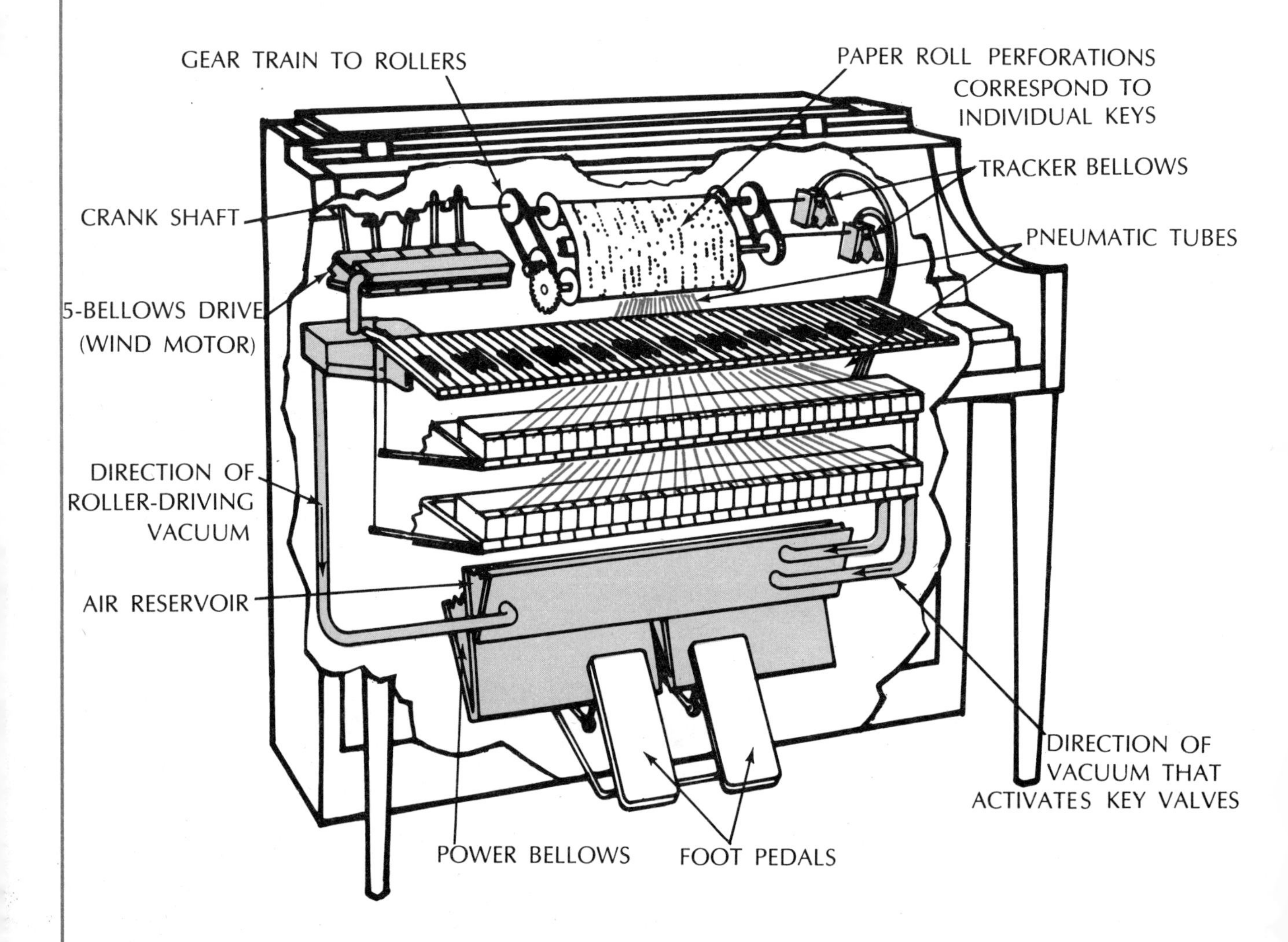

In the spinet-type player piano, action of the foot pedals accomplishes two important tasks: Pedal levers power a gear train that turns the music roll (perforated paper). And pedal action through bellows creates a vacuum under the moving paper. Air from the atmosphere is drawn through the paper's perforated holes into neoprene tubes leading to individual key valves. Each valve connects to a small bellows which raises the back of its key, pushing the hammer forward to strike the string.

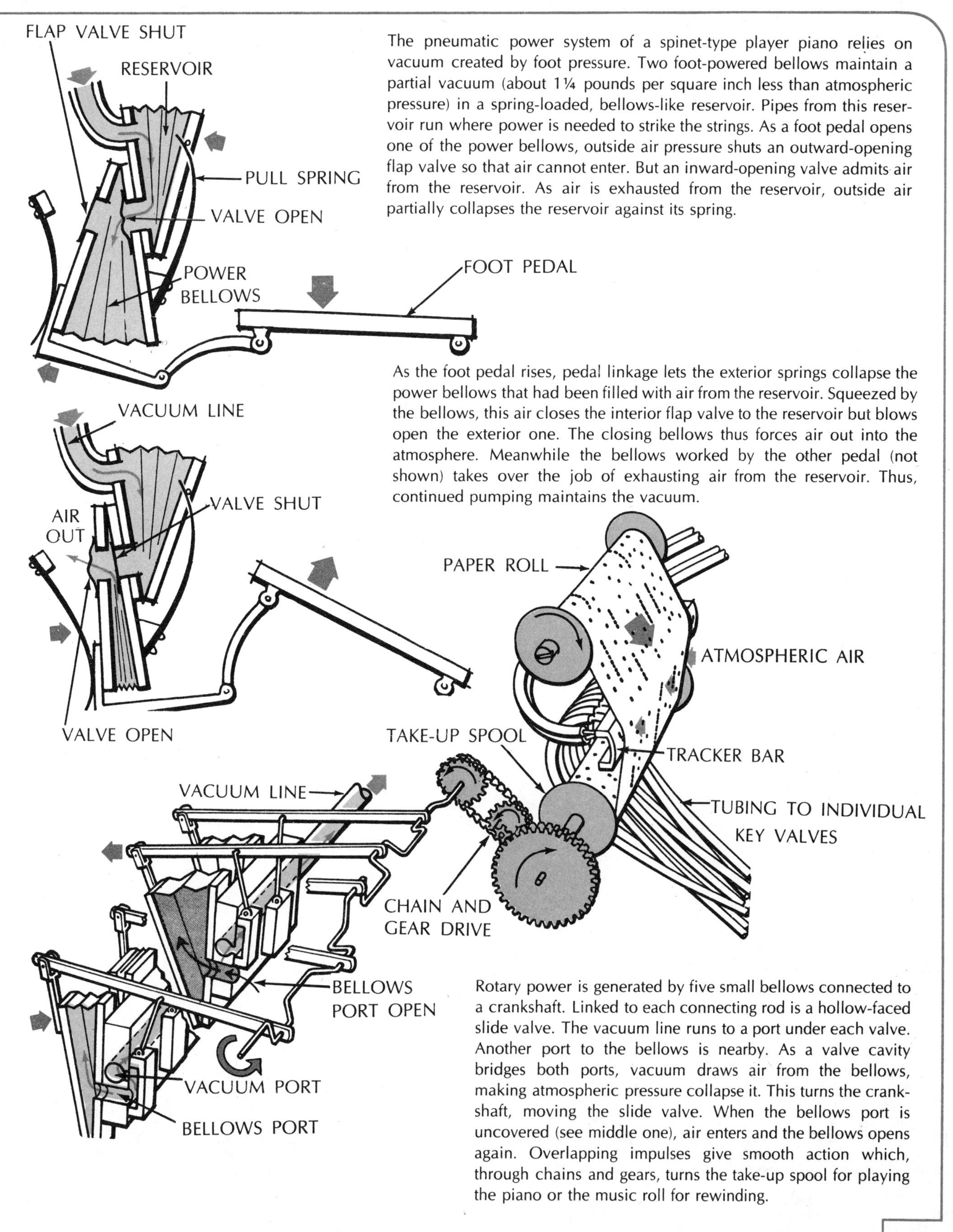

The pneumatic power system of a spinet-type player piano relies on vacuum created by foot pressure. Two foot-powered bellows maintain a partial vacuum (about 1¼ pounds per square inch less than atmospheric pressure) in a spring-loaded, bellows-like reservoir. Pipes from this reservoir run where power is needed to strike the strings. As a foot pedal opens one of the power bellows, outside air pressure shuts an outward-opening flap valve so that air cannot enter. But an inward-opening valve admits air from the reservoir. As air is exhausted from the reservoir, outside air partially collapses the reservoir against its spring.

As the foot pedal rises, pedal linkage lets the exterior springs collapse the power bellows that had been filled with air from the reservoir. Squeezed by the bellows, this air closes the interior flap valve to the reservoir but blows open the exterior one. The closing bellows thus forces air out into the atmosphere. Meanwhile the bellows worked by the other pedal (not shown) takes over the job of exhausting air from the reservoir. Thus, continued pumping maintains the vacuum.

Rotary power is generated by five small bellows connected to a crankshaft. Linked to each connecting rod is a hollow-faced slide valve. The vacuum line runs to a port under each valve. Another port to the bellows is nearby. As a valve cavity bridges both ports, vacuum draws air from the bellows, making atmospheric pressure collapse it. This turns the crankshaft, moving the slide valve. When the bellows port is uncovered (see middle one), air enters and the bellows opens again. Overlapping impulses give smooth action which, through chains and gears, turns the take-up spool for playing the piano or the music roll for rewinding.

Electric Drill

Crude awls and boring tools originated with Neanderthal man sometime after the last glaciers retreated from Central Europe. Egyptians used drills to construct mummy cases. And Caesar's legions employed a tool like the modern brace and bit to make campsites and fortifications. The modern electric drill translates high speed rotation of its universal motor's shaft through speed reducing gears to a bit or accessory that turns at a lower speed but with far greater turning power. Many models offer a trigger control module that can vary bit speed from as low as 10 rpm to over 1,000 rpm. In addition to turning bits designed to bore metal or wood, electric drills can power accessories such as sanding discs, circular saw blades, buffing pads, grinding wheels, and screwdriver heads.

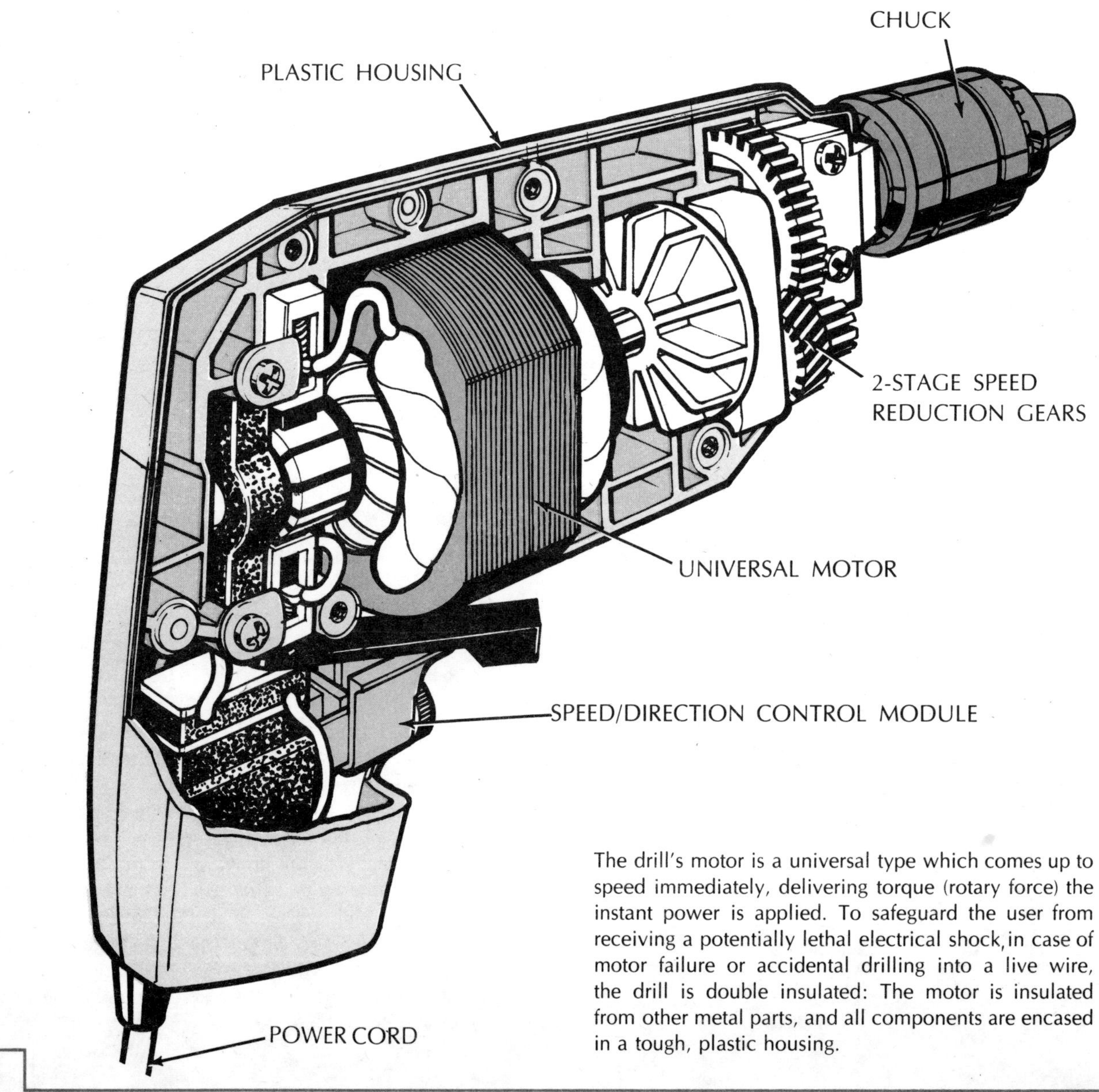

The drill's motor is a universal type which comes up to speed immediately, delivering torque (rotary force) the instant power is applied. To safeguard the user from receiving a potentially lethal electrical shock, in case of motor failure or accidental drilling into a live wire, the drill is double insulated: The motor is insulated from other metal parts, and all components are encased in a tough, plastic housing.

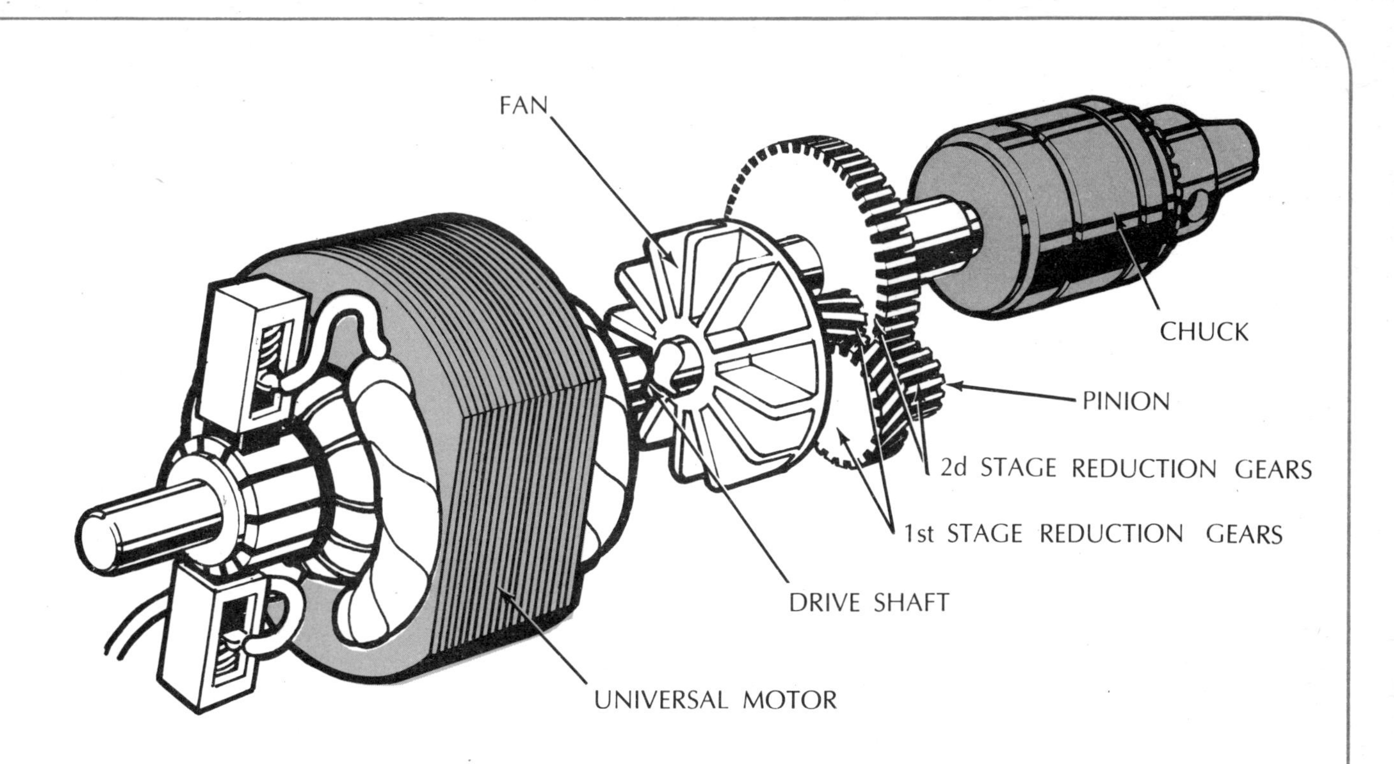

A shaft that does not conduct electricity is fitted to the motor armature and mounted on bearings, fore and aft, that reduce friction drag on the spinning armature. An impeller fan that spins with the shaft, cools the motor by drawing in air through vents in the housing. The high speed of the motor shaft is reduced through a two-stage gear train between the shaft and the chuck. As motor speed is stepped down, torque correspondingly increases in the first-stage of gears. The higher torque applied to the second pinion is further multiplied by the large ratio of the second stage. Thus, the motor shaft turns many times before the chuck completes a single revolution. The result is the exchange of motor speed for torque in the chuck.

Instead of a simple ON/OFF switch, many drills have a built-in speed control module that regulates power to the motor windings so that motor speed becomes proportional to finger pressure. A preset knob in the trigger turns a screw that sets the trigger's travel limit. Thus, a full squeeze of the trigger will produce any desired top speed.

Designed to change rotational direction of the motor, a switch on the module is wired to the two ends of the motor armature winding. Switching the connections to the winding causes the current to reverse the magnetic fields and thereby reverse the direction of the chuck. The directional switch is particularly valuable for backing the bit from tight bores and for removing screws.

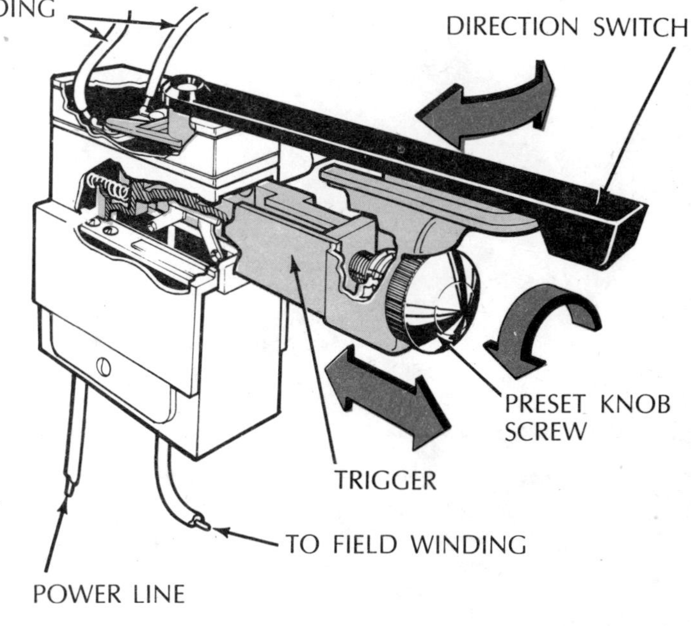

Snow Thrower

Snow throwers, ideal for shoveling out the front walk and the driveway, come in two useful sizes. The smaller ones, which weigh less than 20 pounds, are sufficient in areas of moderate snowfall, while the larger ones are for duty in areas of heavy snowfall. Both operate on the same principle: they have motor-driven rotary, snow-scooping devices that bite into snow, gather it up and then propel it upward and outward, away from the area being plowed. Both expel snow with considerable force. Thus, it is well to keep children and others at a distance in case the thrower picks up a rock or a piece of wood and fires it through the air.

Large snow throwers are self-propelled by a pair of motor-driven wheels. They are equipped with a throttle to adjust engine speed, a traction-control lever for disengaging the drive wheels and a control to alter the direction of the expelled snow.

Light snow throwers have a scoop-shaped housing that contains a cylindrical drum fitted with two paddles. The drum is driven by a two-cycle engine through a sprocket and chain system. The rapidly rotating paddles bite into the snow, compress it against the curved rear portion of the scoop and hurl it upward. Deflector vanes may be angled to direct the stream of snow to the right or left of the thrower's path.

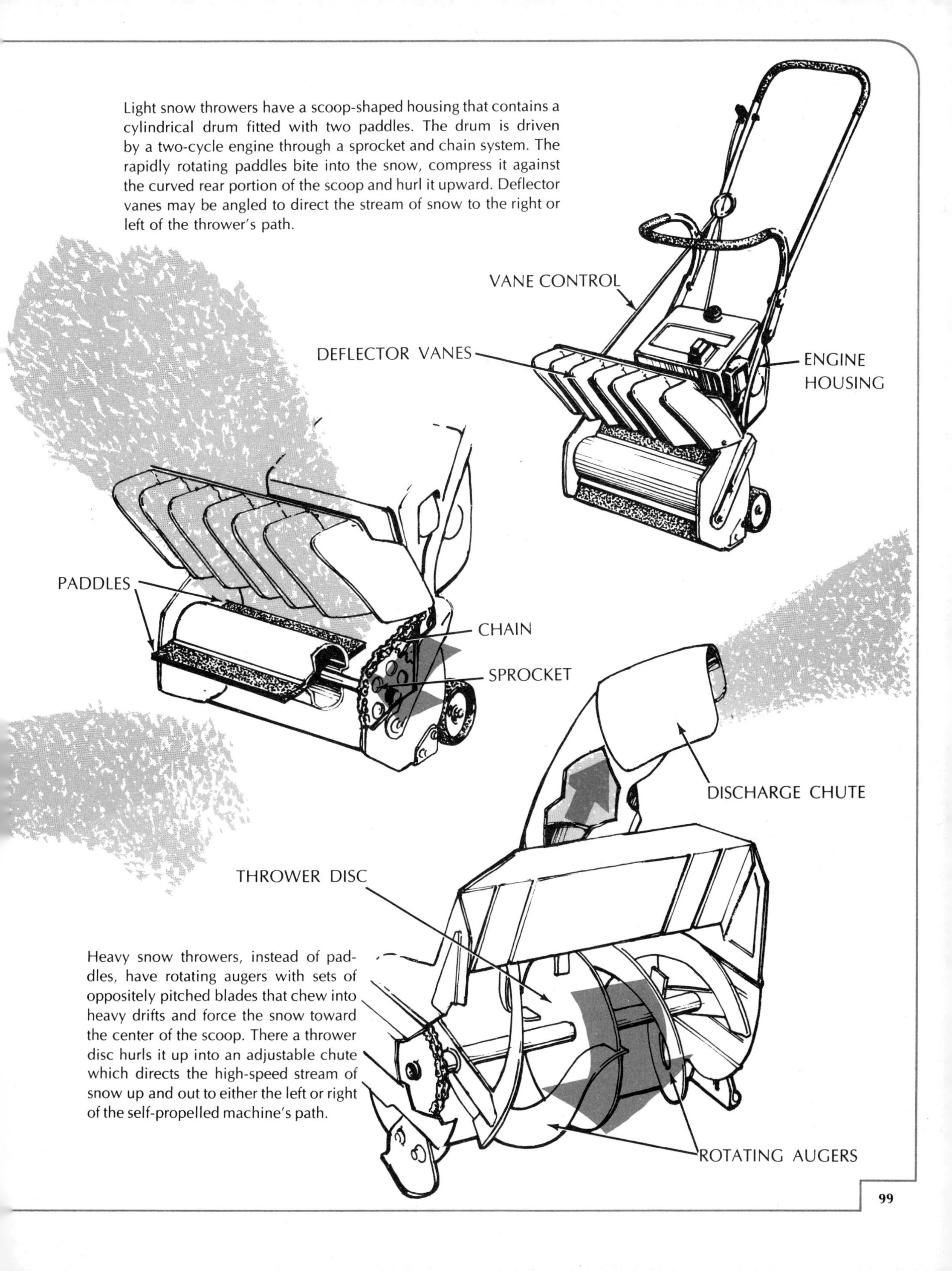

Heavy snow throwers, instead of paddles, have rotating augers with sets of oppositely pitched blades that chew into heavy drifts and force the snow toward the center of the scoop. There a thrower disc hurls it up into an adjustable chute which directs the high-speed stream of snow up and out to either the left or right of the self-propelled machine's path.

Electric Glue Gun

Love may make the world go 'round, but glue holds it together. Around the home shop, and business, adhesives play a vital role. Of the many adhesive products now available, the hot-melt adhesives have been the newest to enter the home craftsman's shop. Strong thermosetting plastic compounds, hot-melts can make neat, permanent bonds between nearly all surfaces. The bond sets in less than a minute and cools to a resilient, waterproof finish. Hot-melt have been a favorite in industry for years, but since they had to be applied at about 350° F., they were considered impractical for home use. The new compact glue gun makes application of hot-melt adhesives and caulking compounds as easy and safe as writing with a ball-point pen.

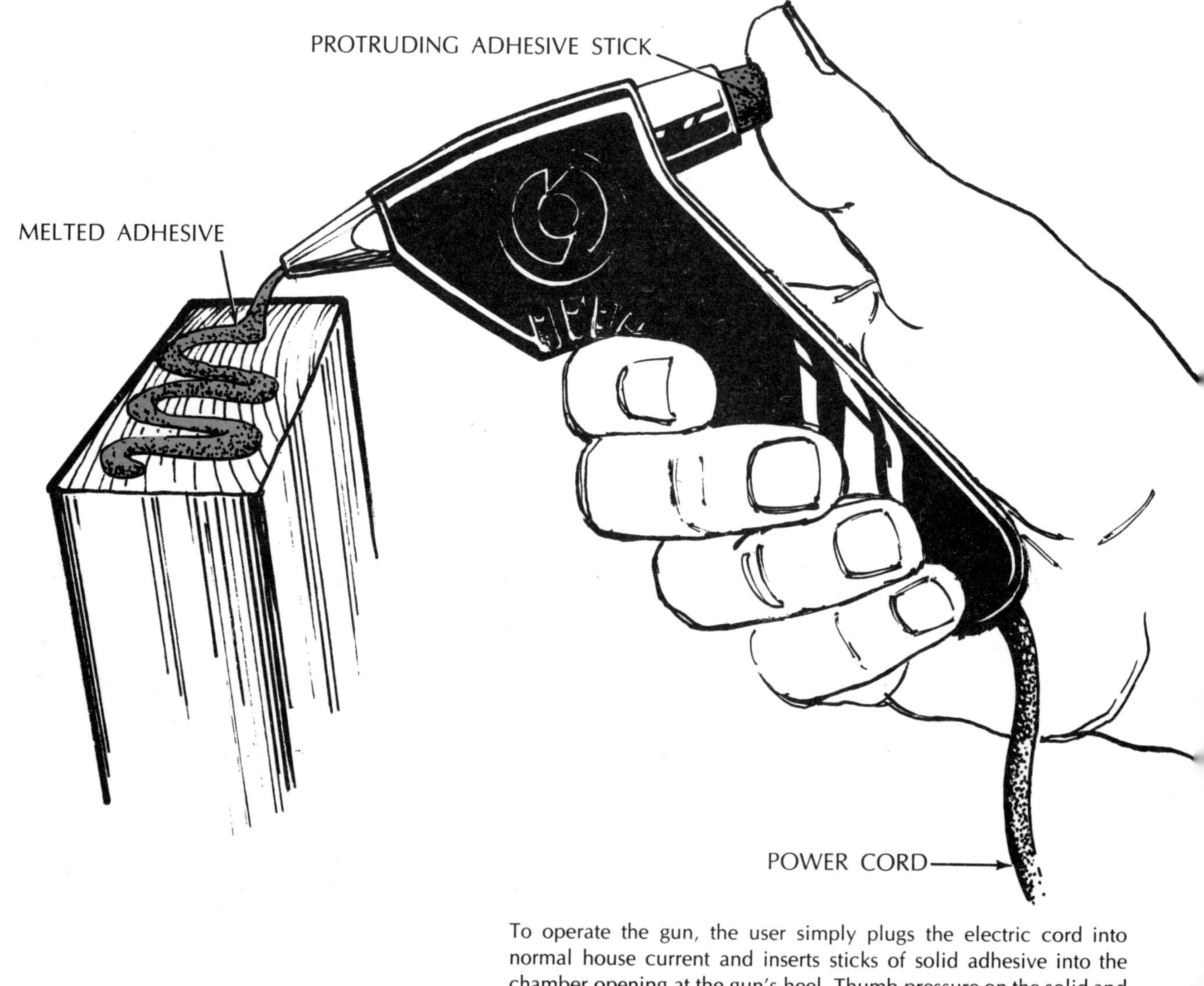

To operate the gun, the user simply plugs the electric cord into normal house current and inserts sticks of solid adhesive into the chamber opening at the gun's heel. Thumb pressure on the solid and cool stick of adhesive protruding from the gun forces the now hot, melted adhesive inside the gun's chamber out through the nozzle. The hot-melt sets firmly as soon as it has given up its heat to the surfaces being bonded.

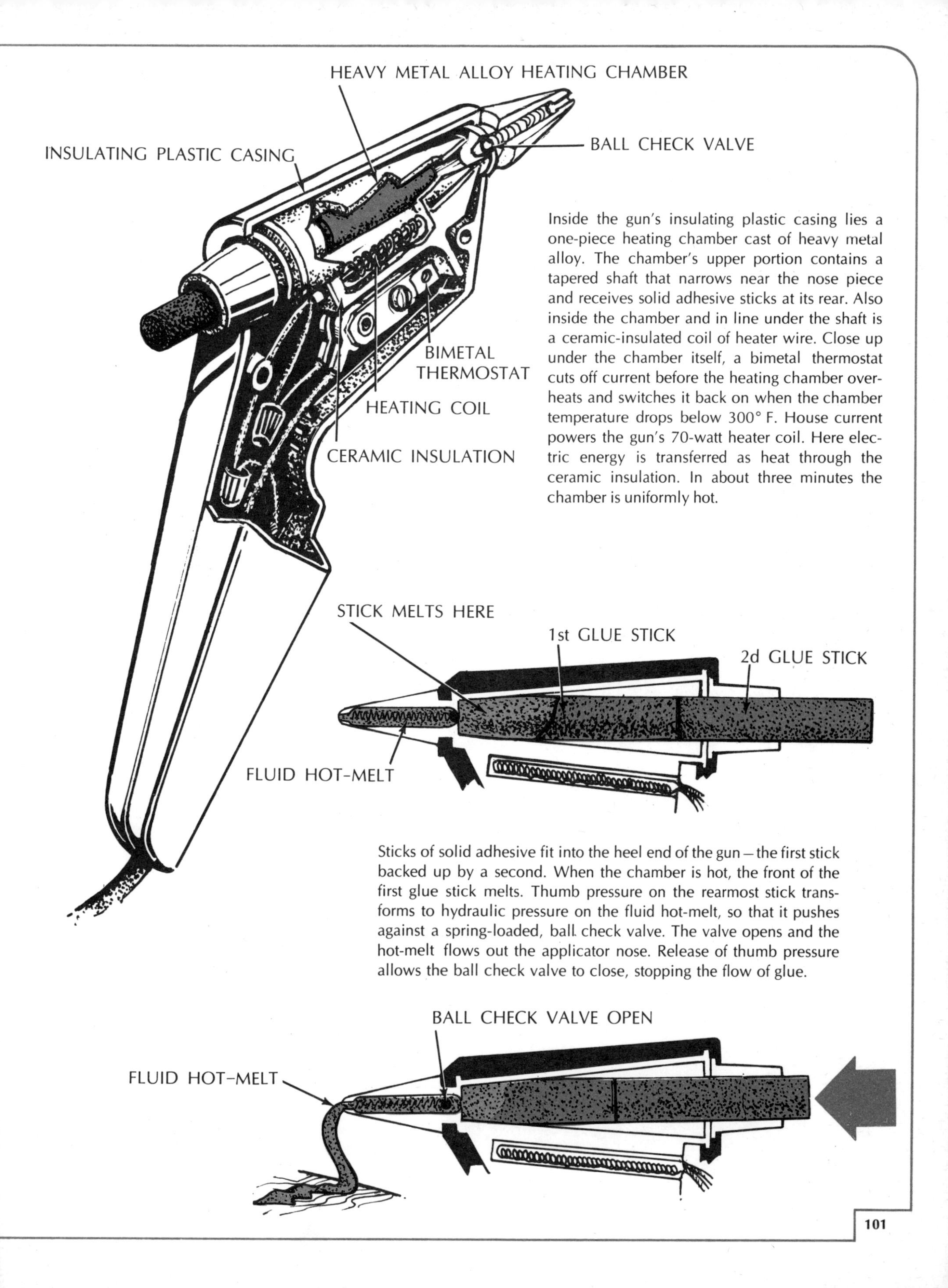

Inside the gun's insulating plastic casing lies a one-piece heating chamber cast of heavy metal alloy. The chamber's upper portion contains a tapered shaft that narrows near the nose piece and receives solid adhesive sticks at its rear. Also inside the chamber and in line under the shaft is a ceramic-insulated coil of heater wire. Close up under the chamber itself, a bimetal thermostat cuts off current before the heating chamber overheats and switches it back on when the chamber temperature drops below 300° F. House current powers the gun's 70-watt heater coil. Here electric energy is transferred as heat through the ceramic insulation. In about three minutes the chamber is uniformly hot.

Sticks of solid adhesive fit into the heel end of the gun – the first stick backed up by a second. When the chamber is hot, the front of the first glue stick melts. Thumb pressure on the rearmost stick transforms to hydraulic pressure on the fluid hot-melt, so that it pushes against a spring-loaded, ball check valve. The valve opens and the hot-melt flows out the applicator nose. Release of thumb pressure allows the ball check valve to close, stopping the flow of glue.

Electric Door Chime

Multiple- and two-tone door chimes employ step-down transformers to reduce household current to a level suitable for chime components. Both types of chime employ solenoid assemblies, consisting of spool-like bobbins, each wound with a coil of fine wire. Inside the solenoid core rides a soft iron rod, tensioned by a coil spring. When a visitor presses the door button, current in the solenoid winding creates an electromagnetic field that swiftly drives the rod through the bobbin until the rod strikes a tone bar. The tone bar produces a precise number of vibrations per second when struck and thus a precise tone. The bar's faint tone is then resonated and amplified in a chamber of air, tuned to the same tone as the bar.

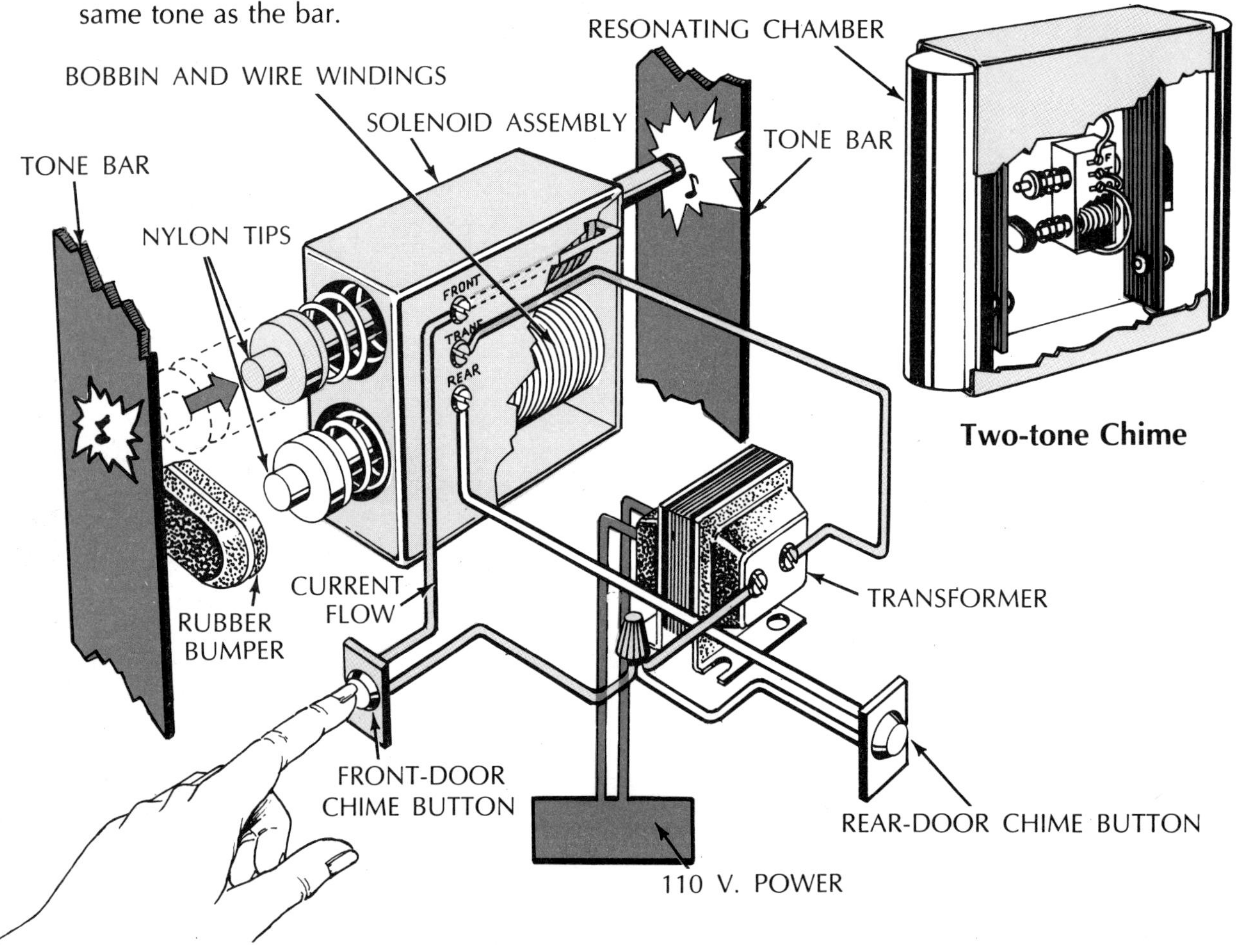

Two-tone Chime

The two-tone chime sounds both tones when a visitor presses the front door button and sounds a single tone for the rear door. Each of the rods passing through the solenoids is tipped with nylon strikers and fitted with a return spring. The separate buttons at the front and rear doors close separate circuits between the front- or rear-door solenoid and the small transformer. A rubber bumper on the end of the rear-door solenoid rod cushions its striker's rearward excursion, thus allowing only one tone to sound.

Current passing through the front door circuit creates an electomagnetic field in the solenoid that quickly pulls the rod through the bobbin. This compresses the return spring, but the inertia of the moving rod is great enough to cause its striker to hit the tone bar and rebound slightly. When the visitor removes his finger from the door button, he breaks the circuit, collapsing the magnetic field and allowing the spring to yank the rod backwards. Inertia causes the rod tip to impact. Spring tension returns the rod.

Multiple-tone Chime

The more complex, multi-tone chime contains a built-in, motor-driven switching mechanism and individual striker solenoids for each tone bar. Pressing the door switch activates a one-revolution-per-minute motor. As the motor begins to turn, it closes a set of holding contacts that will continue to supply current to the motor even when the door button is released. After only a few degrees of revolution, the motor switch closes the first contact, sending current to the first striker solenoid, then to the second and third, producing a pleasing musical tone. When the motor has completed one revolution, the holding contacts finally open, shutting off current until the door button is pressed again.

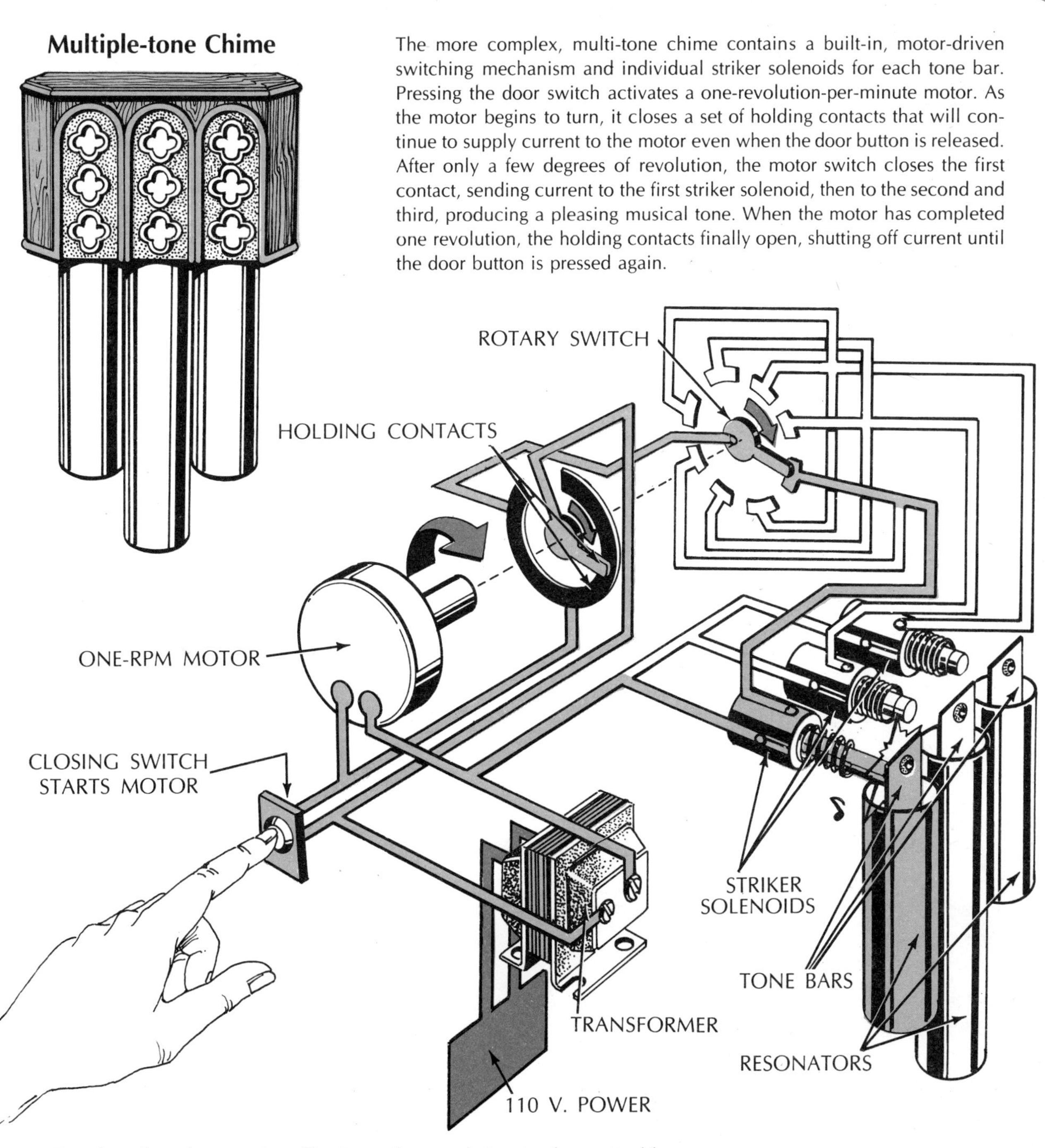

Tone bars flex when struck and begin to vibrate at their natural acoustical frequency. Each bar is supported on rubber grommets fitted to studs on a backing plate. The grommeted holes are located on the bar at nodes which remain stationary even though the rest of the bar is vibrating.

Since tone bars do not create a loud sound when struck, each tone bar is placed next to or inside a tuned resonator. This may be either a cylindrical closed-end pipe or a plastic cube-shaped chamber. The volume of air within the chamber has the same natural rate of vibration as the tone bar. Air is far more elastic than the metal bar. Thus, faint tone-bar vibrations trigger large-scale vibrations in the air column. These vibrations enter into the room air and thence into the ear drum.

Toilet Tank

Credit for the first automatic flush shutoff and other refinements of the modern toilet goes to a London plumber of the late 1800s by the name of Thomas Crapper. The most apparent difference between the early tank and the one shown here was a chain pull (for extra leverage) and a large water-filled chamber that was raised high enough to allow its water to discharge into the inlet of the flush tube—whereupon water rushed with "considerable velocity" into the bowl. Modern tanks, like early models, rely on a flush mechanism and a refill mechanism that harness the combined effects of water pressure, gravity, and buoyancy to accomplish a complex series of actions.

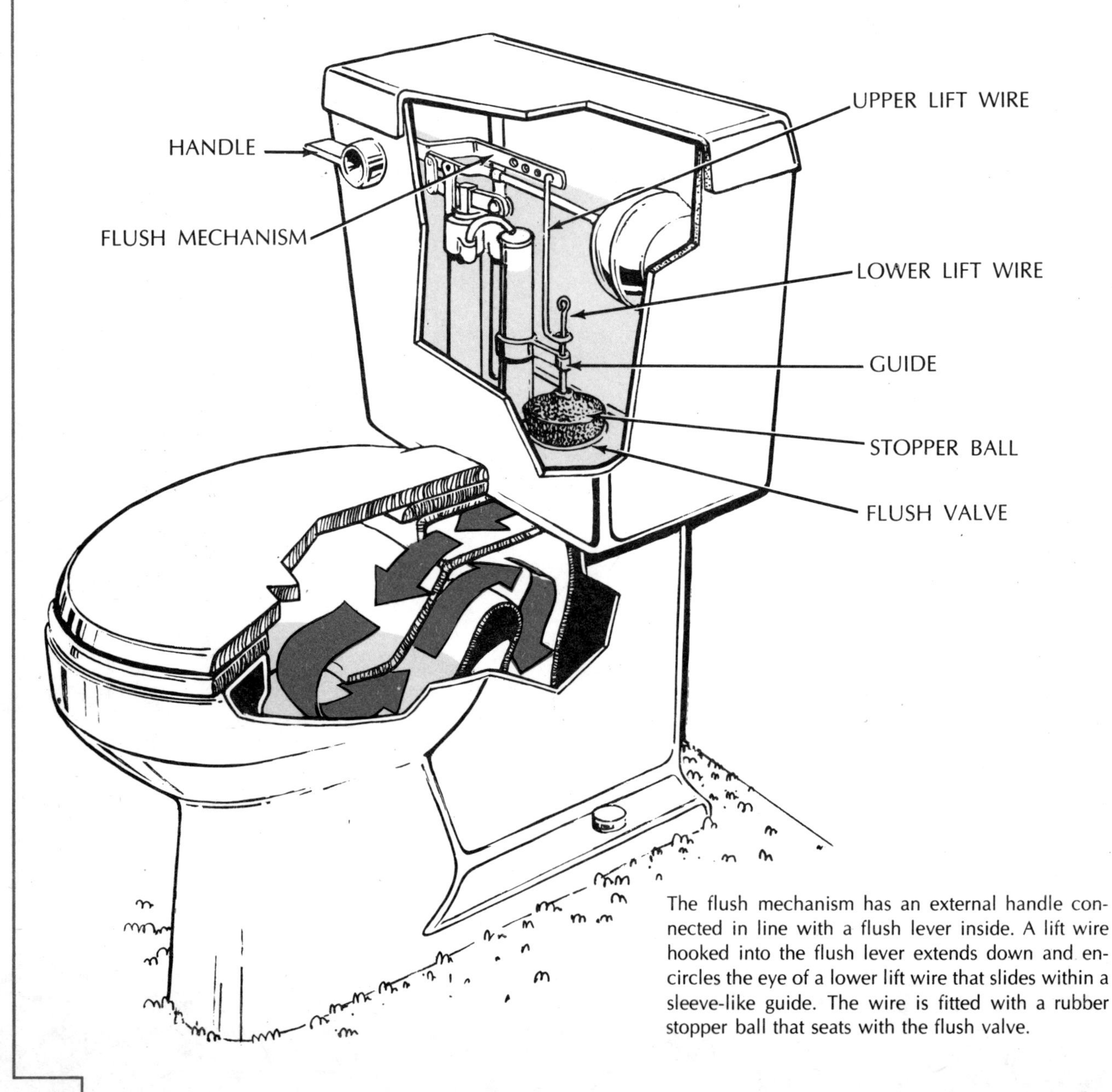

The flush mechanism has an external handle connected in line with a flush lever inside. A lift wire hooked into the flush lever extends down and encircles the eye of a lower lift wire that slides within a sleeve-like guide. The wire is fitted with a rubber stopper ball that seats with the flush valve.

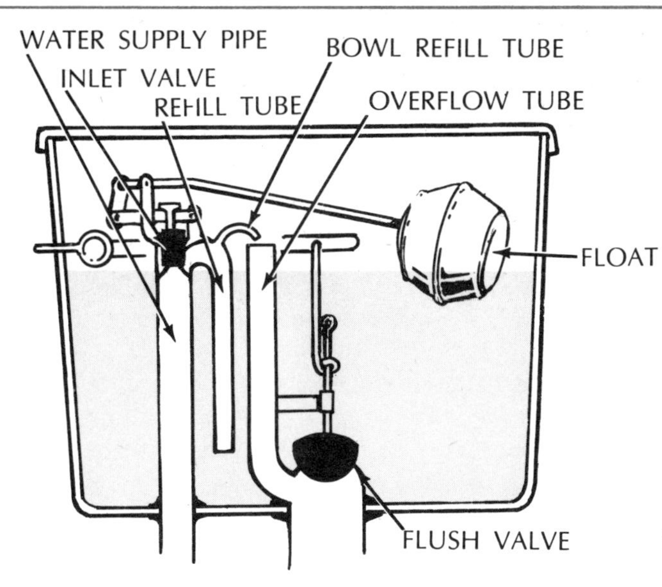

The refill mechanism consists of an incoming water supply pipe leading to an inlet valve, commonly called a ballcock, that admits water into the tank through the tank filler tube. The inlet valve also replenishes water in the bowl through an overflow tube and a bowl-refill tube. The inlet valve plunger is actuated by a compound lever assembly that responds to an air-filled float.

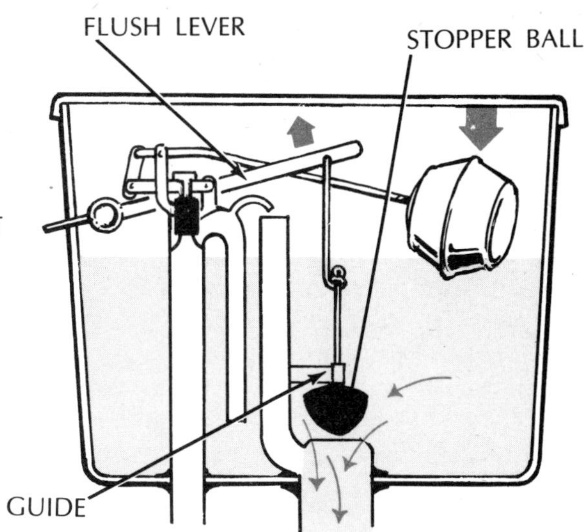

Normally the tank is full of water to within an inch of the top of the overflow tube. When the handle is depressed, the flush lever causes the stopper ball to be pulled out of the flush valve. Gravity cascades the several gallons of water down into the bowl. Meanwhile, the buoyant stopper ball floats up until it is arrested by the guide strapped to the overflow tube.

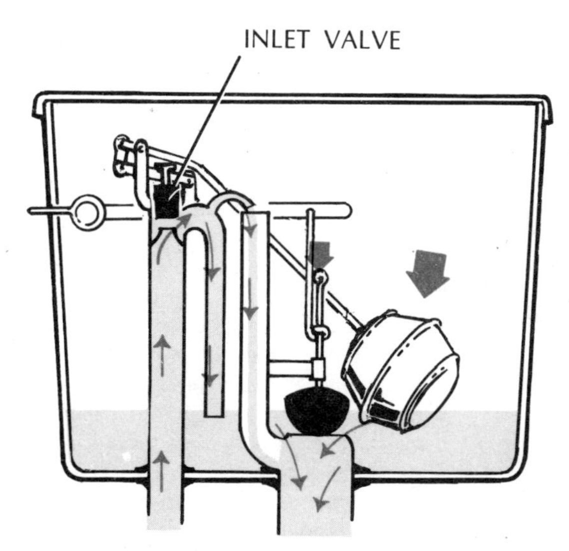

As the water level drops in the tank, the floating stopper drops with it, following a direct path toward the flush valve because the guide directs the movement of the stopper's lift wire. When the stopper ball sinks into the vicinity of the flush valve seat, suction of the escaping water draws the ball firmly into the seat. This shuts off water flow into the bowl and allows the tank to refill.

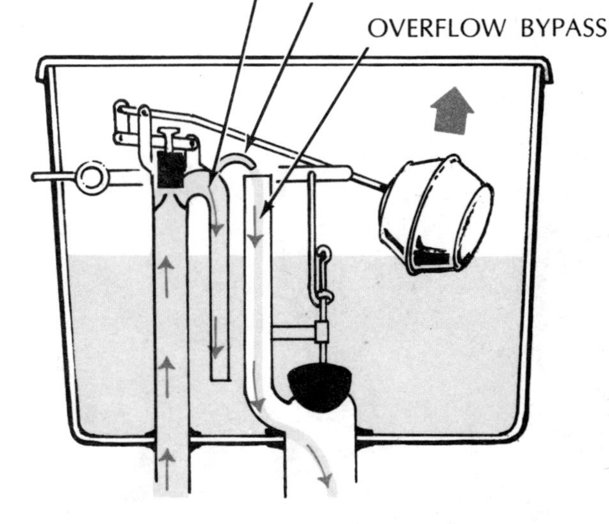

With the stopper seated, the float reaches its lowest level. This causes the float levers to open the inlet valve fully. Water flows into the tank via the filler tube and into the bowl via the bowl-refill tube as well as through the overflow bypass around the closed flush valve. The float rises as the tank fills and, through its levers and valve, gradually stops the flow of water.

Incinerating Toilet

For many years the choice of toilet system has been an easy one—between an indoor flush toilet and an outdoor privy. But water systems are sometimes impractical or inadequate: Faulty septic tanks and sewage systems can lead to dangerous pollution of water resources. Winter freeze-ups take their toll. Recreational and public transport vehicles are subject to dumping restrictions. And environmental legislation on marine discharges imposes strict waste sterilization standards. Today, incinerating toilets are available at prices that easily compete with the compact water-flush systems such as septic tanks and cess pools. Incinerating toilets employ either electricity, natural gas, or other heating fuels to flash off liquid waste as steam and reduce solid matter to a well-sterilized ash.

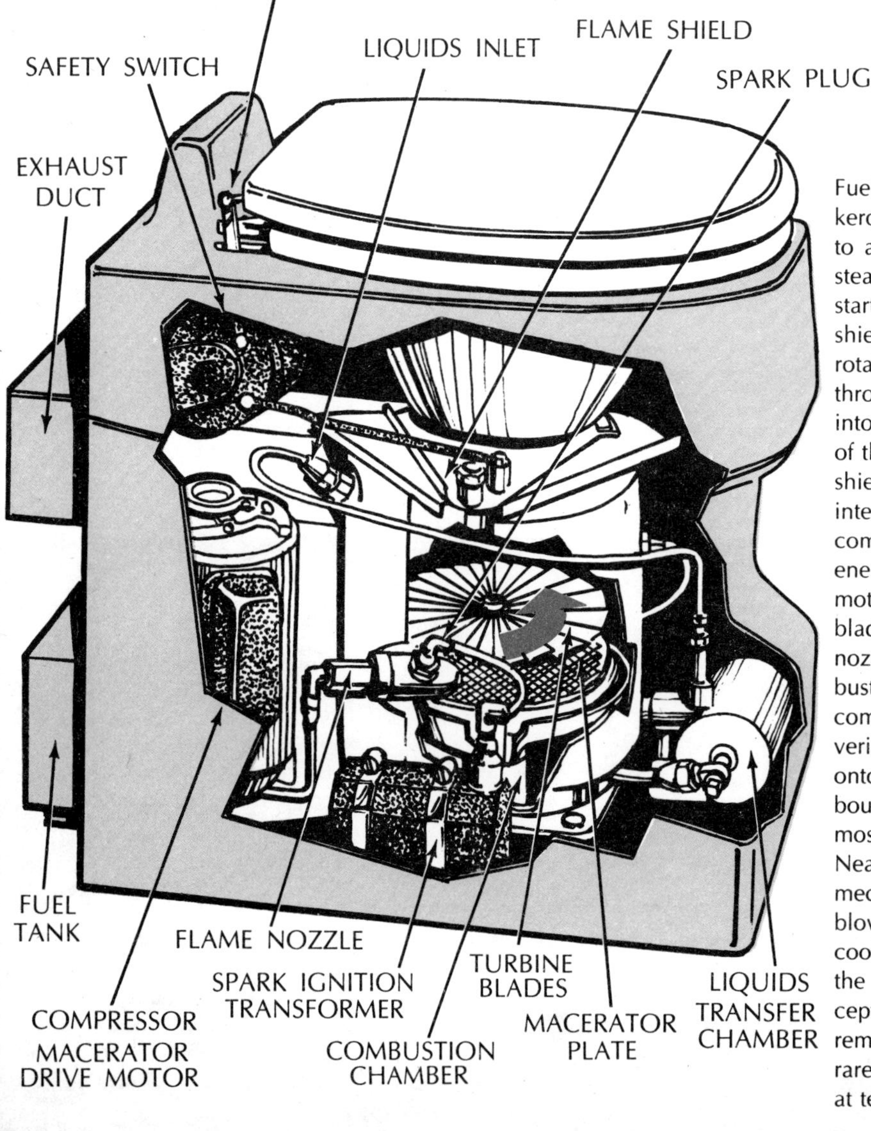

Flame-type

Fueled by either natural gas (LPG), gasoline, kerosene or diesel oil, this toilet reduces solids to ash and flashes off liquids as super-heated steam in four to six minutes. Raising the lid starts the exhaust blower and opens the flame shield. During use, the solids remain on the rotating components while the liquids drain through the turbine blades and macerator plate into a holding tank for treatment near the end of the cycle. Lowering the lid closes the flame shield and activates a safety switch designed to interrupt cycling if the lid is opened before completion. Pushing the "flush" start lever energizes a 110-volt timer motor and a drive motor that powers the compressor, turbine blades, and macerator. The pressurized flame nozzle provides high speed tangential combustion around the entire inner wall of the combustion chamber. The turbine blades pulverize solids and paper, casting them down onto the macerator plate where heat and bouncing reduce waste particles to a fine ash, most of which is drawn into the exhaust duct. Near the end of the combustion cycle, all mechanisms shut down except the exhaust blower which continues during a 3-minute cooling period. Ash that does not exit through the exhaust duct is deposited in a small receptacle near the exhaust duct and can be removed for disposal. The chamber, itself, rarely needs cleaning since combustion occurs at temperatures from 1800 to 2000° F.

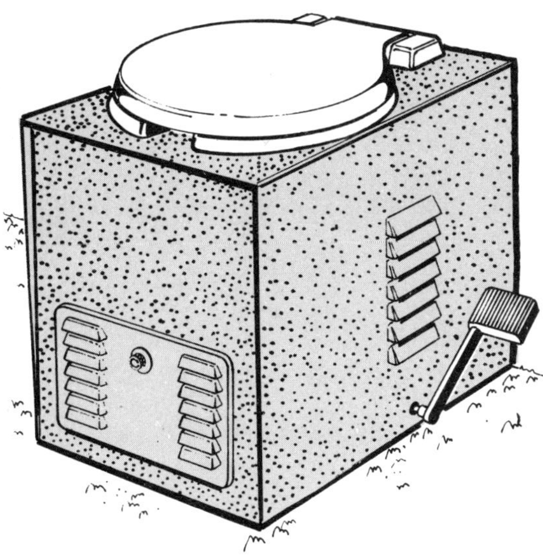

Powered from a 220-volt system, like the home range or clothes drier, this toilet uses about one kilowatt-hour of electricity during its 30-minute incinerating cycle. Before use, a sanitary liner of wax paper is laid into the "bowl." After use, stepping on the mechanically-linked spring-return pedal makes the rear portions of the bowl and the insulated incinerator lid draw back together, allowing wastes to drop into the stainless steel ash pan. A 1300° F. tubular heater in the base of the pan then starts to incinerate the solids and evaporate the liquids. An exhaust fan in the rear of the enclosure pulls air through the incinerator chamber. Most of this air is drawn in through louvres in the casing, serving to cool the outer casing before being blended with the 800 to 1000° F. air from the incinerator chamber. In this way the high temperature of the chamber is reduced to about 120° F. by the time it leaves the exhaust fan. After the heater shuts off, a thermostat keeps the fan *on* until the unit has cooled adequately. The ash pan and heater simply slide out through a front access door, thereby disconnecting the heater's prong-type plug. The heat-sealed toilet bowl can be used during the incineration cycle. After use and closure of the lid, depressing the foot pedal resets the timer for another 30 minutes to ensure complete incineration. For a family of six, the ash pan must be emptied about once a month.

Electric-type

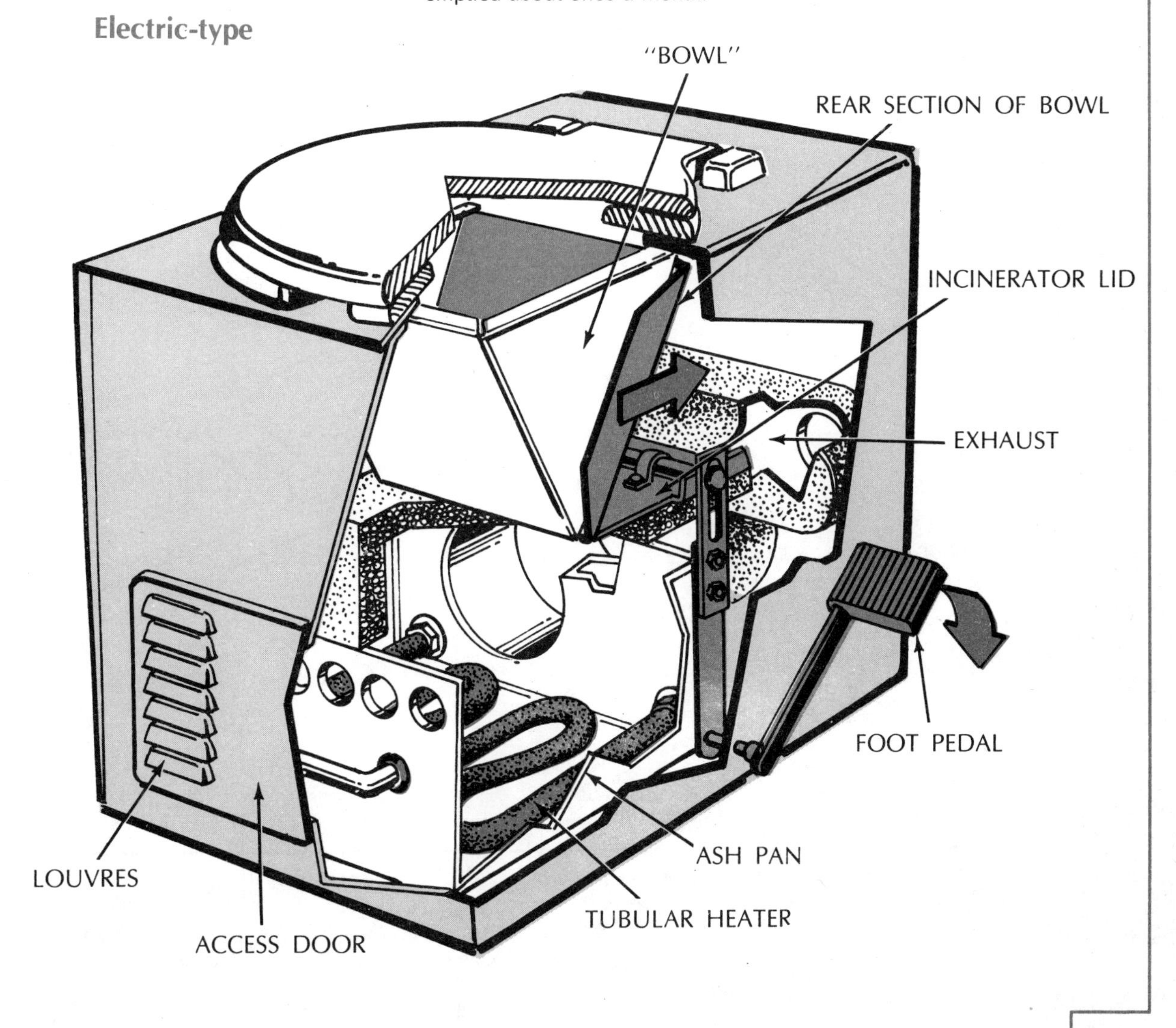

Cylinder Lock

The first pin-tumbler lock was invented by the Egyptians over 4,000 years ago. It had a bolt secured by concealed wooden pins. A key with pegs spaced like the pins raised the pins flush with a bolt, which could then be drawn open. Down through the centuries, locksmiths attempted to bolster the security of weak locks with booby traps that cut off fingers and with devices that trapped or shot would-be lock pickers. Other contrivances included fake key holes and multiple key movements. In the 1860s Linus Yale Jr. patented improvements on a cylinder lock, invented earlier by his father, a builder of bank locks. The resulting Yale cylinder lock, in common use today, was the first mass-produced lock that offered good security, plus master-keying.

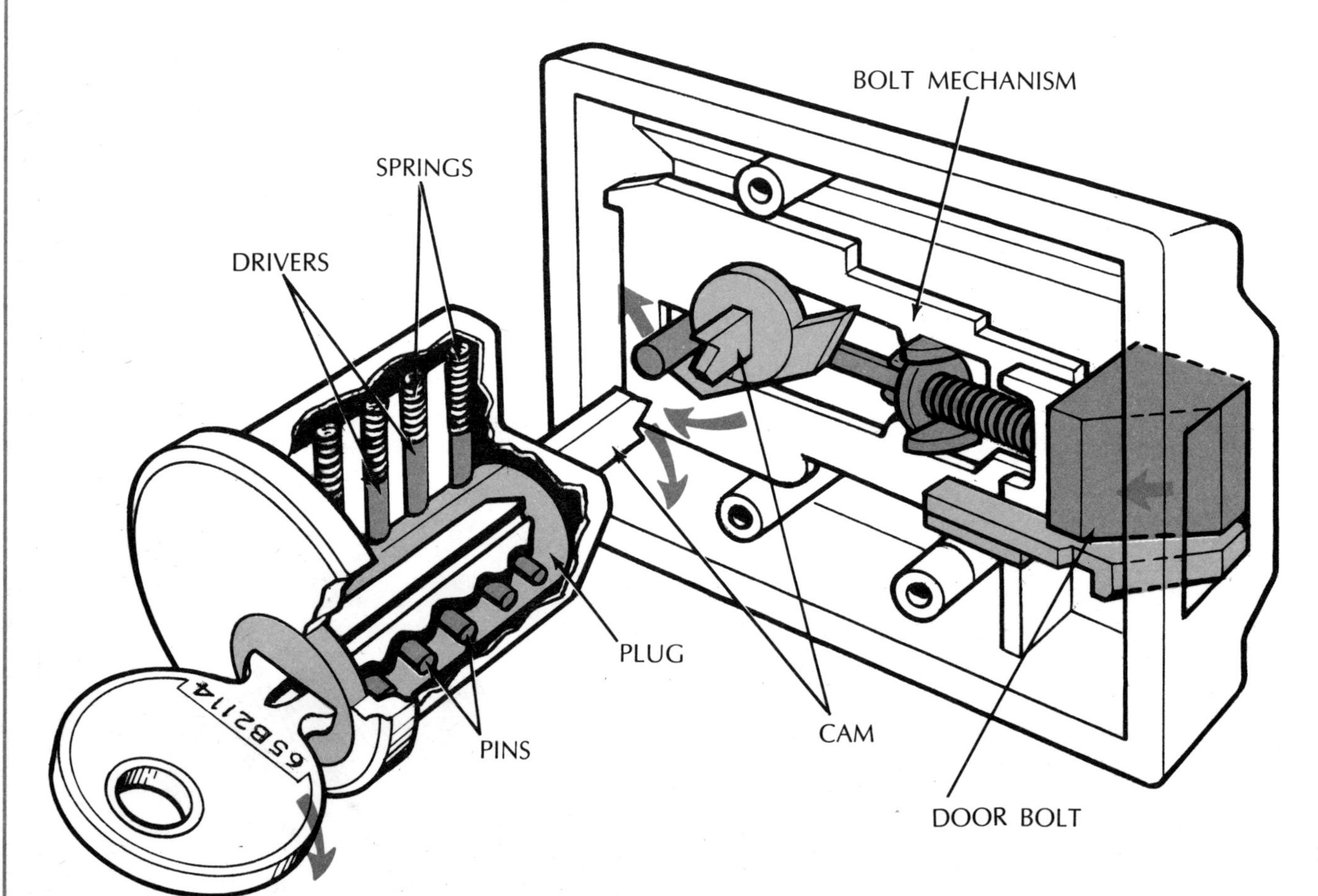

The key receptacle of a pin-tumbler cylinder lock is a cylinder or case with an off-center hole. A plug slotted to receive the key turns in this. Drilled crosswise in both cylinder and plug, and aligned when the key is out, are five small holes. Each contains a round-ended pin and a similar flat-ended driver. When the key is inserted initially (key is shown turned at left), the pins are pushed by small springs into the keyway. The inner end of the plug has a cam or a bar (as shown) that engages the second lock element, the bolt mechanism. As the key turns the plug within the cylinder, this cam or bar operates the bolt mechanism, withdrawing the door bolt.

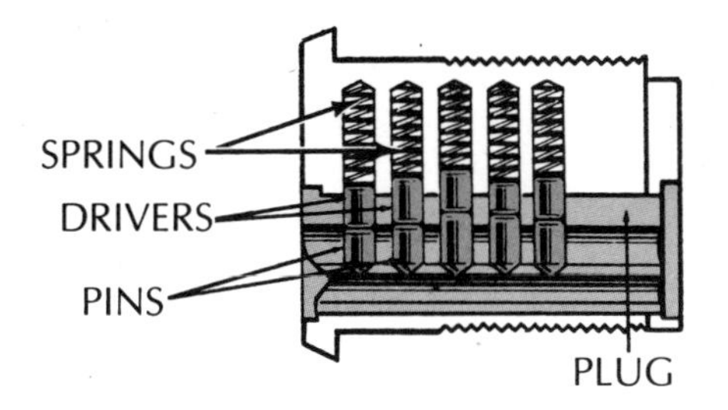

With the key out, all pins are pushed against the lower keyway ridge. Partly in the cylinder, partly in the plug, the drivers keep the plug from being turned and have greater shear strength than anything that can be inserted to force them.

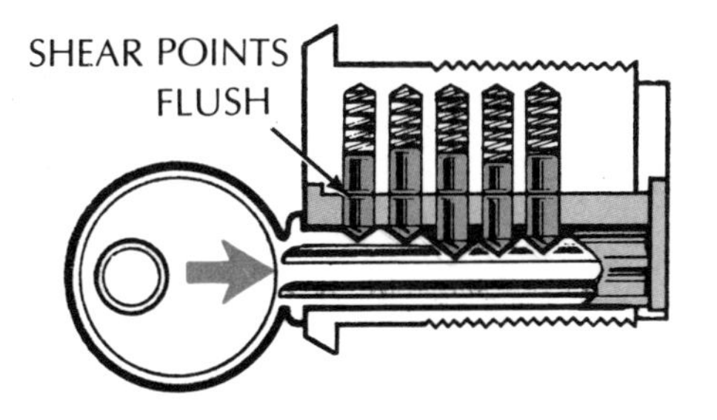

As the key is inserted, the rounded pin ends ride over the key bittings until each pin settles into its own notch. This brings the shear points between drivers and pins exactly to the surface of the plug. The proper key aligns all five shear points.

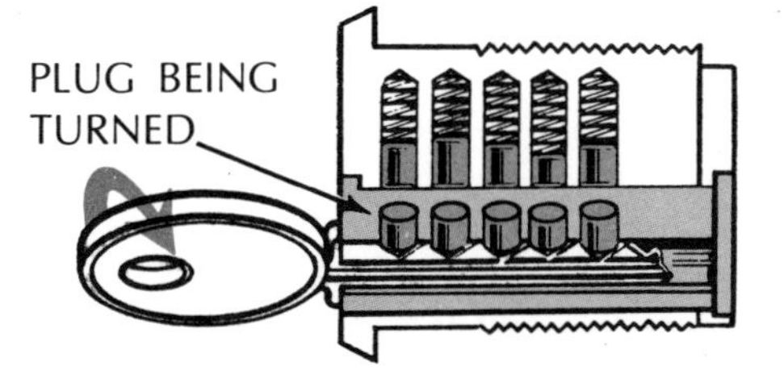

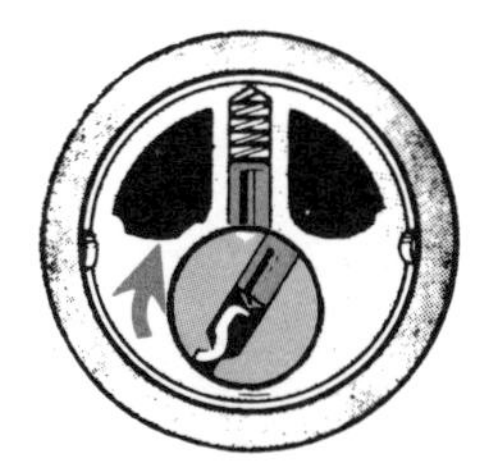

With shear points aligned, the key can turn the plug. A key notch that is either too low or too high will cause a driver or a pin to prevent the plug from turning. Thus a key having but one slightly wrong notch will not free the plug.

Master-keying

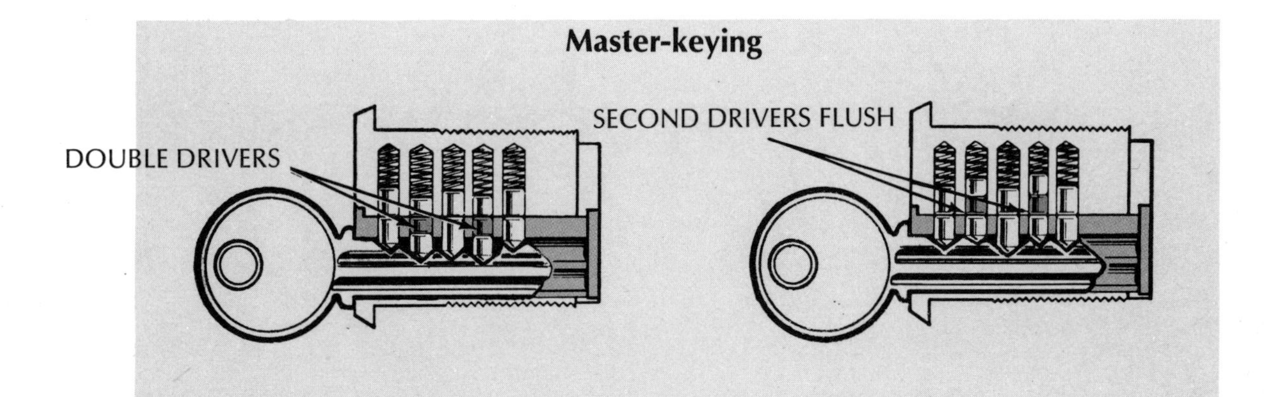

In locks designed for master-keying, some of the drivers are doubled. Here the second and fourth pins have two-part drivers. In this lock a normal key would raise the second and fourth drivers only to their first shear points. But here a master key raises the drivers to the second shear points. Thus this master key could fit two different locks. Increasing the number of pins with double drivers, increases the number of possible key combinations.

Medeco Lock

The best conventional pin tumbler lock can be picked. This may be done by exerting rotational pressure on the plug while lifting each pin individually with a "pick." Some picklocks (legitimate and otherwise) use a lock-picking gun, which is a hand-held mechanism that drives a wave-shaped wire rapidly back and forth beneath the pins. Mathematical chance says there will be an instant when all the pins will be lifted to the correct height. With the gun or a simple pick an expert can open most pin tumbler locks in less than a minute. To frustrate picklocks, MEDECO manufactures a pin tumbler lock which they claim has never been picked. Whereas the standard "pin" need only be lifted the correct distance, each MEDECO pin must be *lifted and turned* the correct amount before the plug can be rotated to open the lock.

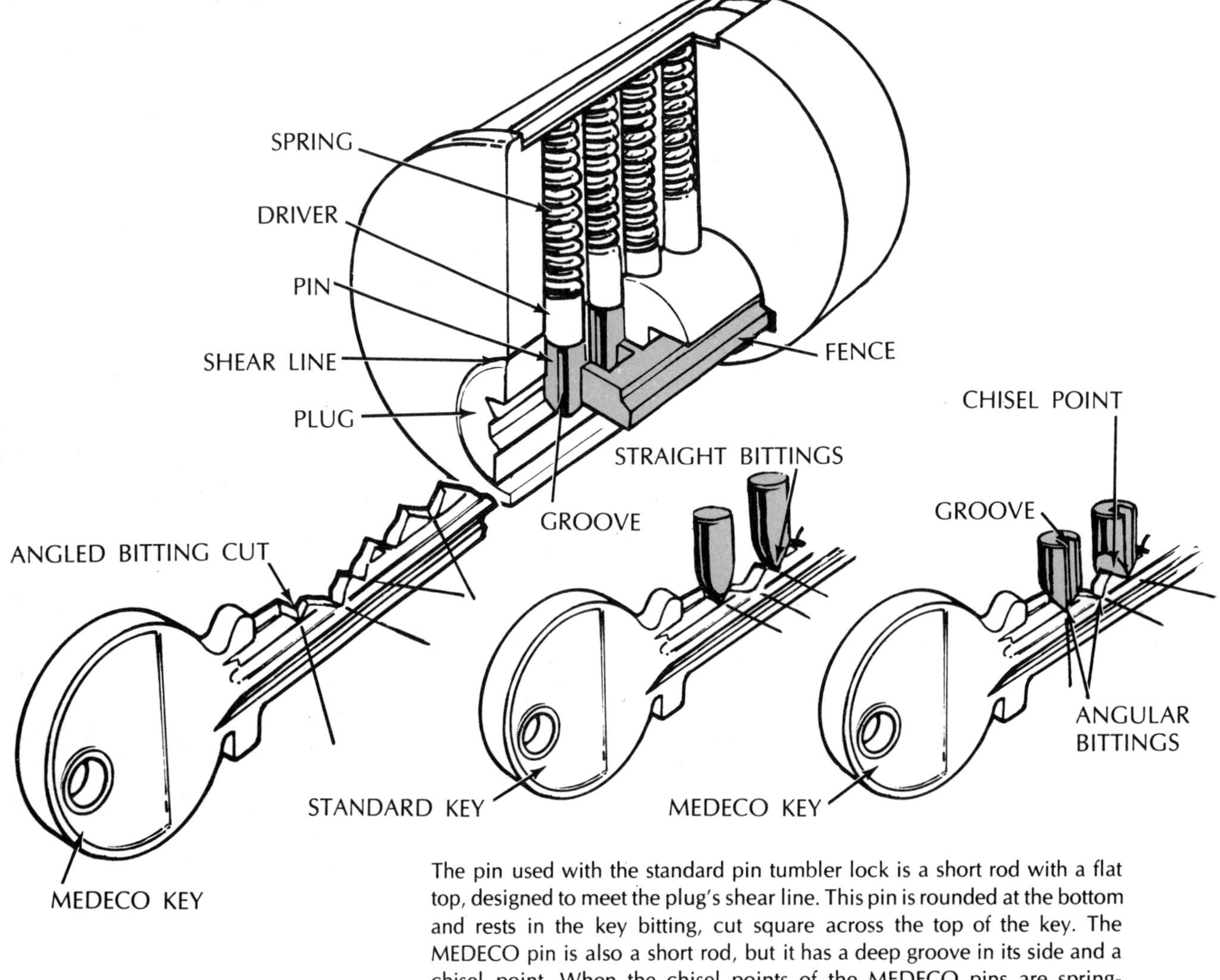

The pin used with the standard pin tumbler lock is a short rod with a flat top, designed to meet the plug's shear line. This pin is rounded at the bottom and rests in the key bitting, cut square across the top of the key. The MEDECO pin is also a short rod, but it has a deep groove in its side and a chisel point. When the chisel points of the MEDECO pins are spring-forced into the angle-cut bittings of the MEDECO key, they rotate the pins in their wells to put the pin grooves in line.

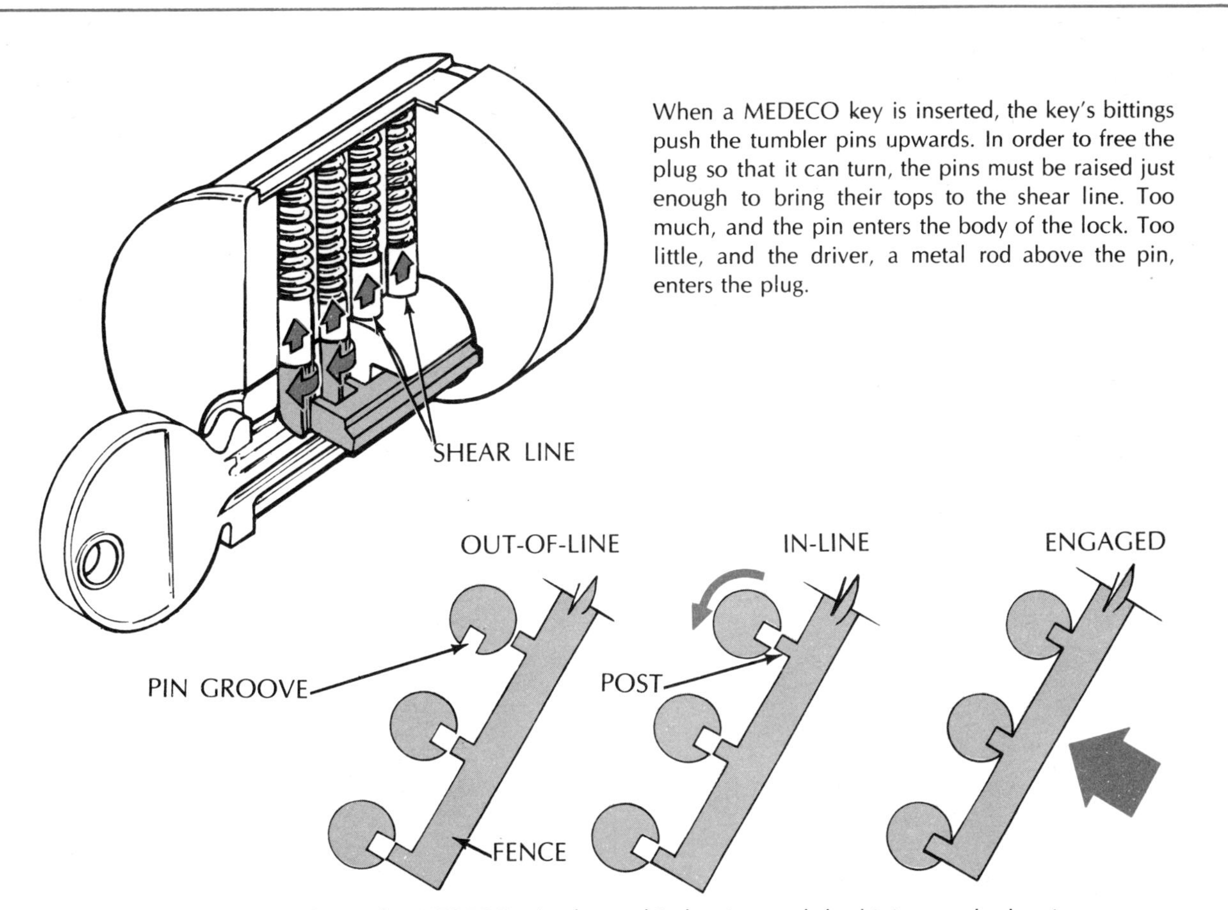

When a MEDECO key is inserted, the key's bittings push the tumbler pins upwards. In order to free the plug so that it can turn, the pins must be raised just enough to bring their tops to the shear line. Too much, and the pin enters the body of the lock. Too little, and the driver, a metal rod above the pin, enters the plug.

Since the MEDECO pins have chisel points and the bitting on the key is cut at an angle, the pins must rotate in their wells a specific amount. When the degree of rotation is correct, the grooves in the sides of each pin line up with the posts on the fence. At left, when one angle on one "cut" on the key is incorrect, the fence cannot enter the plug, and the key cannot be turned. At center, the bitting angles are correct for that lock since the grooves line up with the fence posts. At right, the fence posts have entered all the grooves, and the fence no longer projects into the body of the lock.

When the posts on the fence enter the grooves on the pin and the tops of the tumbler pins are flush with the shear line, the plug can be turned to open the lock. If each pin were to have five possible degrees of rotation and ten different heights, it could assume 10×5 or 50 different positions. Thus, in theory, a 4-pin MEDECO lock (some have more) can be made to accept any one of 50^4 or 6,250,000 different keys. In addition to multiplying the number of key "changes" a lock will take, the rotating pin improvements compound the picklock's problems. Not only must each pin be lifted an unknown distance, it must also be rotated an unknown number of degrees.

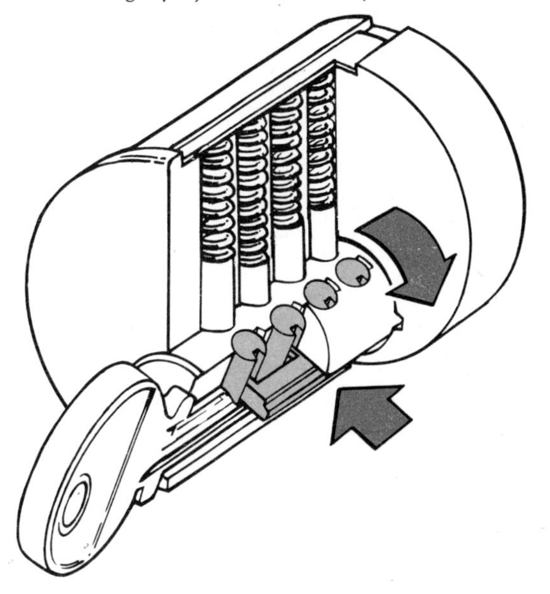

Water Filter

A municipal water treatment plant fights a constant battle against bacterial contamination of its water supplies while trying to cope with rising levels of man-made pollutants. But the water treatment plant does not remove *all* impurities and unpleasant odors from tap water. The best way to improve the water is to use a sink-connected water filter. Tap water originates from rain and snow that seep into the ground or run off into lakes and reservoirs. In this process, water dissolves natural chemical and organic matter as well as man-made chemicals and pollutants that combine to give the water some unwanted properties. Water filters employ either cellulose discs or activated charcoal, an extremely pourous material that has enormous surface area. A mere handful of activated charcoal has about as much absorptive area as the surface of a football field.

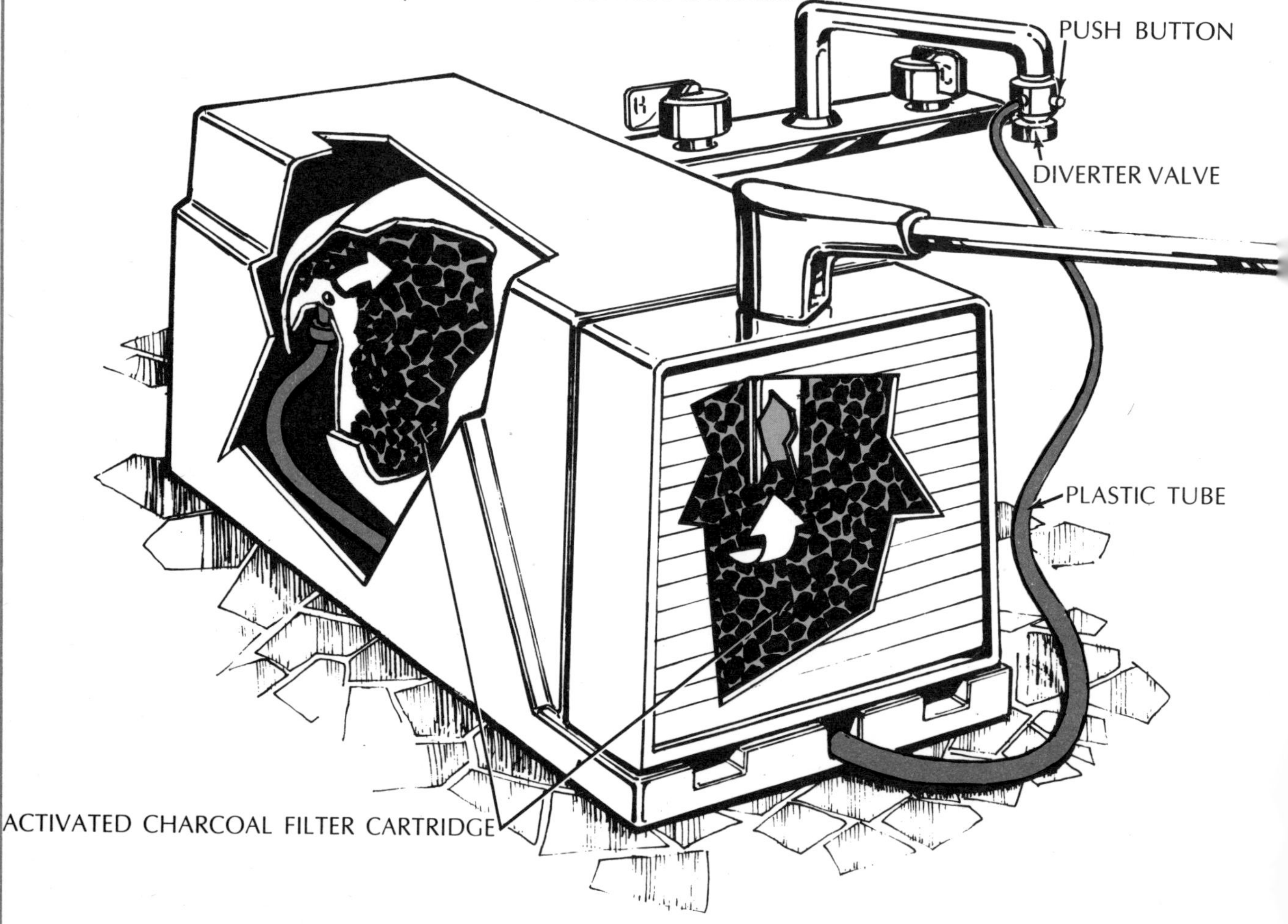

This counter-top filter has a snap-on fitting that attaches to the faucet. The fitting contains a diverter valve that can be switched to direct water flow either to the sink or into the filter through a flexible plastic tube. This tube connects to a plastic container that encloses the charcoal. As water moves slowly through the filter's intricate surface system, molecules and ions of odor and taste-producing pollutants are absorbed by the charcoal granules. Thus, the water flowing out the filter spout is crystal clear and free of odor. (Of course, this filtered water may still contain microscopic and sub-microscopic bacteria and viruses.)

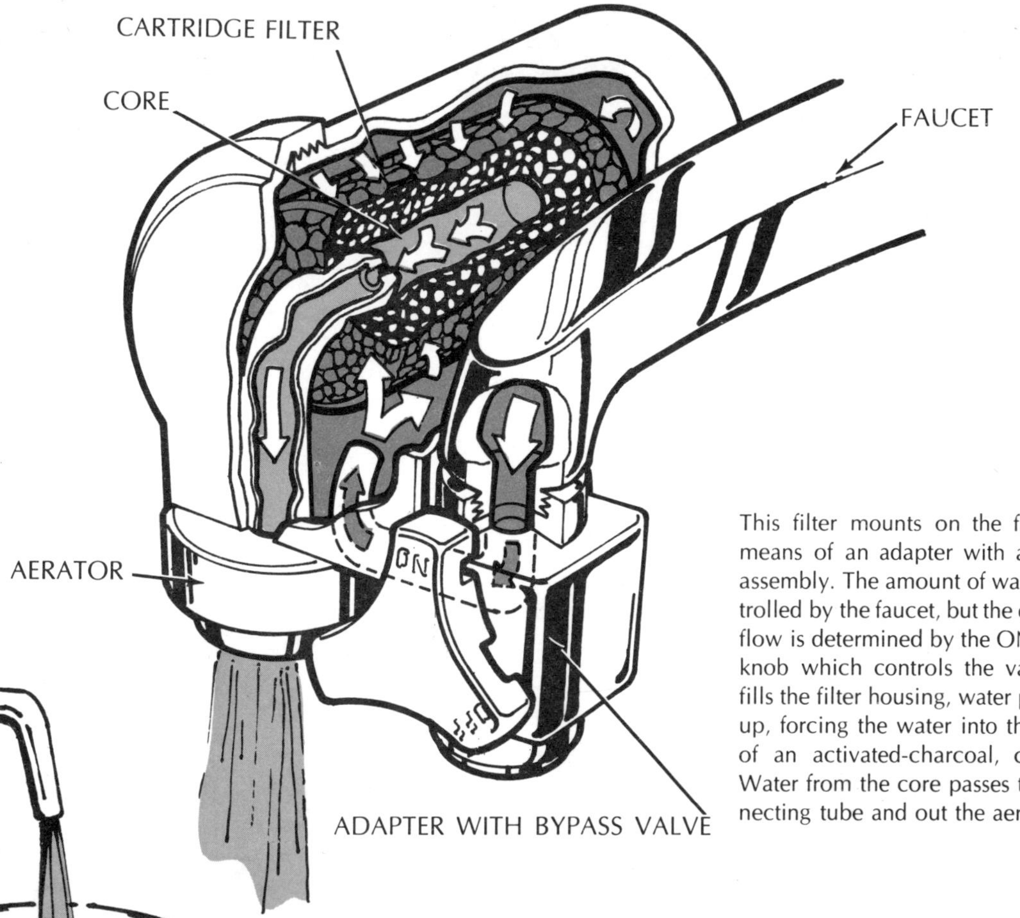

CARTRIDGE FILTER
CORE
FAUCET
AERATOR
ADAPTER WITH BYPASS VALVE

This filter mounts on the faucet itself by means of an adapter with a bypass valve assembly. The amount of water flow is controlled by the faucet, but the direction of the flow is determined by the ON/OFF selector knob which controls the valve. As water fills the filter housing, water pressure builds up, forcing the water into the hollow core of an activated-charcoal, cartridge filter. Water from the core passes through a connecting tube and out the aerator assembly.

All filter manufacturers warn against running *hot* water through filters, since heat deteriorates the holding power of activated charcoal granules and may drive out some of the trapped odor-producing molecules. When used only to filter drinking water, charcoal cartridges normally last up to six months. But cartridges handling water with heavy amounts of pollutants may have to be changed more often.

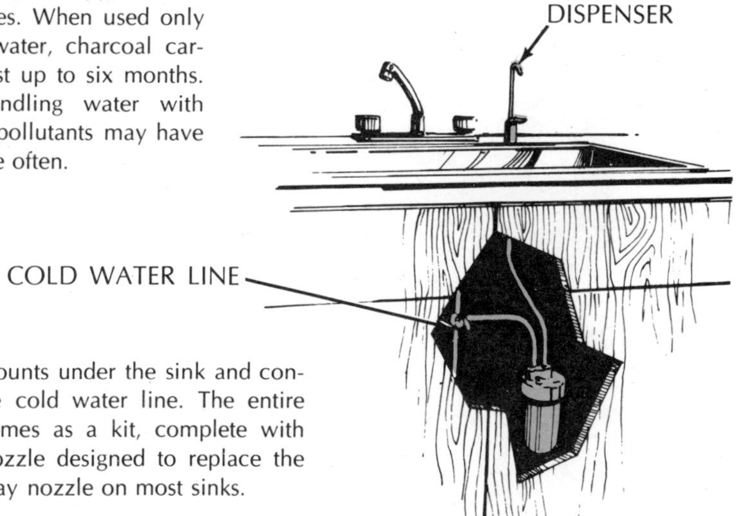

DISPENSER
COLD WATER LINE

This filter mounts under the sink and connects to the cold water line. The entire assembly comes as a kit, complete with dispenser nozzle designed to replace the standard spray nozzle on most sinks.

Xerographic Copy Machine

Tinkering in a darkened room in Astoria, New York, in the mid 1930s, a physicist turned patent attorney by the name of Chester F. Carlson discovered a fundamental copy-making process that has revolutionized office work. Carlson originally set out to develop a new way of reproducing blueprints and other documents. By 1938 he had made the first crude copies by a technique that uses electric charges—similar to static electricity—to duplicate an original on a second sheet of paper. The process acquired the name "xerography" from the Greek words *xeros* (dry) and *graphikos* (to print). Thus, literally, xerography is dry printing.

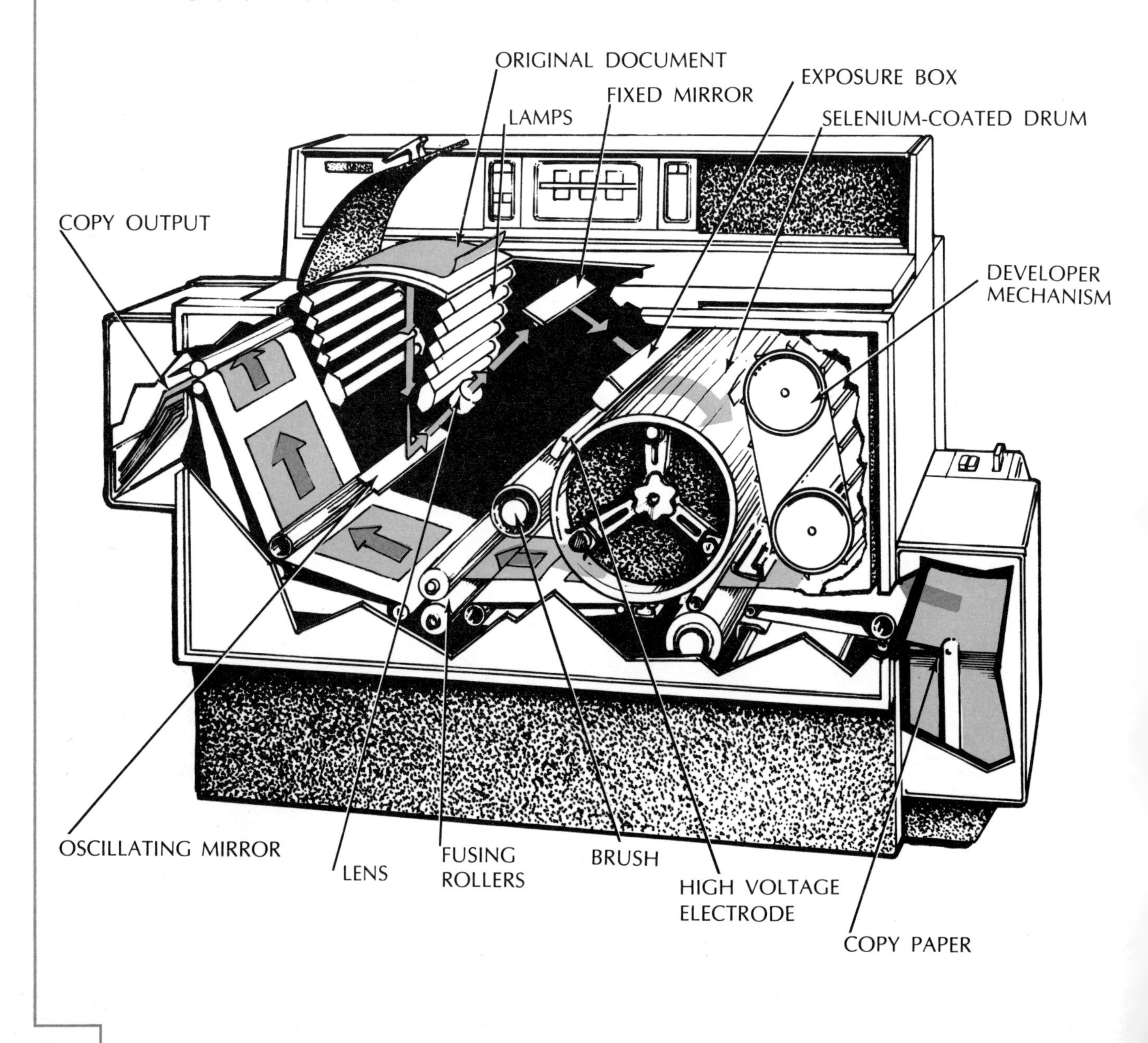

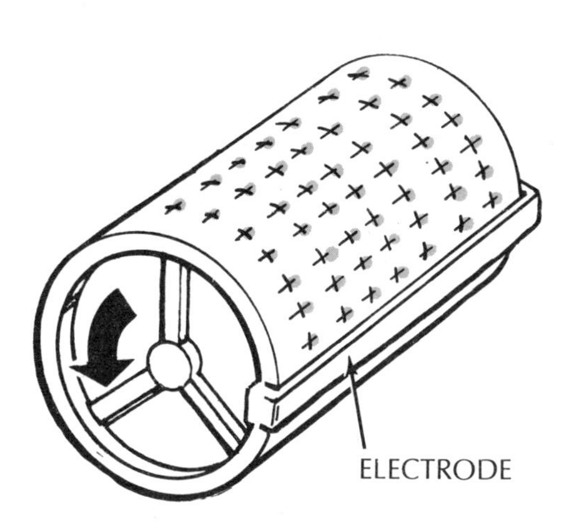

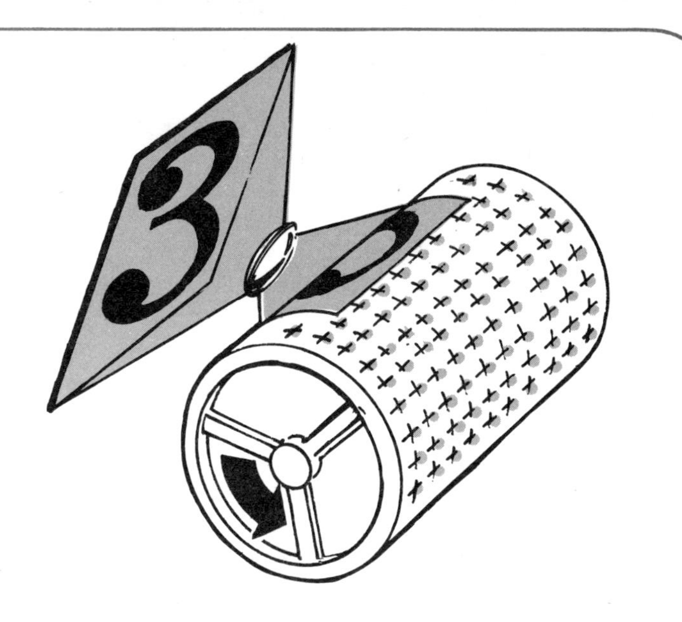

1. In the five-step copying process, a selenium coated, cylindrical metal drum is rotated beneath an electrode that is connected to a very-high-voltage (7,000 volts) power supply. This gives the drum's selenium surface a positive electrical charge.

2. Lenses and mirrors then focus an image of the original document onto the revolving drum. On the drum, light areas of the image, corresponding to white areas of the original, lose their positive charge. But where the image is dark the charge remains.

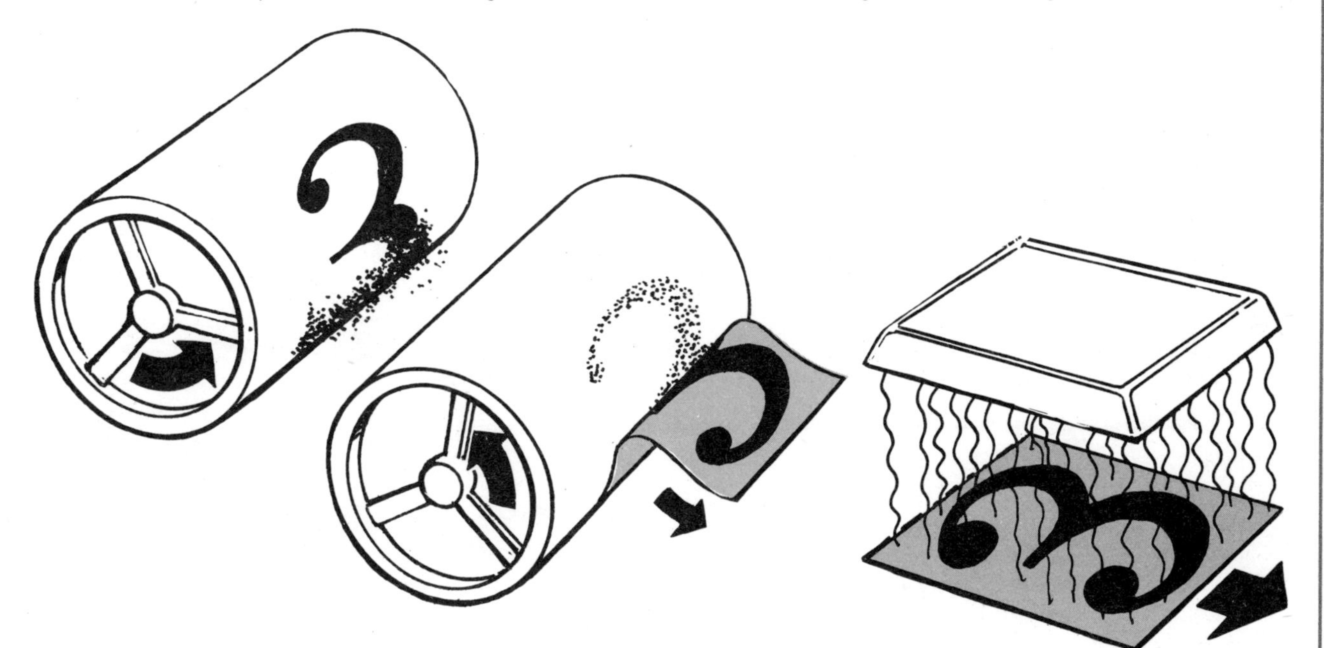

3. The charge on the drum becomes an electrical "mirror image" of the original. Negatively-charged "toner" powder then clings to the positive charges on the drum, forming a replica of the original.

4. A positively-charged sheet of paper is rolled against the revolving drum. The paper attracts most of the negatively-charged toner particles and acquires an image that is a close likeness to that of the original.

5. A heat source then softens the toner particles and fuses them to the paper. This creates a permanent copy of the original which then emerges from the machine. Some machines have collating mechanisms.

Electrostatic Copy Machine

To simplify things, some people refer to xerography as the plain-paper copying process and to electrostatic copying as the coated-paper process. In both processes a uniform negative charge is employed to bring about an image transfer from an original document. However, the electrostatic process employs materials and techniques much like those for photographic printing. First a negative electrical charge is applied to the specially coated copy paper. Then the paper is exposed to a reflected replica of the document. Next the paper passes through a developing solution before being squeegeed and air dried. It is then delivered into the receiving tray.

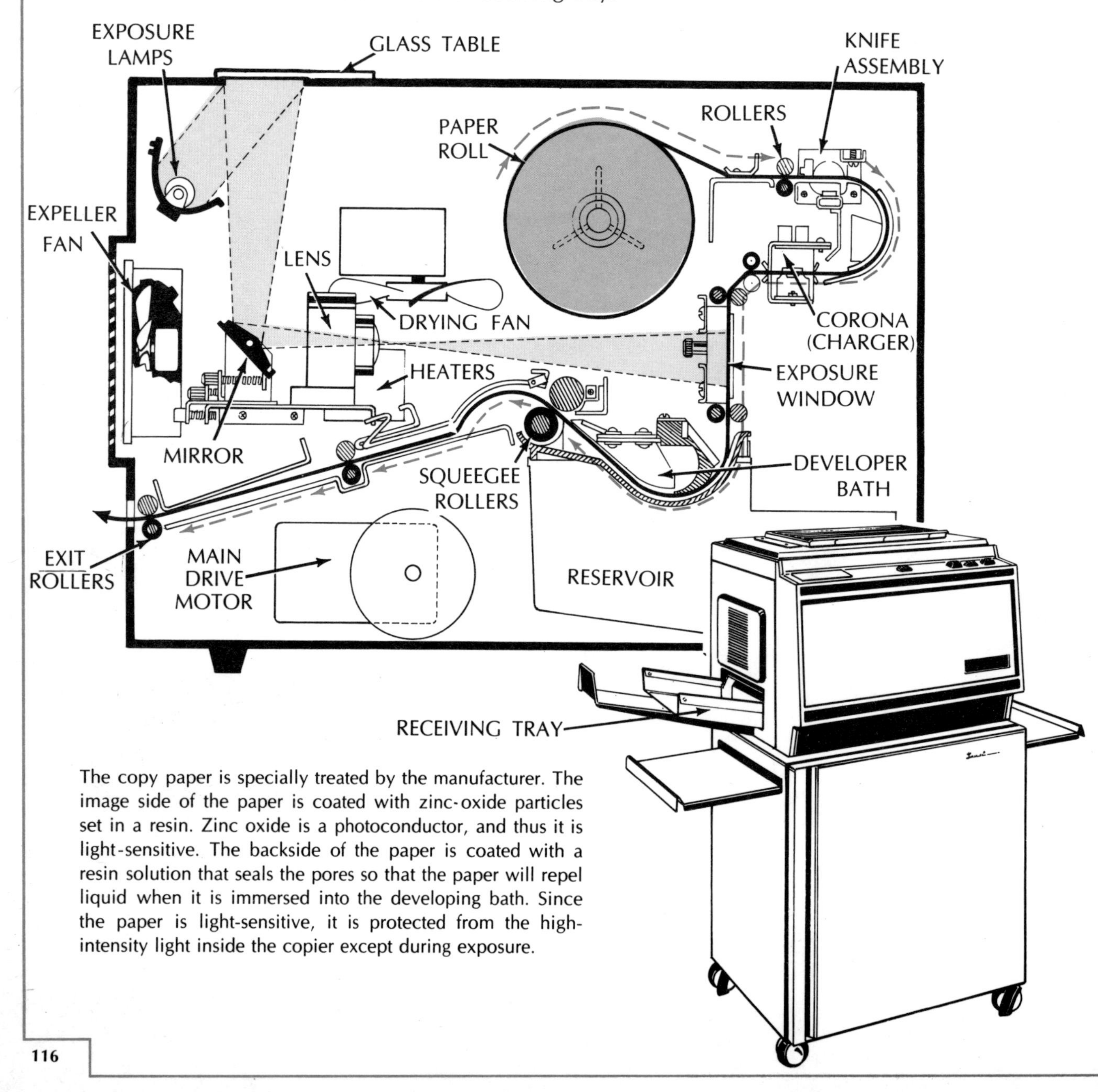

The copy paper is specially treated by the manufacturer. The image side of the paper is coated with zinc-oxide particles set in a resin. Zinc oxide is a photoconductor, and thus it is light-sensitive. The backside of the paper is coated with a resin solution that seals the pores so that the paper will repel liquid when it is immersed into the developing bath. Since the paper is light-sensitive, it is protected from the high-intensity light inside the copier except during exposure.

After the paper drum rolls off the correct length of paper, a knife assembly cuts it to size. Rollers then pass the paper through the electrical fields produced by two sets of charged wires in a high voltage unit called a corona. The positively charged wires deposit a uniform positive charge on the back-side of the paper, while the negatively-charged wires produce a uniform negative charge on the paper's coated image side.

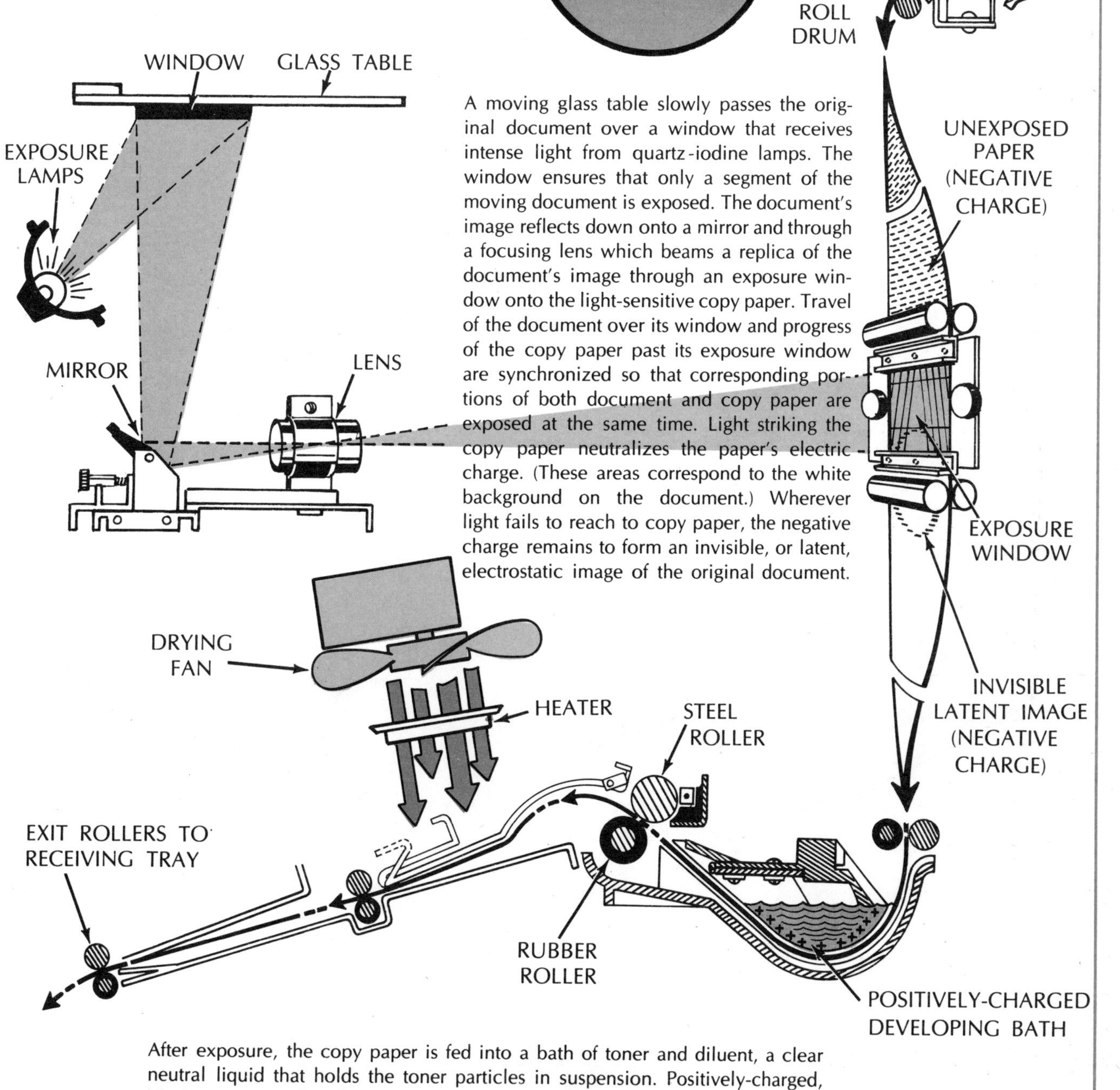

A moving glass table slowly passes the original document over a window that receives intense light from quartz-iodine lamps. The window ensures that only a segment of the moving document is exposed. The document's image reflects down onto a mirror and through a focusing lens which beams a replica of the document's image through an exposure window onto the light-sensitive copy paper. Travel of the document over its window and progress of the copy paper past its exposure window are synchronized so that corresponding portions of both document and copy paper are exposed at the same time. Light striking the copy paper neutralizes the paper's electric charge. (These areas correspond to the white background on the document.) Wherever light fails to reach to copy paper, the negative charge remains to form an invisible, or latent, electrostatic image of the original document.

After exposure, the copy paper is fed into a bath of toner and diluent, a clear neutral liquid that holds the toner particles in suspension. Positively-charged, black toner particles cling to the paper's negatively-charged latent image, thus forming a copy of the original document. As the copy emerges from the developing tank, the image is made permanent as it passes between a steel and a rubber roller. Finally, the copy is dried beneath a heater coil and fan before it is transported into the receiving tray.

Xerographic Color Copier

This copier electrostatically prints color reproductions on ordinary paper by mixing three dry, primary-colored ink powders, called "toners," to create varied hues of yellow, cyan (dark blue), and magenta (deep purplish red). In subtractive combination, these primary toners also produce green, red, blue, and black, for a total of seven basic colors. These allow reproduction in full color, although selector buttons allow single-color and three-color modes as well.

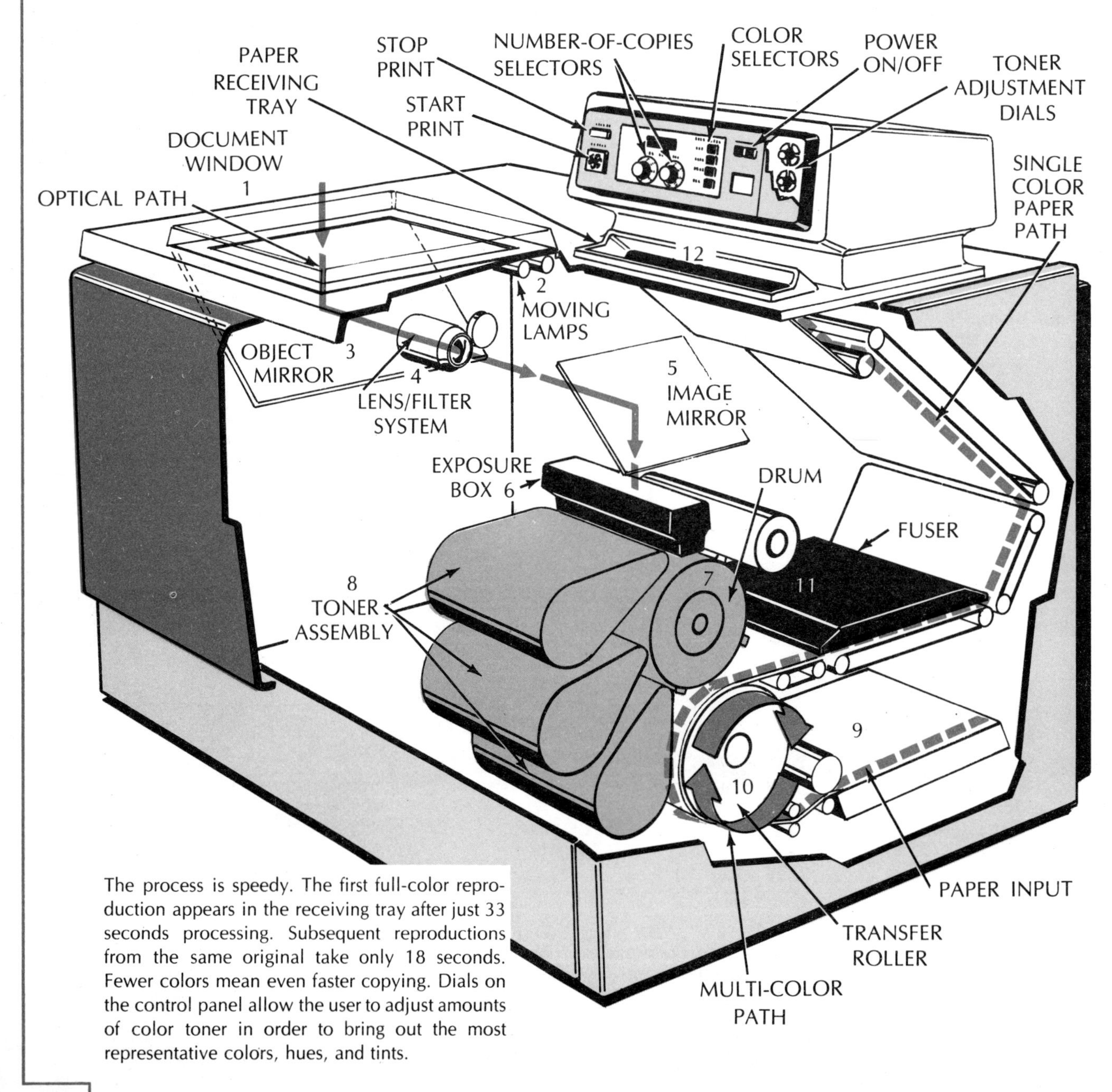

The process is speedy. The first full-color reproduction appears in the receiving tray after just 33 seconds processing. Subsequent reproductions from the same original take only 18 seconds. Fewer colors mean even faster copying. Dials on the control panel allow the user to adjust amounts of color toner in order to bring out the most representative colors, hues, and tints.

First, the original is placed face down on the document platen window (1). Moving lamps (2) scan the document, projecting the image onto an object mirror (3). The image is reflected from the object mirror through a lens and color-separation filter system (4) onto the image mirror (5). This mirror projects the image through the exposure box (6) onto a positively-charged, light-sensitive drum (7) which has a positive electric charge.

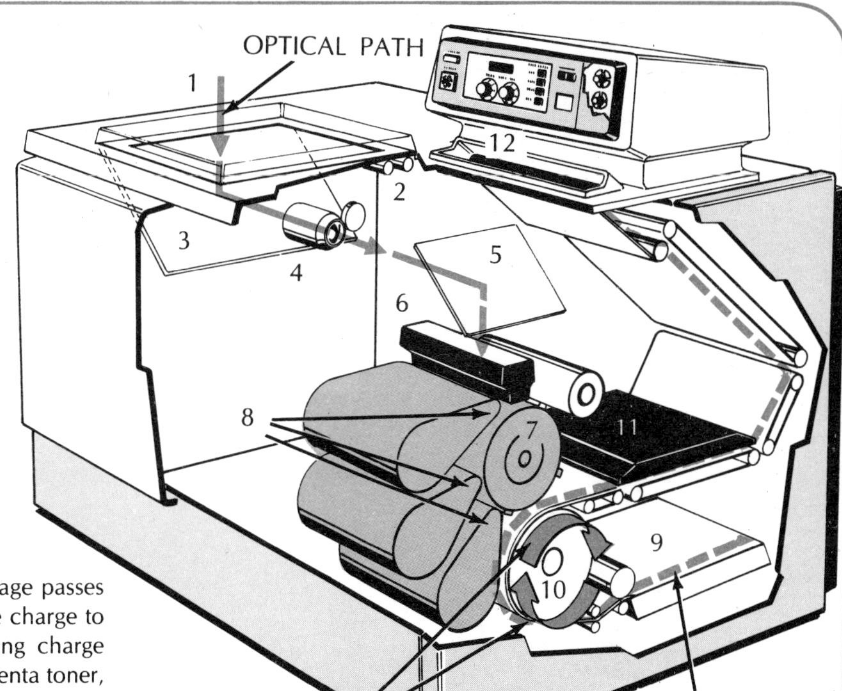

In the first cycle, light reflected from the image passes through a green filter and causes the positive charge to leak off the non-image areas. The remaining charge pattern constitutes a latent image. Next, magenta toner, the filter's complementary color, is applied by magnetized rollers (8) to the "charge image" remaining on the drum. Negatively-charged toner, which coats magnetic-carrier particles, is taken to the rollers by paddles and is attracted to the positively-charged areas on the drum, developing the image.

Paper from the input tray (9) is held on a charged transfer roll (10) which attracts the toner image from the alloy drum and holds it on the paper. The paper remains on the transfer roll for up to three sequential revolutions depending on the number of colors selected. In the second cycle, light from the original passes through a blue filter, and its complementary color, yellow toner, is applied in exact alignment over the magenta toner. In the third and final cycle, light reflected from the original passes through a red filter, resulting in application of cyan toner over the already deposited magenta and yellow. The paper is then transported to the fuser (11) where heat melts the color toners, fusing them to the paper. The fused toners merge to create the color copy, and the finished reproduction appears in the receiving tray (12).

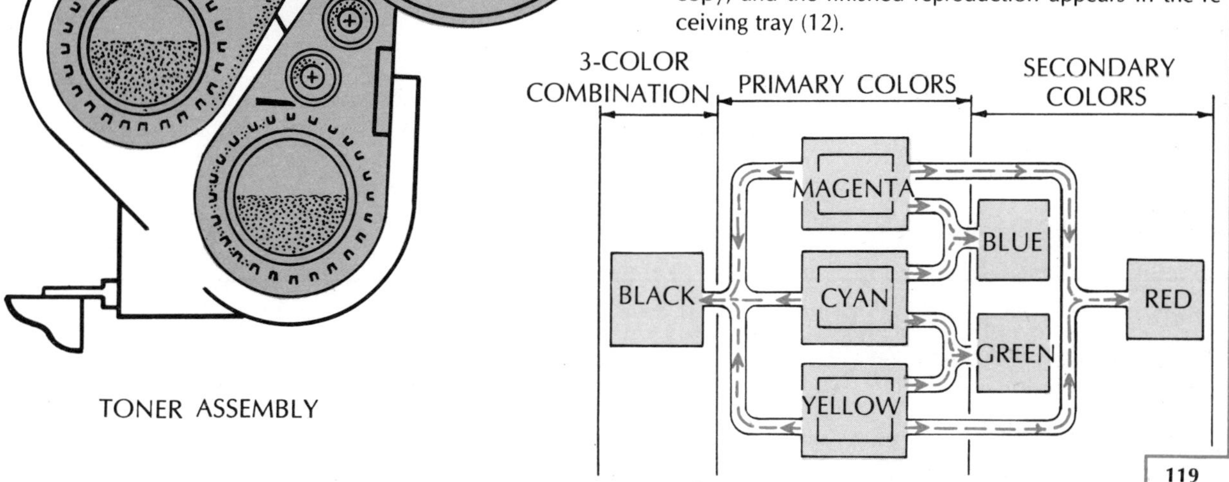

TONER ASSEMBLY

Scuba

Until the 1930s, man had dived either as a breath-holding swimmer or as a helmeted diver with cumbersome equipment. But the clouds of World War II triggered interest in "frogmen" who could perform underwater reconnaissance and demolition. Vest-like "rebreathers" that had been used in World War I for gas attacks and submarine escapes came into use. But the rebreather's chemically re-circulated oxygen was hazardous at depths over 30 feet. Effective face masks, swim fins, and insulating suits appeared. In 1943, two Frenchmen, Jacques-Yves Cousteau and Emile Gagnan, developed an air cylinder which fed compressed air into a demand-type regulator. Today similar devices are widely used by commercial and sport divers and are called "scuba" (self-contained underwater breathing apparatus).

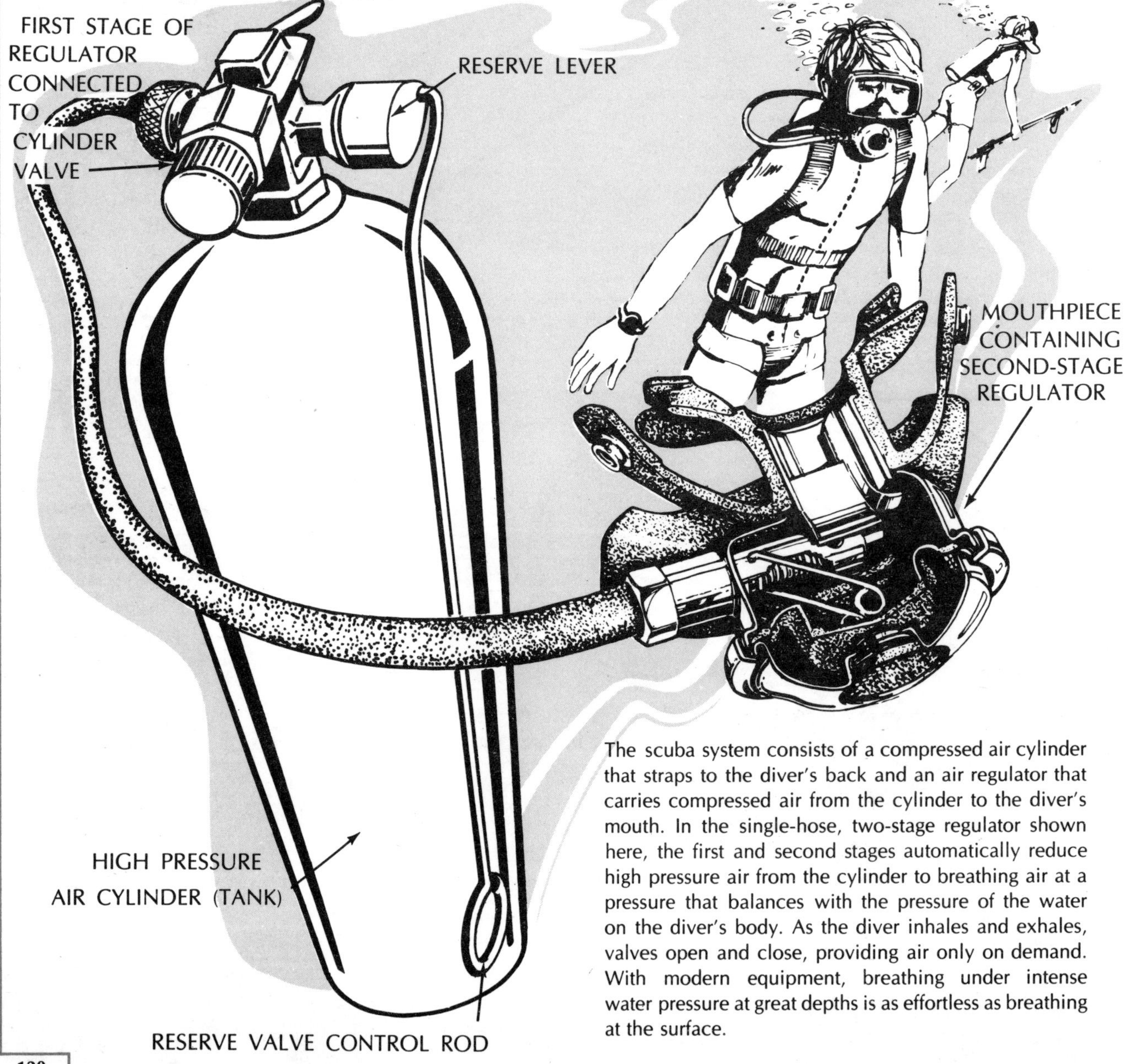

The scuba system consists of a compressed air cylinder that straps to the diver's back and an air regulator that carries compressed air from the cylinder to the diver's mouth. In the single-hose, two-stage regulator shown here, the first and second stages automatically reduce high pressure air from the cylinder to breathing air at a pressure that balances with the pressure of the water on the diver's body. As the diver inhales and exhales, valves open and close, providing air only on demand. With modern equipment, breathing under intense water pressure at great depths is as effortless as breathing at the surface.

The three-dimensional drawing at right shows the regulator's first stage in position to be fitted over the cylinder valve and seated on the valve's "O" ring. In the line drawing below, the cylinder valve's screw-type ON-OFF control is turned on and thereby opens the main valve seat, allowing high pressure air to exit via the valve opening. Air pressure in a freshly filled tank varies from about 2250 to 3000 psi (pounds per square inch). As air is consumed, the tank's internal pressure decreases. When air pressure drops near 300 psi, the reserve valve spring forces the reserve valve seat to greatly restrict the flow of air. This signals the diver that the tank has only 300 psi, or three to five minutes of air remaining. He then pulls the reserve lever open by means of a control rod. Pulling the rod down rotates the cam-action reserve seat open, permitting the diver to make a safe and leisurely ascent.

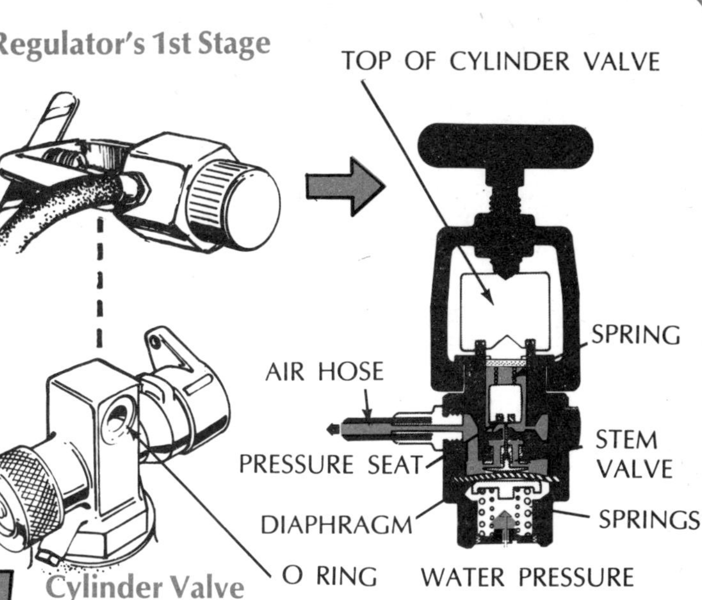

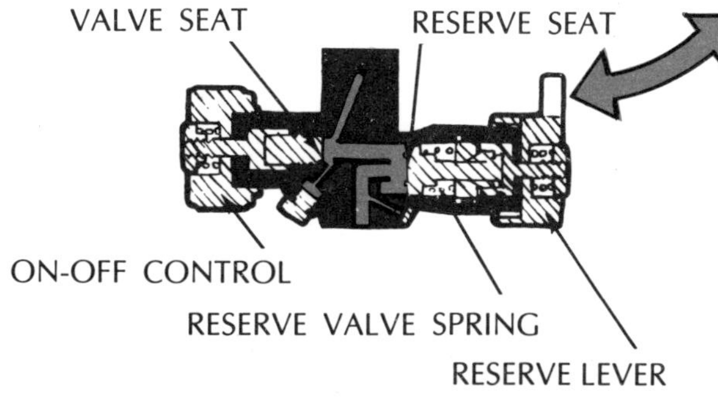

Within the regulator's first stage, opposing springs cause a diaphragm to dimple inward any time pressure within the regulator drops below about 130 psi over outside water pressure. Movement of the diaphragm is transmitted to the stem valve which forces the high pressure seat to open. This allows high pressure cylinder air to flow into the first stage and then through the hose to the second stage. The high pressure valve remains open until the diver stops inhaling or exhales, causing pressure in the first stage to climb back to at least 130 psi over outside water pressure. This allows the stem valve springs to close the high pressure seat and check the flow of high pressure air from the cylinder.

The regulator's second stage, located within the diver's mouthpiece, has a demand diaphragm that balances internal air pressure with outside water pressure. The diver's slightest inhalation lowers pressure under the diaphragm, allowing water pressure to force the diaphragm toward the lower pressure. The diaphragm depresses a demand valve lever that unseats the second stage valve. Air flows in as long as the diver inhales or manually depresses a purge button, designed to purge the mouthpiece of water so that the diver can insert it into his mouth and resume breathing—all under water. When inhalation stops, incoming air continues to flow until the air pressure under the diaphragm equals outside water pressure. The diaphragm thus balances. The demand valve lever causes the second stage valve to reseat. And air flow stops. The diver's exhaled air passes through the non-return valve in the mouthpiece into the water. Thus, the breathing cycle is completed, and the regulator is again ready to respond to demand and another inhale-exhale cycle.

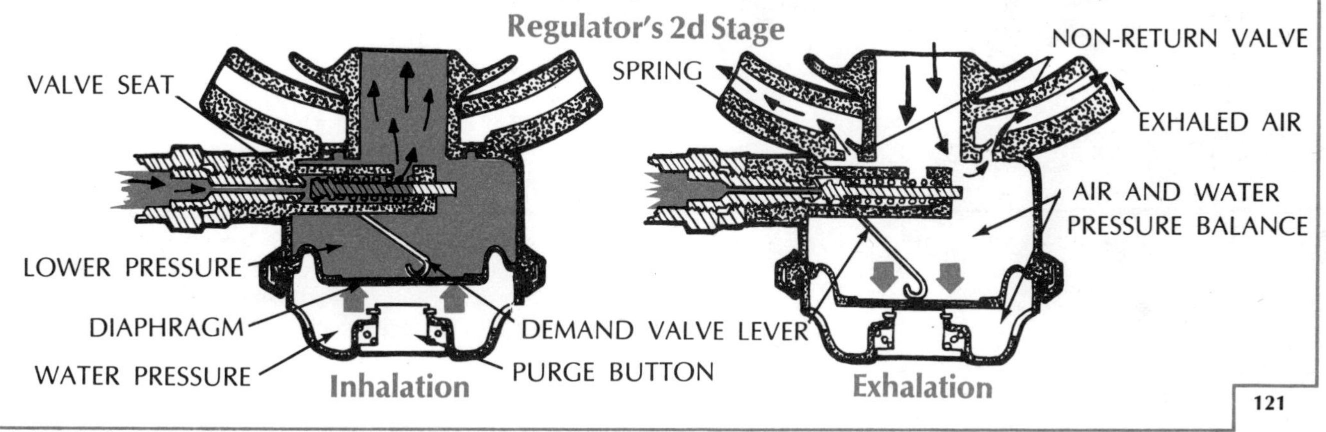

Electric Carving Knife

If fault were to be assigned for the ragged meat carving of Stone Age Man, it would go less to lack of culinary skills than to poor cutlery. Yet many modern dinner hosts with quality cutlery parcel out more tattered morsels per meal than did their flint-wielding ancestors. Few people today have the time or interest in the culinary arts to study carving and maintain well-honed knives. But, using an electric knife, anyone with a little practice can develop the essentials for coaxing handsome slices from the most "difficult" meats. The electric knife employs removable serrated blades that reciprocate rapidly side-by-side in ½-inch strokes. The knife's smooth blade action eliminates the need for manual saw action and requires little more from the user than a watchful eye and good intentions.

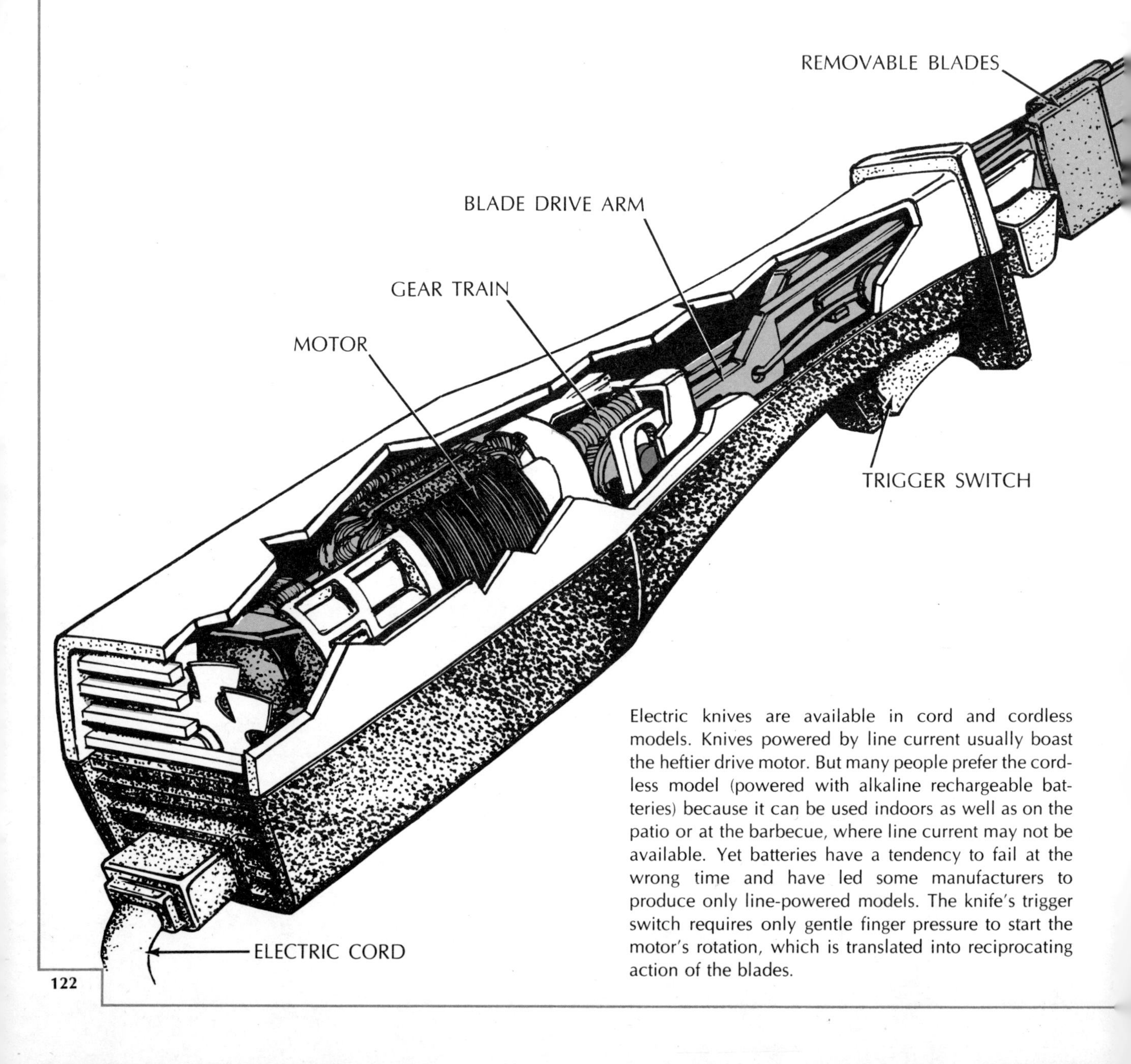

Electric knives are available in cord and cordless models. Knives powered by line current usually boast the heftier drive motor. But many people prefer the cordless model (powered with alkaline rechargeable batteries) because it can be used indoors as well as on the patio or at the barbecue, where line current may not be available. Yet batteries have a tendency to fail at the wrong time and have led some manufacturers to produce only line-powered models. The knife's trigger switch requires only gentle finger pressure to start the motor's rotation, which is translated into reciprocating action of the blades.

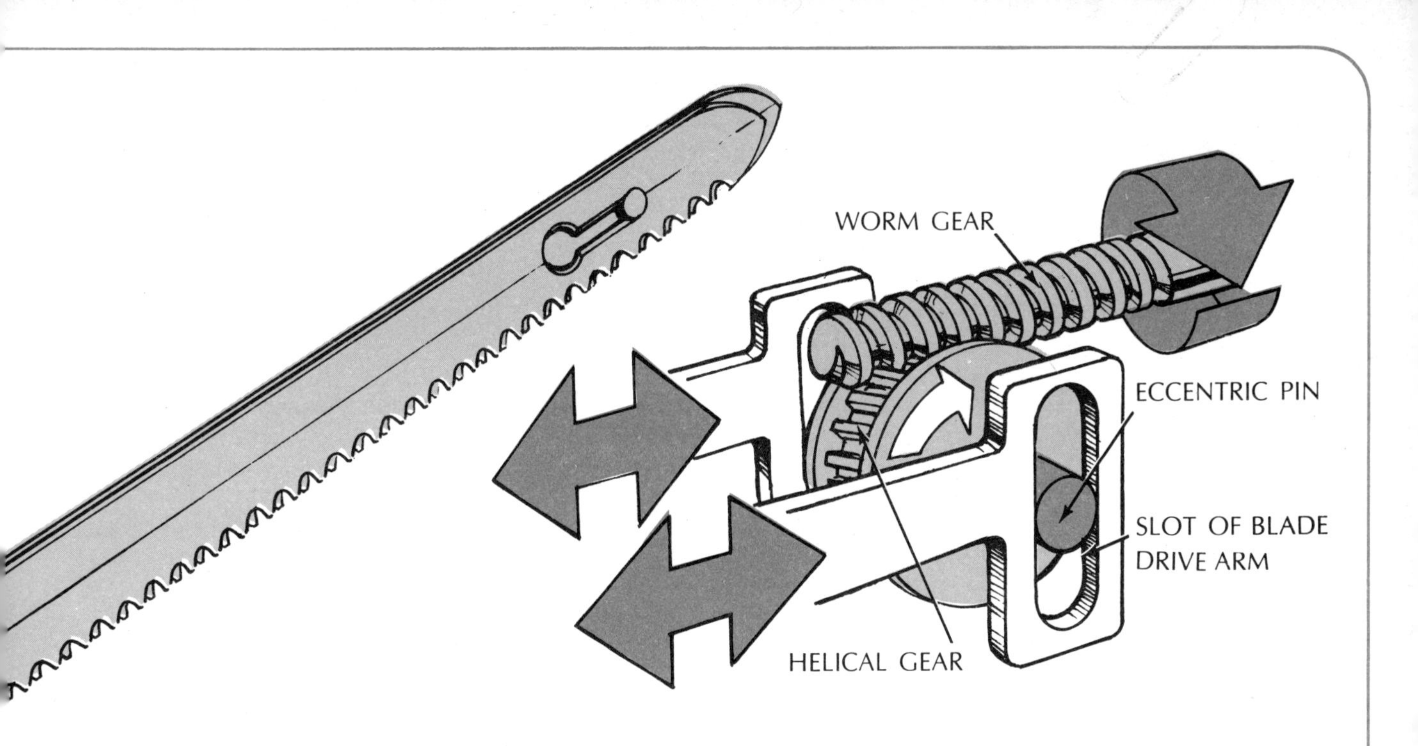

The motor shaft is fitted with a worm gear that meshes with a large-diameter helical gear. The helical gear is fitted with two eccentric (off-center) pins which fit into the oblong slots of the blade drive arms. As the worm gear, extending from the motor shaft, rotates, its speed is exchanged for torque (turning power) in the helical gear. The rotation of the larger diameter helical gear is converted to a linear reciprocating action of the blade arms as the eccentric shafts rotate within their oblong slots. The drive arms, thus, move in opposition to one another, driving the blades in a saw-like fashion.

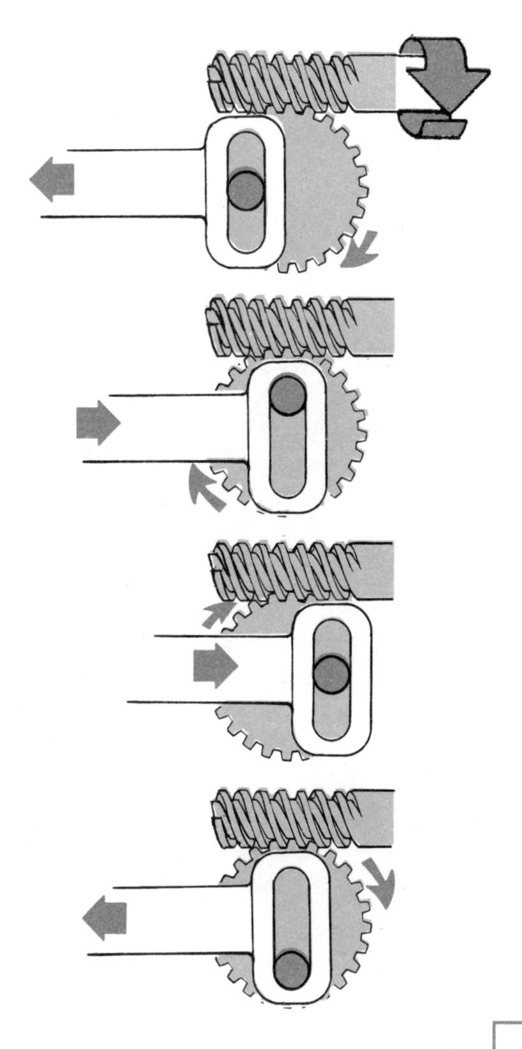

Cordless Models
After each use, a battery-powered knife handle assembly must be deposited into the mating receptacle of a charging unit. A low-voltage charger circuit built into the charging stand reduces house current to a trickle that recharges the batteries.

Electronic Watch

Man has tried for millennia to record time accurately. More than 4,000 years ago, Egyptians kept time with a crude sun stick that cast a time-approximating shadow. Greeks and Romans later improved on the shadow principle with a sun dial. Some societies have kept time with falling sand, dripping water, and burning candles. It wasn't until the 15th century in Europe that clockwork mechanisms appeared. At that time, town watchmen carried crude clocks in cage-like cabinets as they made their rounds. These hand-carried mechanisms had only an hour hand and became known as the watchman's "watch." Pocket watches were introduced about a hundred years ago. Since then, improvements have ranged far beyond the limits of the traditional spring, lever, and gear.

One of the greatest time-keeping improvements came about with the development of the "electronic" watch that employs the miniature battery and transistor to supply its motive power. But the key to the electronic watch's accuracy lies in a tool formerly associated only with musicians and piano tuners—the tuning fork. Electronic watches can accurately record time to within a 2-second tolerance each day.

The tuning fork has long been recognized as a precision vibrational standard. Striking or twanging a tuning fork causes it to vibrate (oscillate) at an exact acoustical rate (natural frequency) that is dependent only upon the fork's dimensions. The electronic tuning fork is vibrated by magnetic impulses delivered at exact intervals by the watch's tiny battery and a transistor circuit. The watch's fork is designed to hold a natural frequency of 360 vibrations per second, or slightly above middle C on the musical scale.

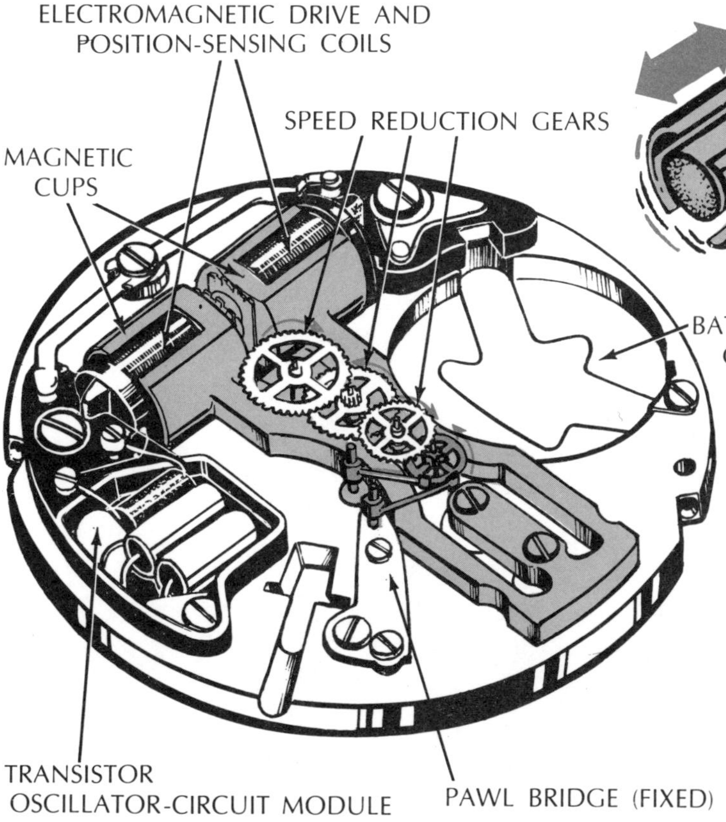

Tuning Fork

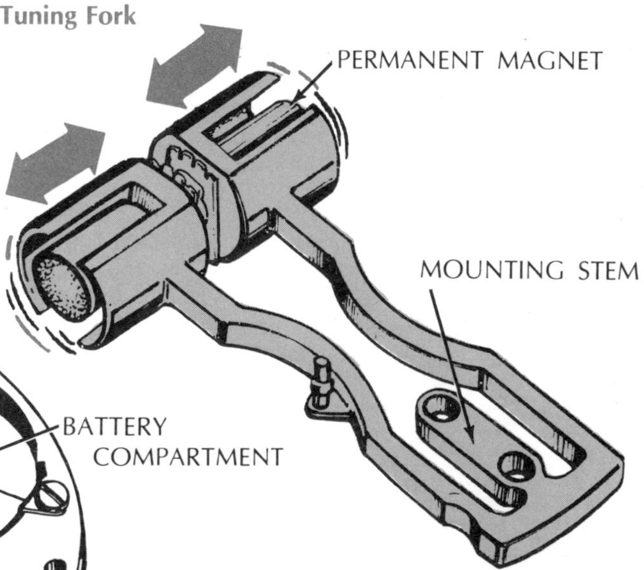

Indexing Mechanism

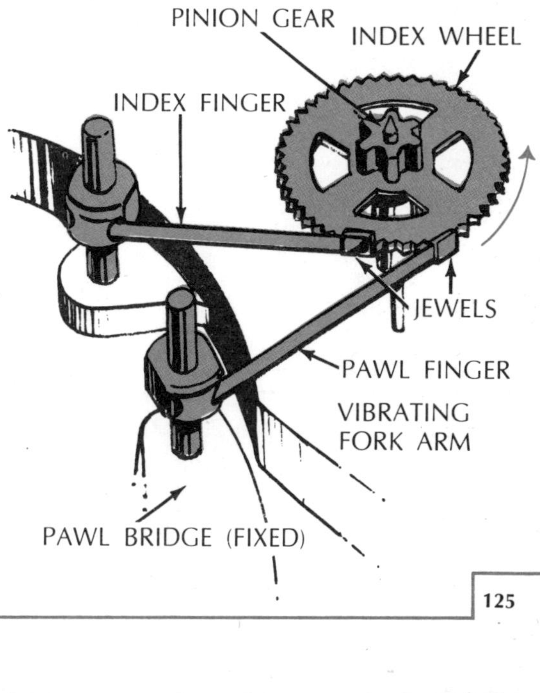

The vibratory motion of the fork is converted into rotary motion by a pawl and ratchet mechanism. One tine of the tuning fork is attached to a straight index finger, tipped with a jewel that engages ratchet teeth on an index wheel. The index wheel advances one tooth for each complete back-and-forth cycle of the tuning fork. A pawl finger attached to the stationary pawl bridge holds the index wheel in position and prevents its slipping backward during the return stroke of the fork-driven index finger. The shaft of the index wheel has a pinion which turns the second, minute, and hour hands through a train of speed-reduction gears. The precision of these gears and the stability of the tuning fork mechanism ensure optimum reliability of the time-piece.

Self-winding Watch

Today's self-winding watch dates back to 1752 and the appearance of a pedometer that was used to measure walking distances. In the pedometer a pivoted weight is so balanced by a hairspring that each step of the walker swings it down momentarily. A spring finger advances a fine-toothed count wheel as the weight swings back. A click or detent holds the wheel during the downswing. Around 1800 a Swiss watchmaker applied the pedometer's action to a watch, and during the next 20-odd years many self-winders were made. In 1893, the Swiss mass-produced a self-winder with a reserve-power indicator. Englishman Thomas Harwyn invented the self-winding *wrist* watch in 1929.

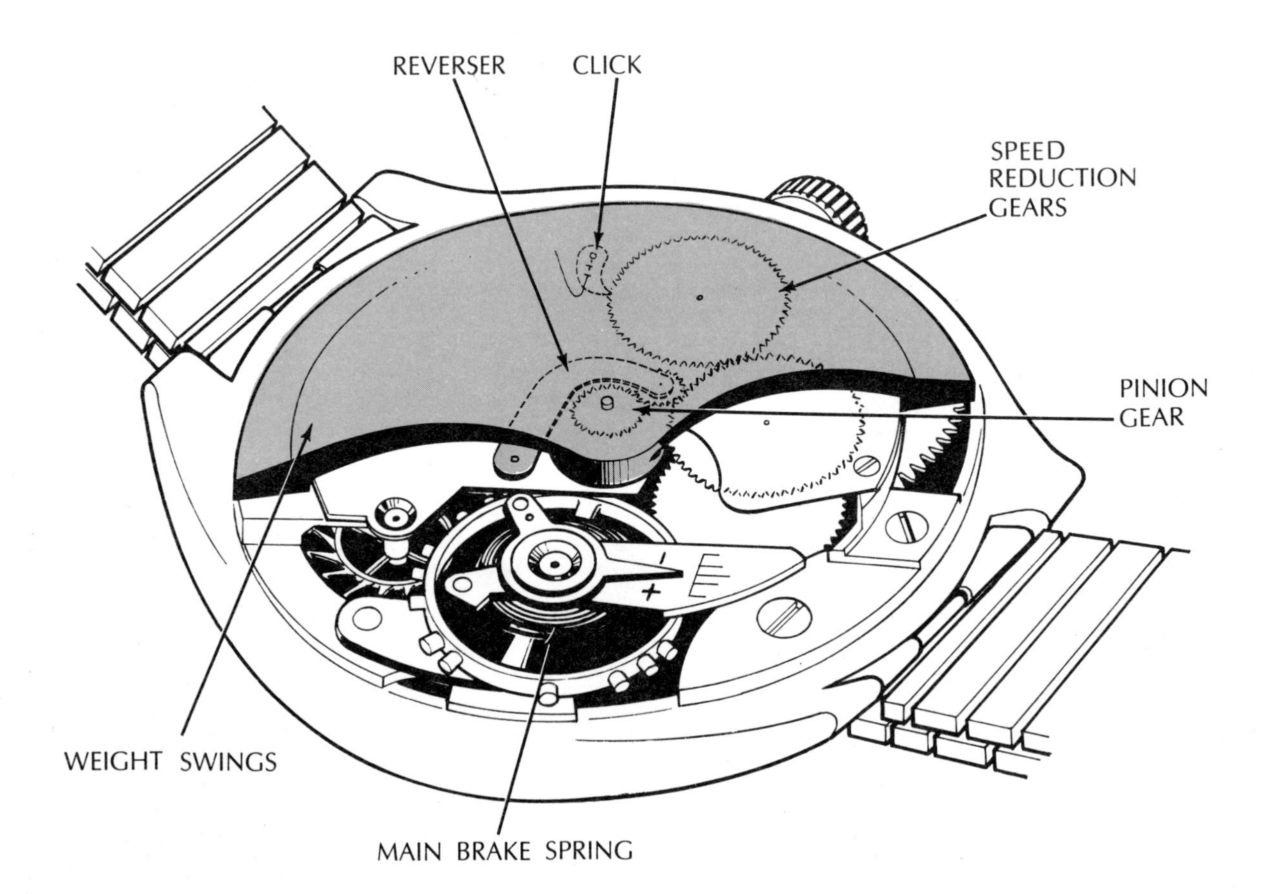

Beneath the mechanism shown here lies a weighted disc that winds the clockspring that sets the precision gears into motion. In earlier actions, the weighted disc hit springs at each end of its swing and wound only when turning in one direction. In modern types, the weight can turn completely around. Thus, it winds going either way, restoring mainspring energy quickly and reliably.

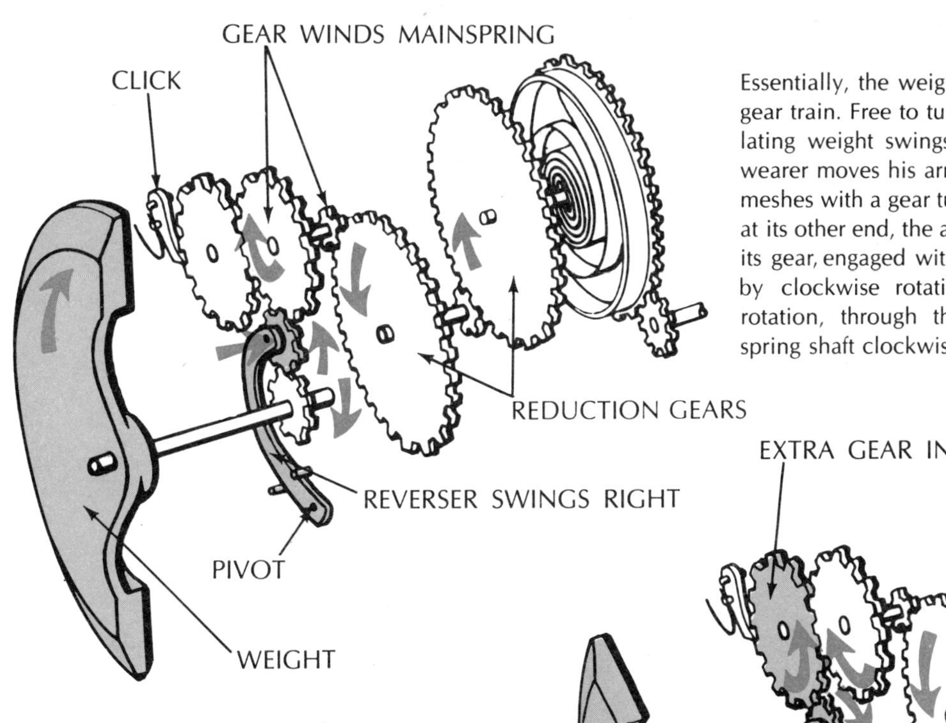

Essentially, the weight winds the spring through the gear train. Free to turn around completely, the oscillating weight swings one way or the other as the wearer moves his arm. A pinion on the weight shaft meshes with a gear turning on a curved arm. Pivoted at its other end, the arm is held toward the right, and its gear, engaged with the right-hand one of the pair by clockwise rotation of the weight shaft. This rotation, through the gear train, turns the mainspring shaft clockwise and thus winds the watch.

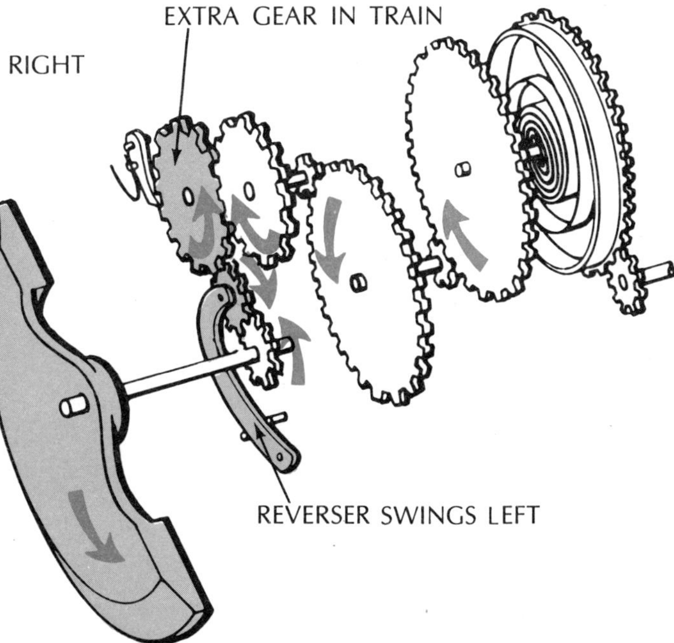

When the weight stops, a click on the left-hand wheel holds the winding train. As the weight swings the other way, the click prevents rotation, so the pinion rolls the reverser off the right-hand gear into mesh with the left-hand one. Now it turns this left-hand gear in the direction permitted by the click. Driving through an extra gear, counterclockwise rotation of the weight turns the mainspring shaft clockwise, the same way as before, and so again winds the watch despite its reversed swing.

Overwinding can't break the springs

Coiled up in a shallow barrel and wound by the shaft on which it is fixed, the mainspring has a short, stiffer, brake spring fastened to its outer end. This presses against the smooth inner wall of the barrel. Acting like a friction clutch, it drives the barrel and the watch movement. But when almost fully wound, the mainspring's constricting coil pulls the brake spring off the barrel wall. It then slips around until the mainspring unwinds enough to let the brake spring again apply driving friction.

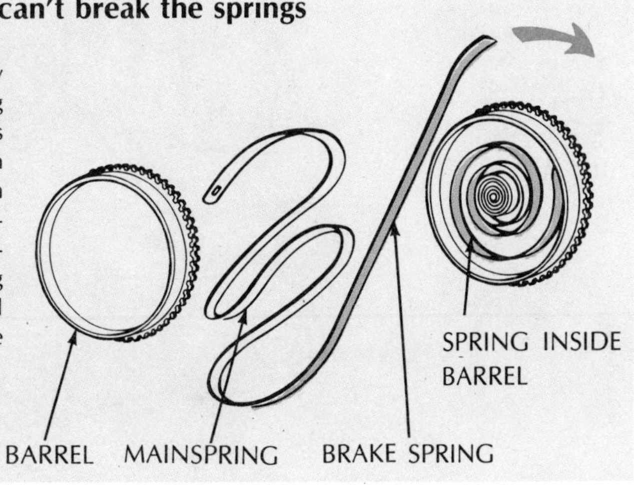

Stopwatch

Most people associate the word "stopwatch" with athletic events. But it is industrial time-study experts—and not sports officials—who give the stopwatch its heavy use, timing a great variety of processes and reactions down to minute fractions of a second. The typical stopwatch has seven jewels. Its balance wheel oscillates five times a second when at work. The more complex timers may run for hours with the balance wheel alternating direction 100 times a second. A good stopwatch has rate-regulating parts made of an alloy that is nonmagnetic and unaffected by temperature. The internal mechanism compensates for the watch's being held in any position — vertical, horizontal, or at any angle. For many years the industry standard for the stopwatch was that it deviate no more than six seconds in six hours.

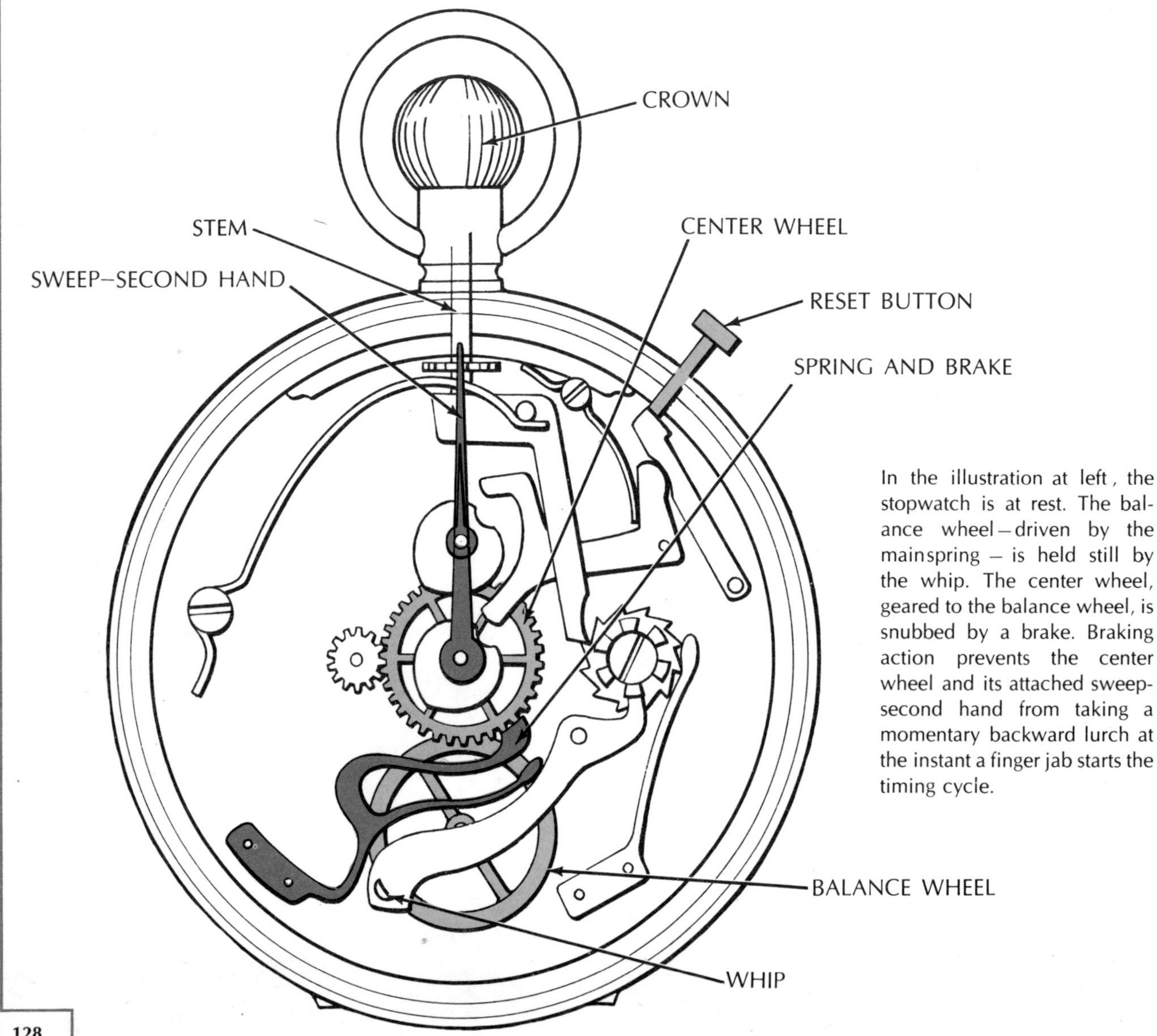

In the illustration at left, the stopwatch is at rest. The balance wheel—driven by the mainspring — is held still by the whip. The center wheel, geared to the balance wheel, is snubbed by a brake. Braking action prevents the center wheel and its attached sweep-second hand from taking a momentary backward lurch at the instant a finger jab starts the timing cycle.

A push on the crown starts the watch. The stem presses the spring-loaded main push-piece down. (It slides but doesn't pivot.) This rotates the star wheel, advancing it one tooth. As the star wheel moves, the whip lever slips off a column on the star wheel and falls between columns. This pivots the whip away from the balance wheel, freeing it, the center wheel, and sweep-second hand. Simultaneously, the whip lever's movement releases a brake. Now the center wheel and second hand turn.

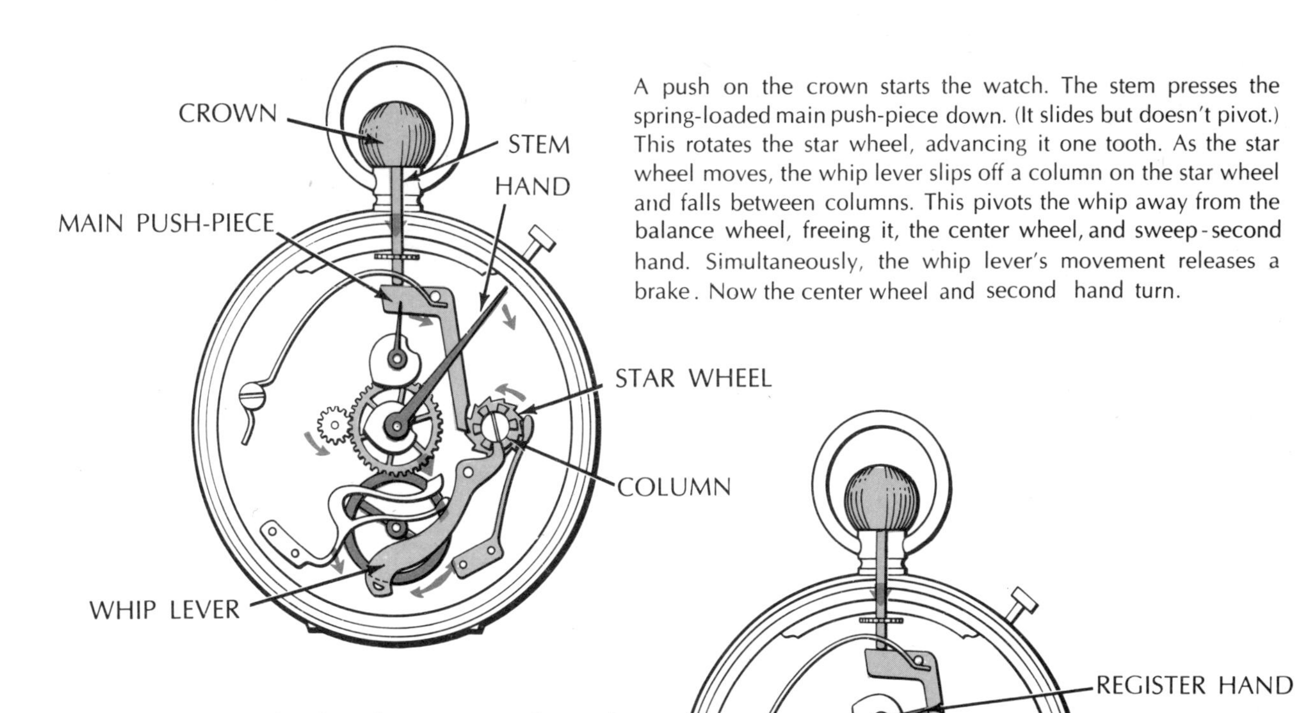

A second push on the crown stops the watch for time out. The hands stay where they were when stopped. Forced down by the stem, the push-piece rotates the star wheel another tooth. The whip lever slides out of its resting space between columns and rests on the outside of one column. Pivoting, the whip again stops the balance wheel while the brake snubs the center wheel.

Pressing the side button returns the hands to zero. The button's stem pushes down a spring-loaded, side push-piece. This pivots the hammer, whose yoke-shaped end rides on heart-shaped cams—one cam for each hand. In pivoting, the hammer jabs the two cams and spins them around. The cams stop when a notch in each of them reaches the hammer. Notches are exactly positioned so they come to rest against the hammer when both hands point to zero.

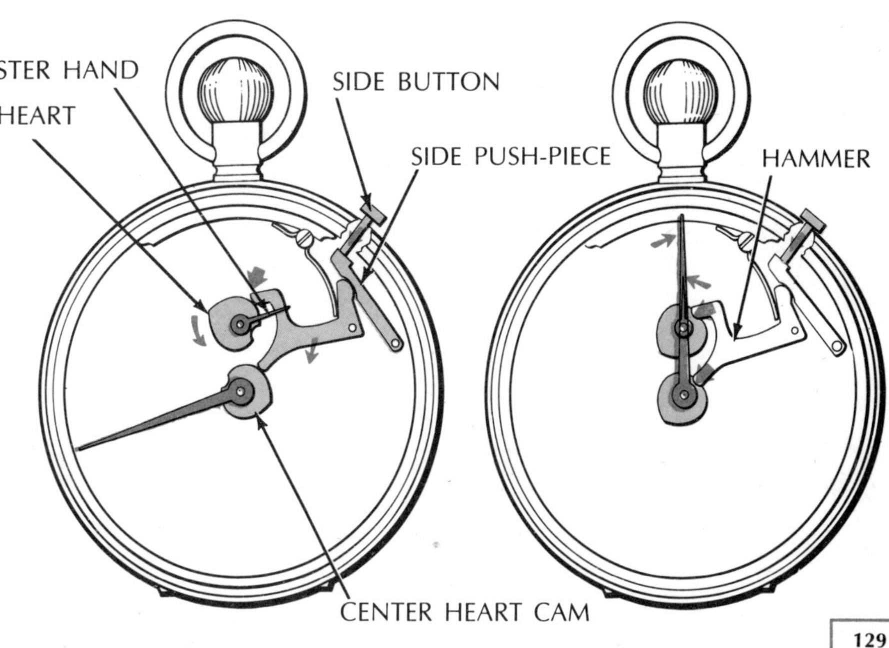

Taxi Meter

Travel meters in vehicles for hire date back to ancient Rome. These earliest meters employed a compartmented wheel driven by a road wheel that dropped pebbles from a hopper into a box. Counted at the end of the ride, the pebbles set the fare. Modern taxi meters can be adjusted to charge established rates for mileage as well as for waiting time, minimum fare, and extras such as luggage in the trunk. A fast ride with few delays may cost less than the same trip in stop-and-go traffic. Wheel revolutions determine the mileage charge. Thus winter chains, tires with high air pressure, and oversize tires may provide extra travel per wheel revolution and save the customer a few cents on long trips. Low tires work to the customer's disadvantage. But in most communities, the meters are inspected regularly to ensure that they are accurate to within 100 feet per mile.

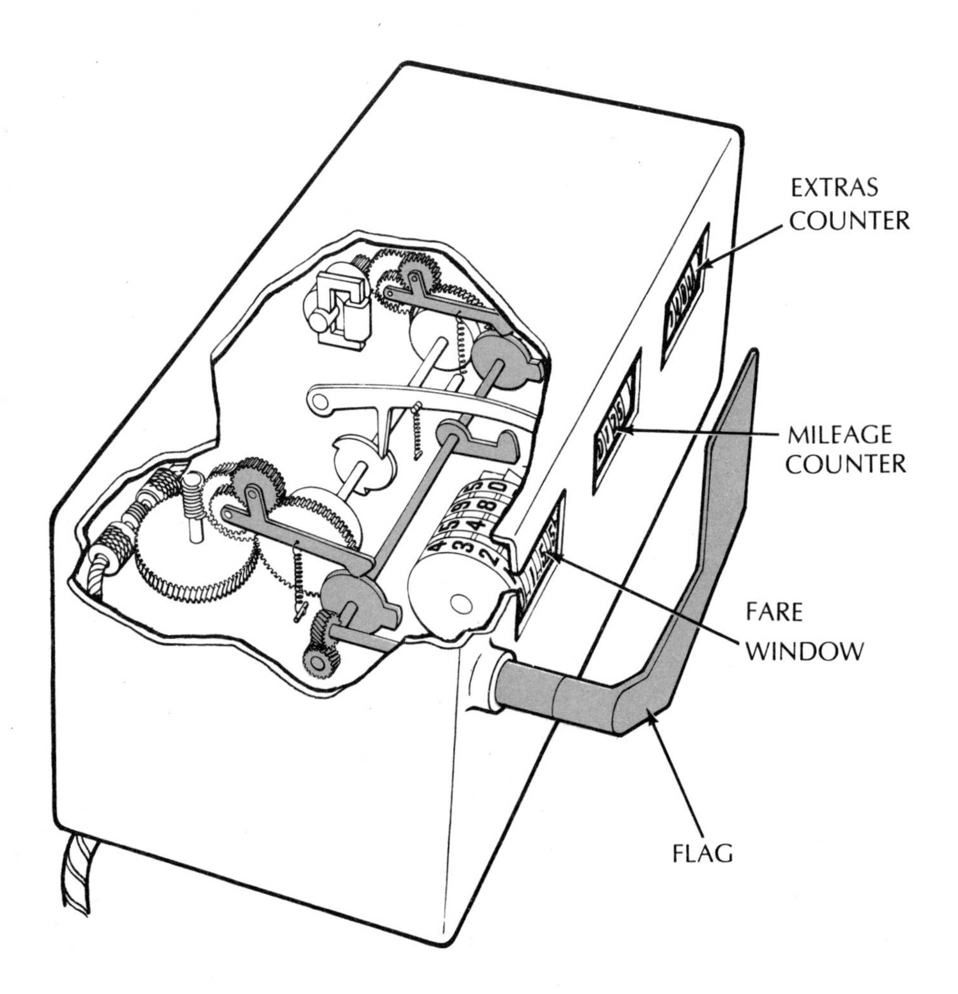

The fare is probably fair because meters are usually checked and sealed by city authorities. The meters are geared to suit tire size and rear-axle ratio, to charge only the legal rate for waiting, and to bill riders at advertised rates. The complex box not only prices each ride but registers the fare, paid and unpaid mileage traveled, number of trips, and extras.

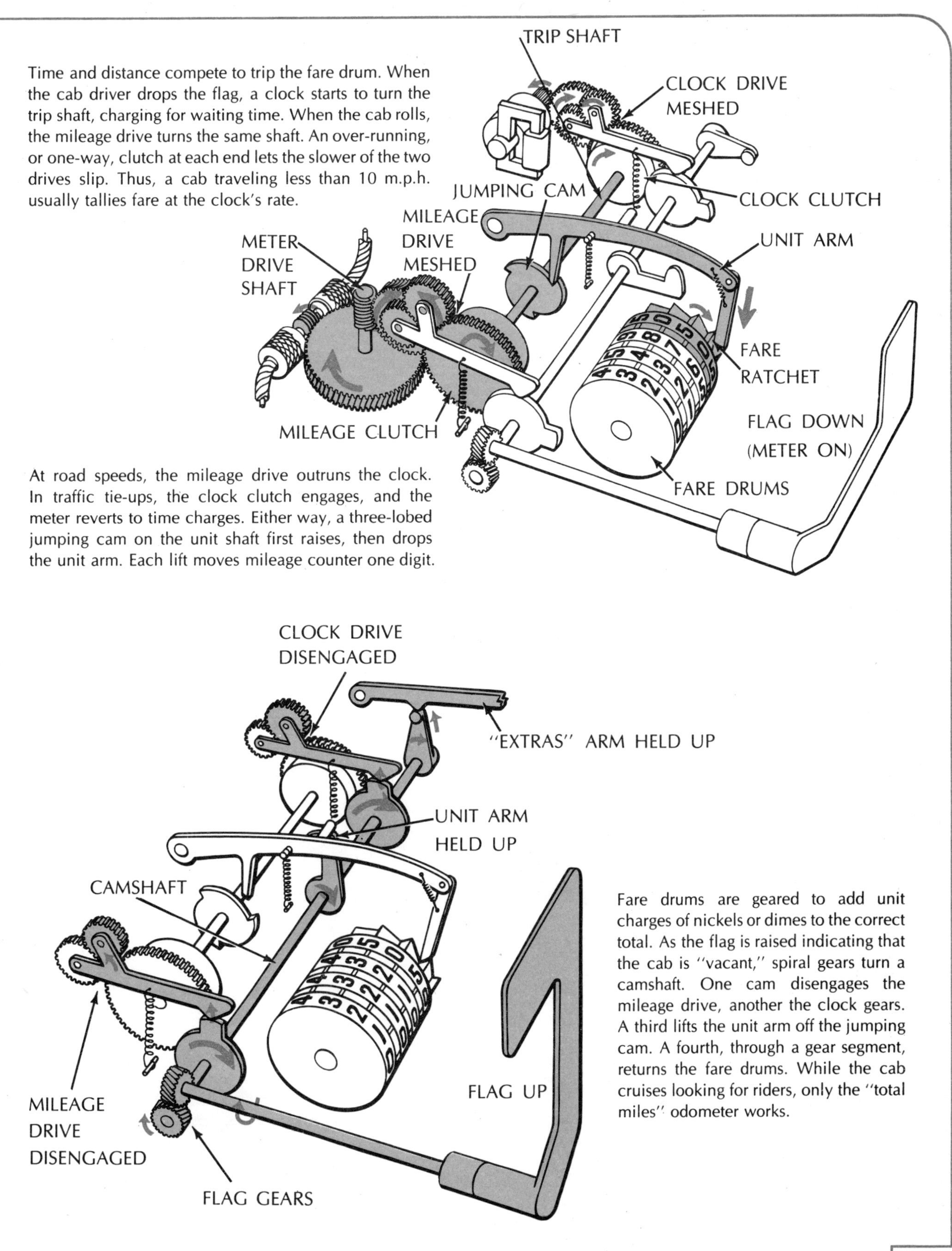

Time and distance compete to trip the fare drum. When the cab driver drops the flag, a clock starts to turn the trip shaft, charging for waiting time. When the cab rolls, the mileage drive turns the same shaft. An over-running, or one-way, clutch at each end lets the slower of the two drives slip. Thus, a cab traveling less than 10 m.p.h. usually tallies fare at the clock's rate.

At road speeds, the mileage drive outruns the clock. In traffic tie-ups, the clock clutch engages, and the meter reverts to time charges. Either way, a three-lobed jumping cam on the unit shaft first raises, then drops the unit arm. Each lift moves mileage counter one digit.

Fare drums are geared to add unit charges of nickels or dimes to the correct total. As the flag is raised indicating that the cab is "vacant," spiral gears turn a camshaft. One cam disengages the mileage drive, another the clock gears. A third lifts the unit arm off the jumping cam. A fourth, through a gear segment, returns the fare drums. While the cab cruises looking for riders, only the "total miles" odometer works.

Electric Engraver

Until recently, electric engravers were found mostly in the hands of jewelers skilled in etching sweet sentiments into rings and charm bracelets. But the steady rise in crime in the United States has led to a new role for this instrument. Many people are using them to permanently mark some identification, such as their Social Security numbers or initials, on valuables that range from watches to television sets. This has made the police's task of tracing stolen goods easier, a fact that might well deter many a thief.

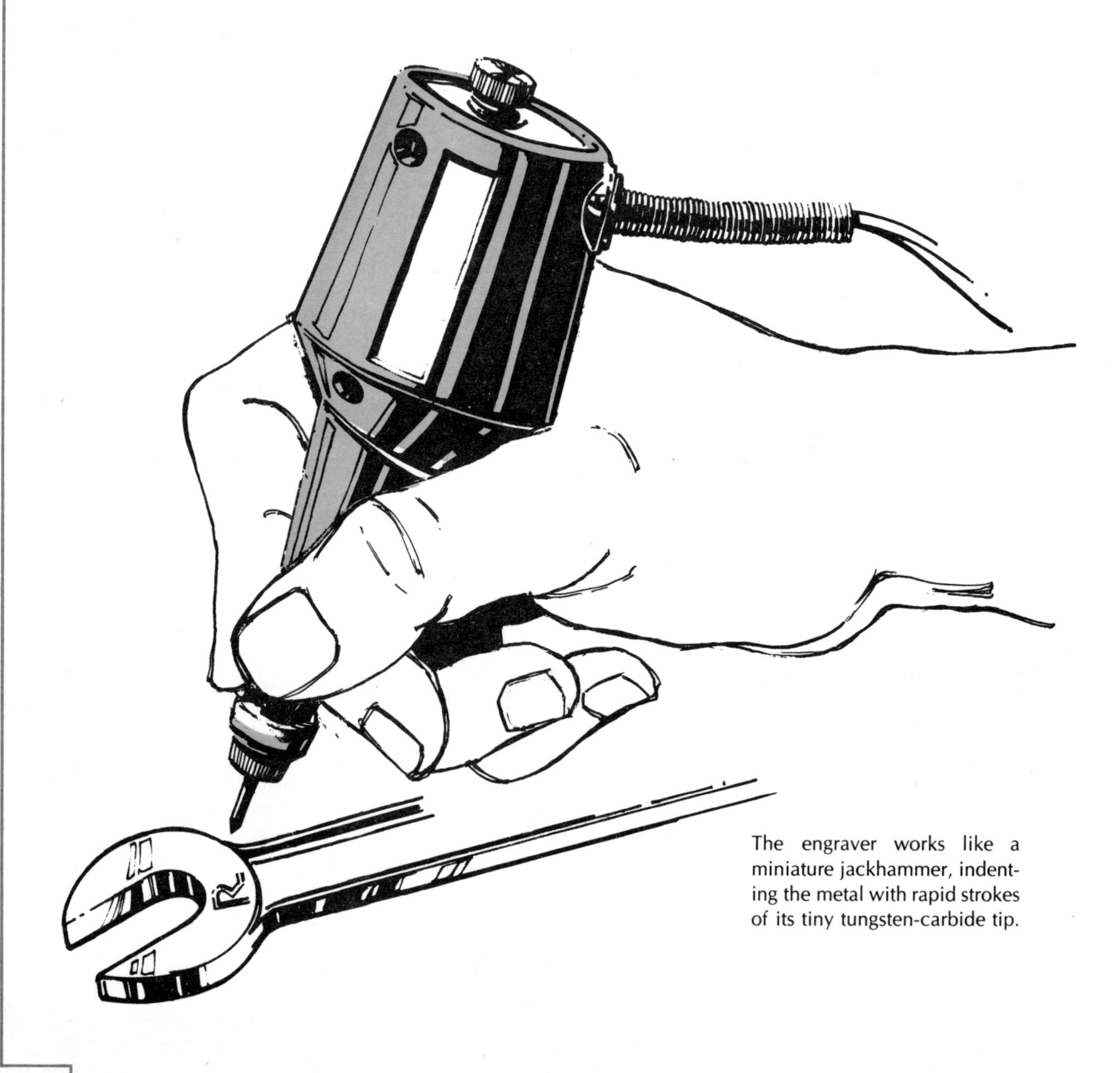

The engraver works like a miniature jackhammer, indenting the metal with rapid strokes of its tiny tungsten-carbide tip.

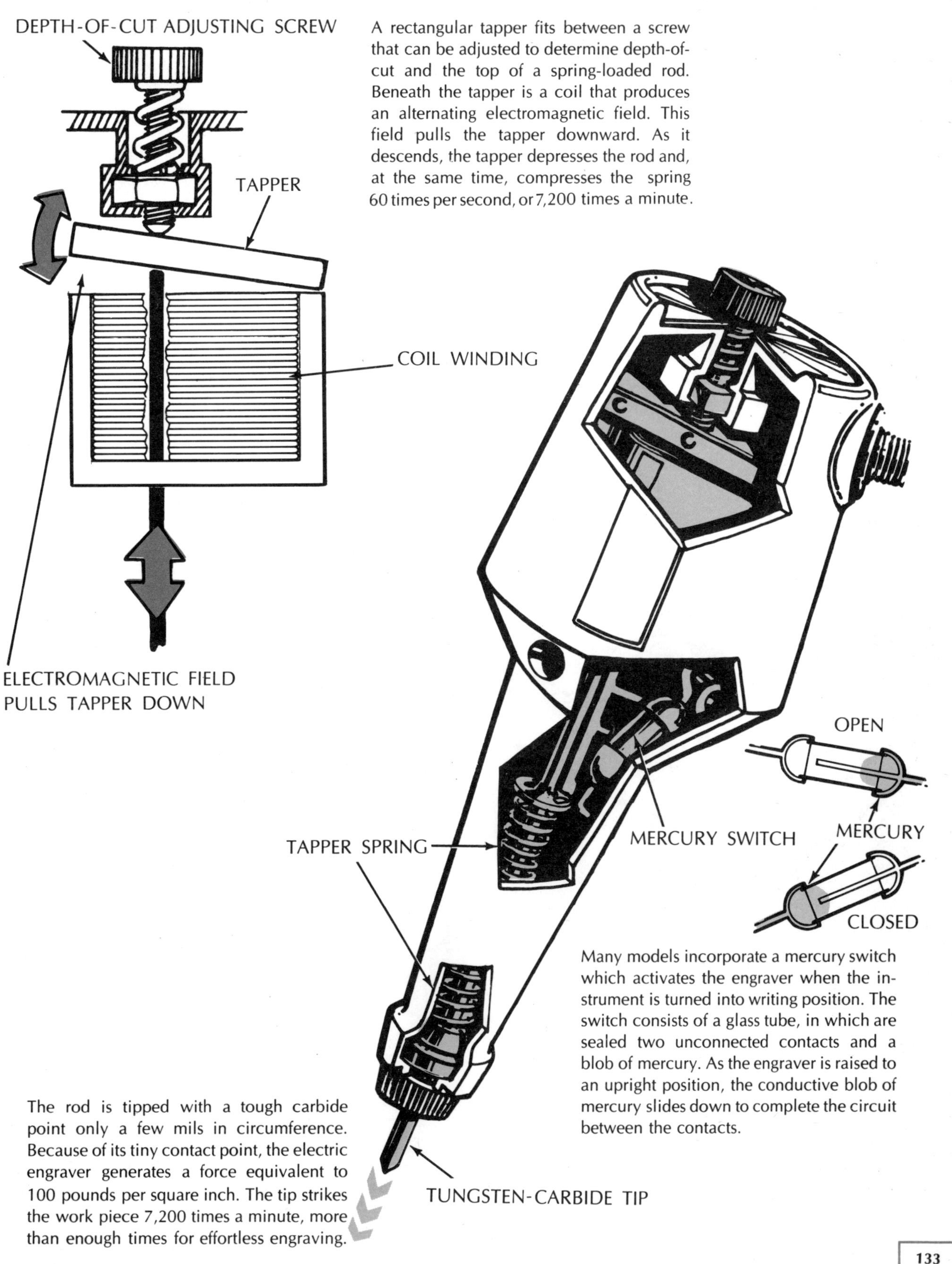

A rectangular tapper fits between a screw that can be adjusted to determine depth-of-cut and the top of a spring-loaded rod. Beneath the tapper is a coil that produces an alternating electromagnetic field. This field pulls the tapper downward. As it descends, the tapper depresses the rod and, at the same time, compresses the spring 60 times per second, or 7,200 times a minute.

Many models incorporate a mercury switch which activates the engraver when the instrument is turned into writing position. The switch consists of a glass tube, in which are sealed two unconnected contacts and a blob of mercury. As the engraver is raised to an upright position, the conductive blob of mercury slides down to complete the circuit between the contacts.

The rod is tipped with a tough carbide point only a few mils in circumference. Because of its tiny contact point, the electric engraver generates a force equivalent to 100 pounds per square inch. The tip strikes the work piece 7,200 times a minute, more than enough times for effortless engraving.

Bathroom Scale

The sophistication of the common bathroom scale belies its modest cost. The scale gauges downward force—weight—and converts it into the horizontal motion that is needed to rotate the indicator dial. But to keep weight watchers from getting motion sickness, the top platform's drop must be limited. To achieve this stability, the mechanism is sprung so that a 100-pound load depresses the platform only 1/100 of an inch. Yet, this tiny movement is enough to register an accurate reading.

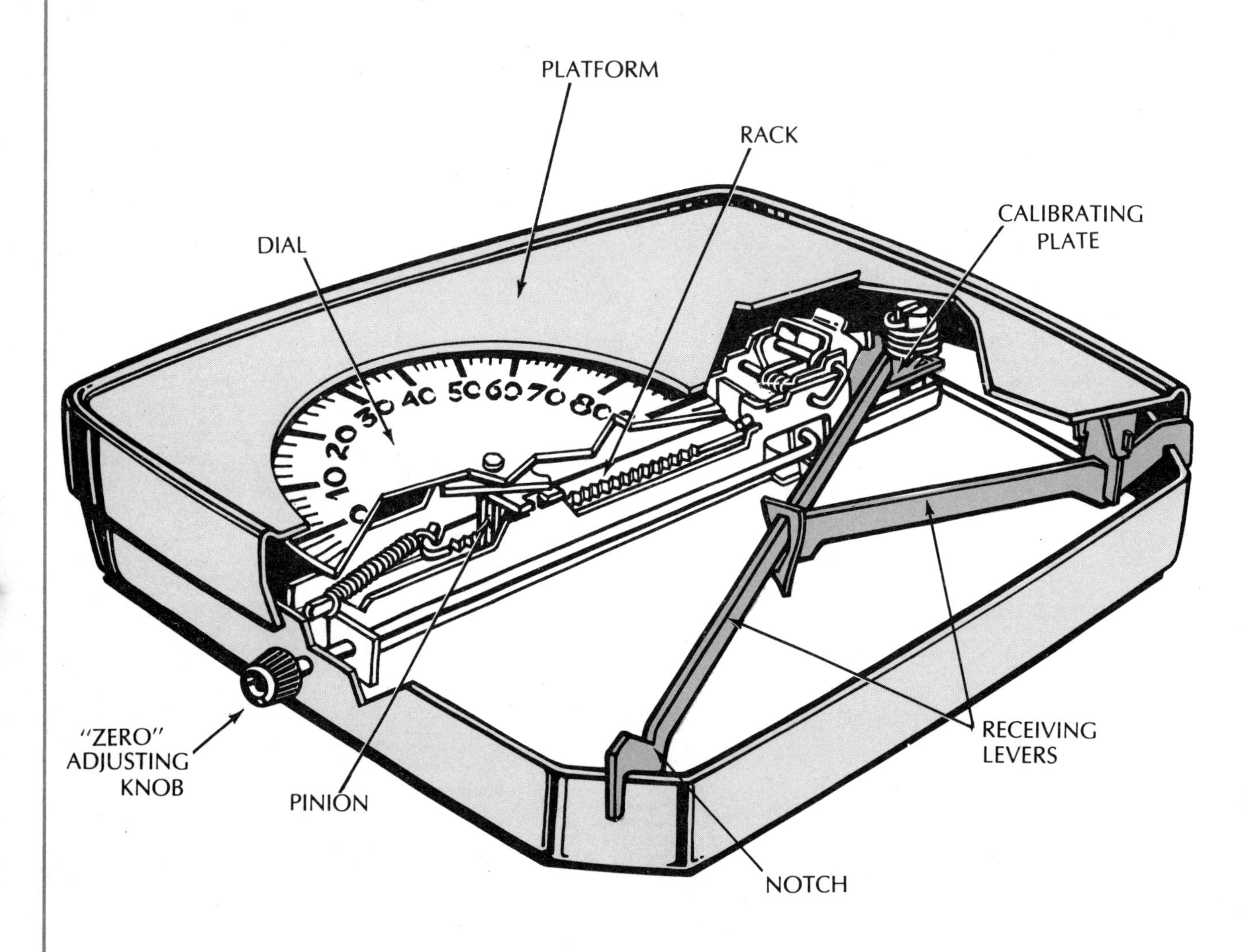

The independent top platform has on its underside four ridges that fit into the notches at the ends of the coordinated receiving levers. When weight is placed on any part of the platform, that force is transmitted through the ridges to the levers. The levers in turn focus the entire force on the spring-loaded calibrating plate.

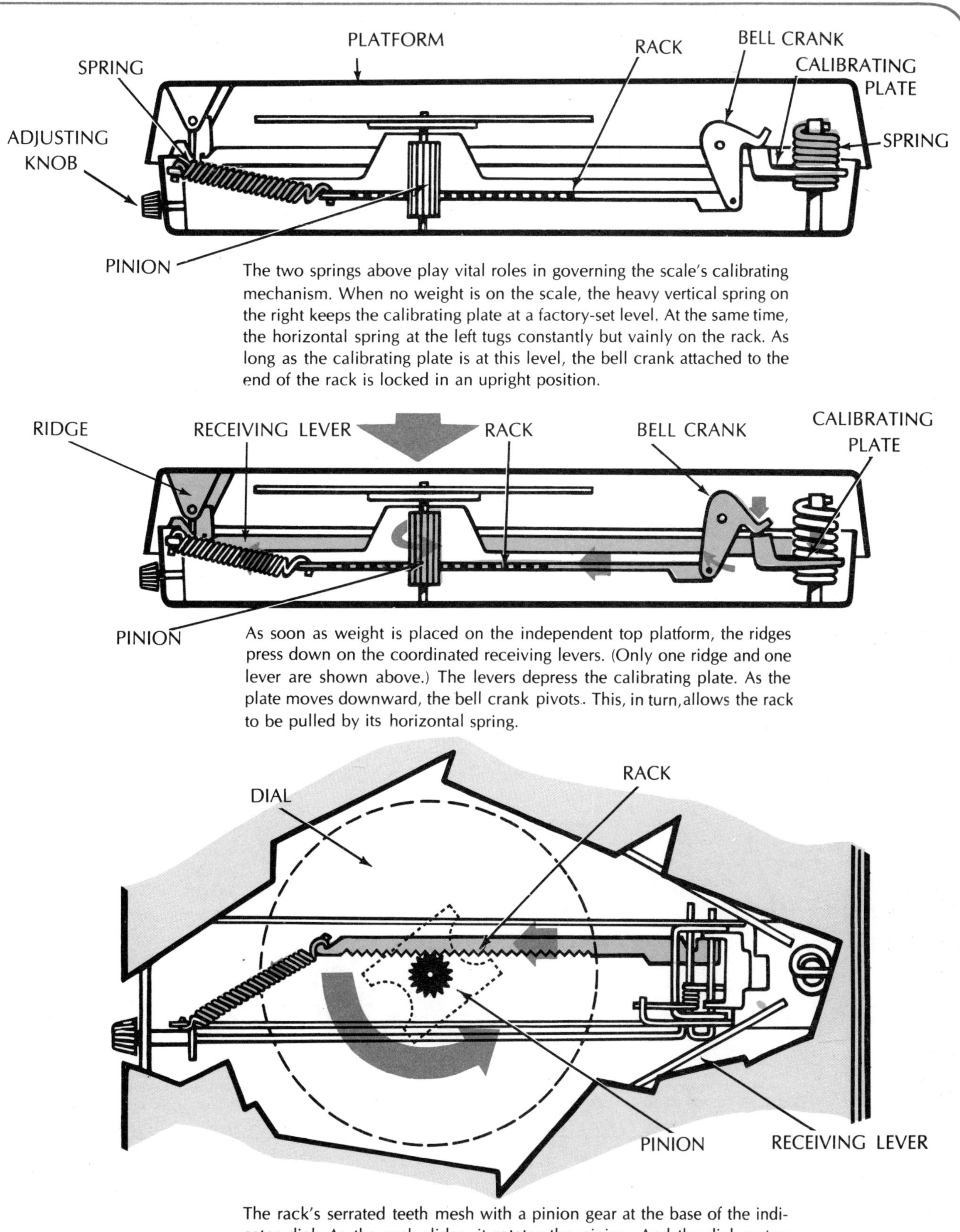

The two springs above play vital roles in governing the scale's calibrating mechanism. When no weight is on the scale, the heavy vertical spring on the right keeps the calibrating plate at a factory-set level. At the same time, the horizontal spring at the left tugs constantly but vainly on the rack. As long as the calibrating plate is at this level, the bell crank attached to the end of the rack is locked in an upright position.

As soon as weight is placed on the independent top platform, the ridges press down on the coordinated receiving levers. (Only one ridge and one lever are shown above.) The levers depress the calibrating plate. As the plate moves downward, the bell crank pivots. This, in turn, allows the rack to be pulled by its horizontal spring.

The rack's serrated teeth mesh with a pinion gear at the base of the indicator dial. As the rack slides, it rotates the pinion. And the dial on top rotates to the correct weight.

Indoor/Outdoor Heating Control

An ideal heating system controls indoor temperature so that occupants don't feel outdoor temperature changes. For this, a steady indoor temperature is not sufficient. Even the best indoor thermostats respond late to chilling winds. Then, after getting a late start from the thermostat, the furnace labors for a time to catch up with demand. As outside walls cool, they absorb body heat from the occupants even though room temperature remains near the thermostat setting. This phenomenon is often called the "cold seventy" feeling and is combatted by turning the thermostat up. An indoor-outdoor control system puts a second thermostat outside the house. The second one alerts the system to the need for heat before the indoor thermostat can detect it and reduces furnace output as outside temperatures rise.

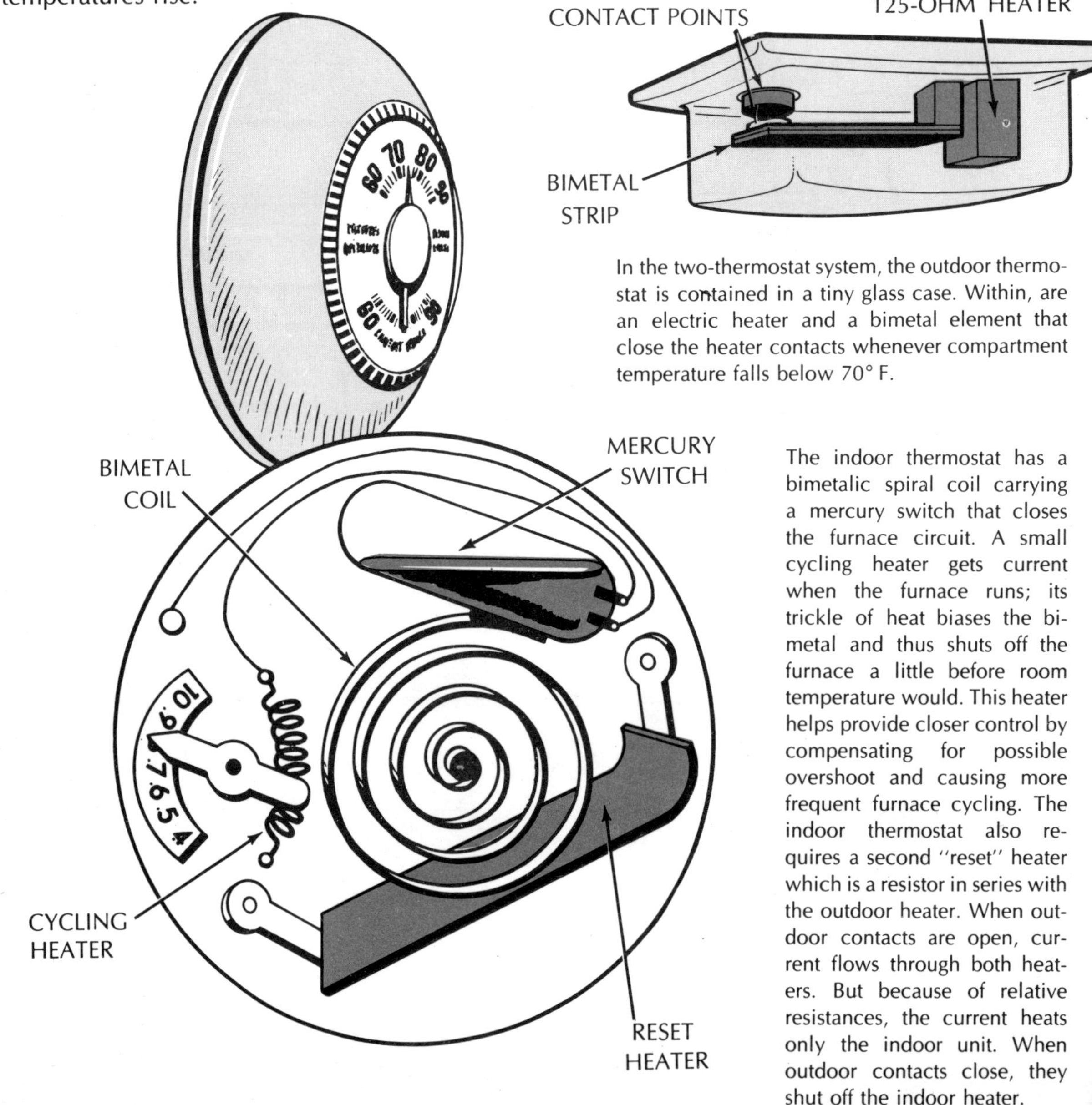

In the two-thermostat system, the outdoor thermostat is contained in a tiny glass case. Within, are an electric heater and a bimetal element that close the heater contacts whenever compartment temperature falls below 70° F.

The indoor thermostat has a bimetalic spiral coil carrying a mercury switch that closes the furnace circuit. A small cycling heater gets current when the furnace runs; its trickle of heat biases the bimetal and thus shuts off the furnace a little before room temperature would. This heater helps provide closer control by compensating for possible overshoot and causing more frequent furnace cycling. The indoor thermostat also requires a second "reset" heater which is a resistor in series with the outdoor heater. When outdoor contacts are open, current flows through both heaters. But because of relative resistances, the current heats only the indoor unit. When outdoor contacts close, they shut off the indoor heater.

In mild temperatures of 70 degrees or more, the outdoor bimetal keeps the points open. The outdoor heater carries too little current to affect this unit. The same current in the indoor reset heater makes the indoor thermostat sense a higher-than-true temperature. The control point at which the furnace starts is in this way kept close to the actual setting, such as 72° F.

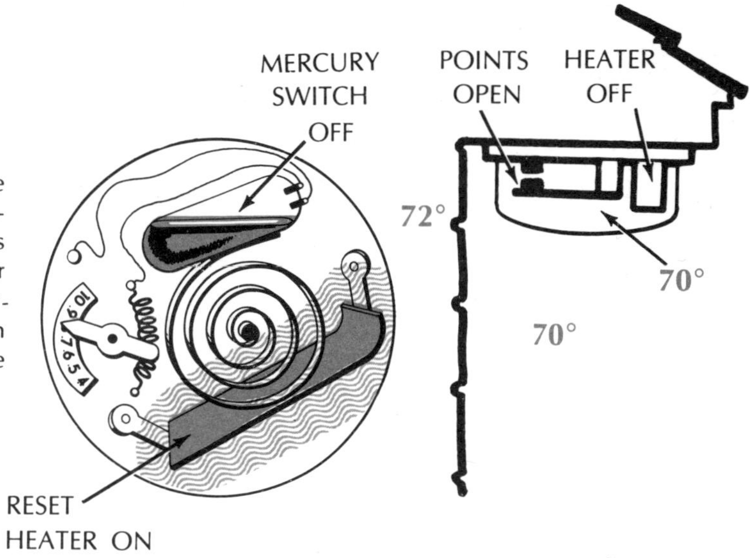

MERCURY SWITCH OFF
POINTS OPEN
HEATER OFF
72°
70°
70°
RESET HEATER ON

Wind, cold, or both affect the outdoor bimetal, as they eventually do indoor temperature. As outdoor contacts close, the heater begins warming the air inside the glass case. Closed contacts also detour current around the indoor heater, alerting the thermostat to the cold outside temperature. This raises the furnace control point so that the furnace starts putting out heat.

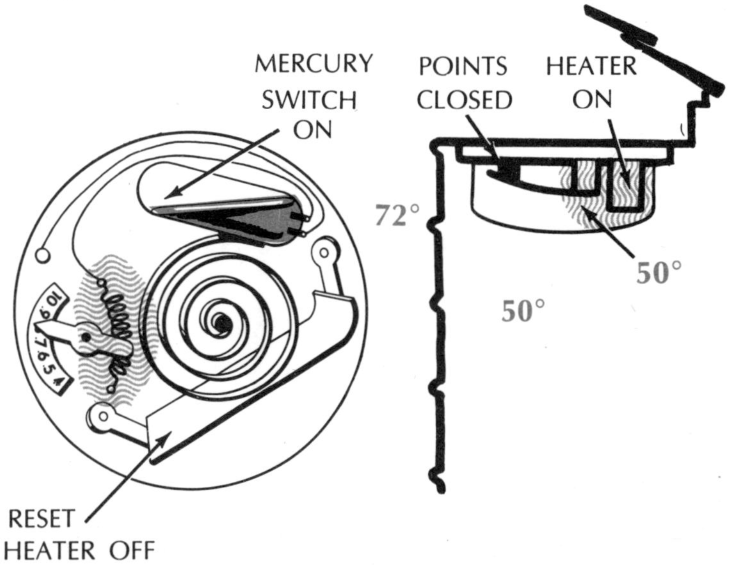

MERCURY SWITCH ON
POINTS CLOSED
HEATER ON
72°
50°
50°
RESET HEATER OFF

Outdoor contacts open when the heater has restored temperature within the case to 70° F. The indoor heater is once more on. But the outdoor case will cool fast or slowly depending on wind and weather, and repeat the cycle. How long outdoor contacts remain closed and how long the heater remains out of circuit determine the shift of the furnace control point. This raises room temperature about 1° F. for every 20° drop in outdoor temperature.

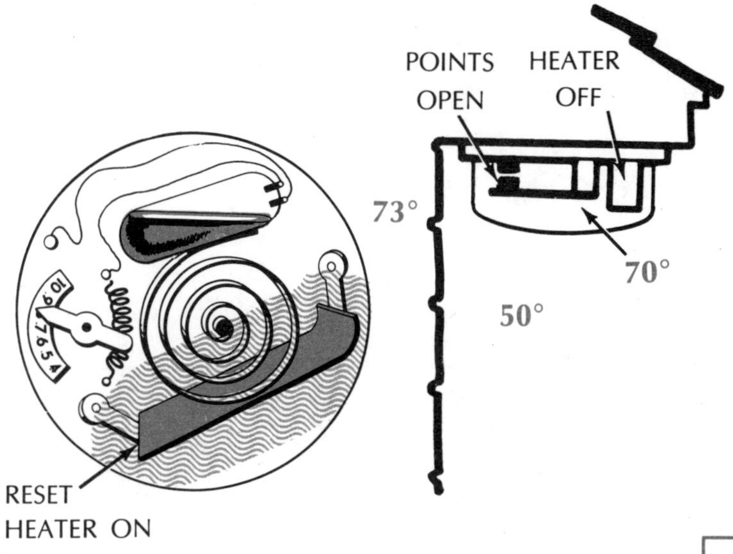

POINTS OPEN
HEATER OFF
73°
70°
50°
RESET HEATER ON

Jackhammer

Known also as air hammer, pneumatic drill, rock drill, and widow maker, the jackhammer hits its steel bit 2,000 times a minute while delivering a one-minute cumulative wallop of 40,000 foot-pounds and turning its bit with every blow. Inside the jackhammer, compressed air is routed through various chambers and ports and thereby drives a piston up and down. When the piston impacts the bit shaft at the bottom of its downstroke, the bit strikes rock with a shattering force. In the U.S. large power drills took their first bite in Massachusetts in 1868. The power drill's rotating, spiral-grooved rifle bar was invented by John Leyner in 1897. He also invented the first one-man drills and a hollow bit, through which air and water could flush out drilling dust.

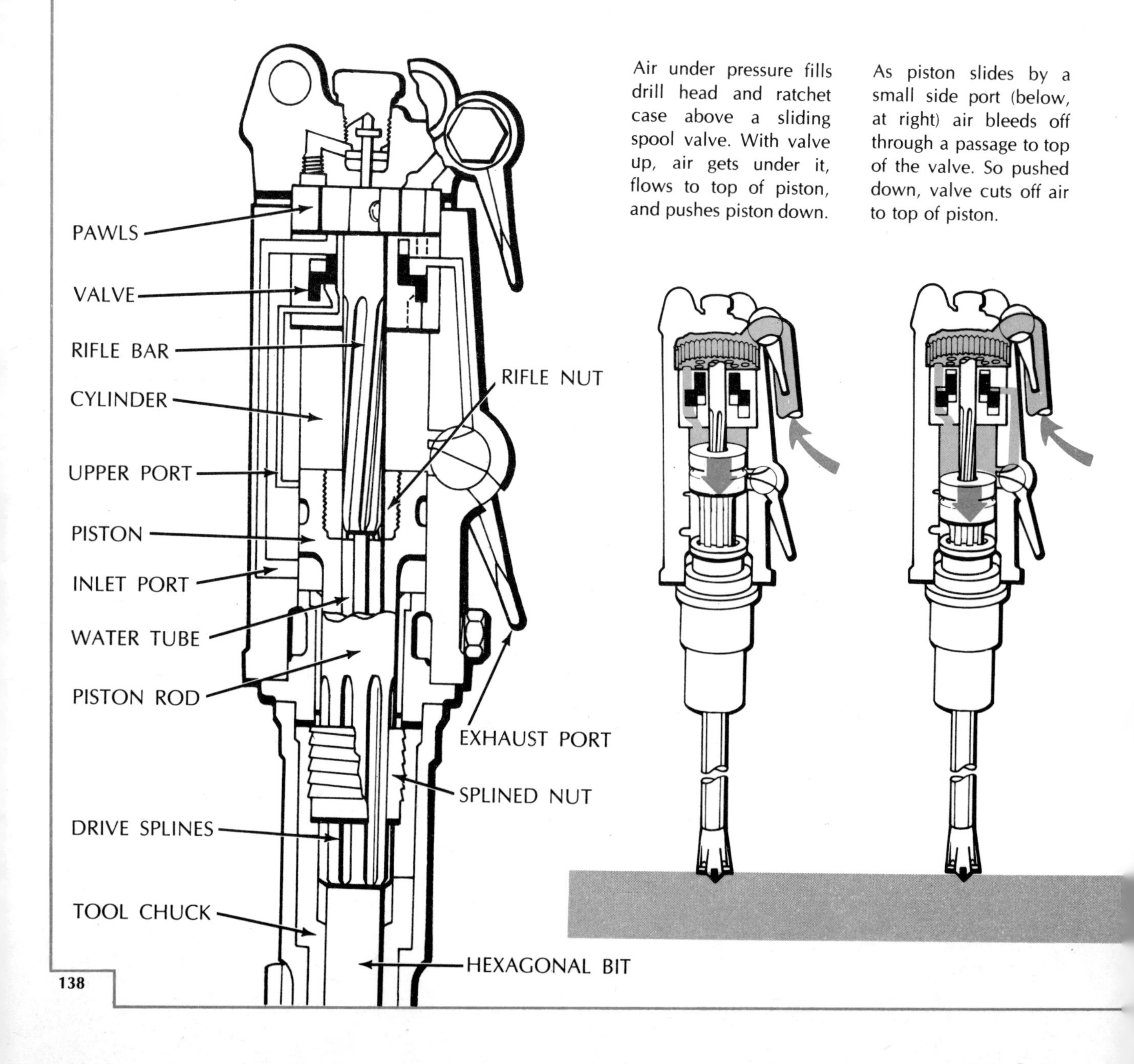

Air under pressure fills drill head and ratchet case above a sliding spool valve. With valve up, air gets under it, flows to top of piston, and pushes piston down.

As piston slides by a small side port (below, at right) air bleeds off through a passage to top of the valve. So pushed down, valve cuts off air to top of piston.

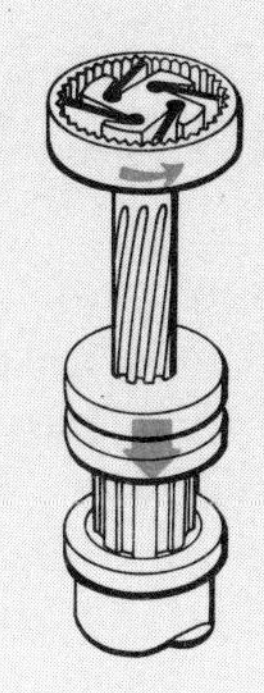

Pawls at the top of a spiral-grooved bar can click past ratchet teeth in one direction but not the other. As piston goes down (far left), pawls let bar turn counterclockwise. On the upstroke, pawls lock, hold bar immovable, so that the piston must turn instead.

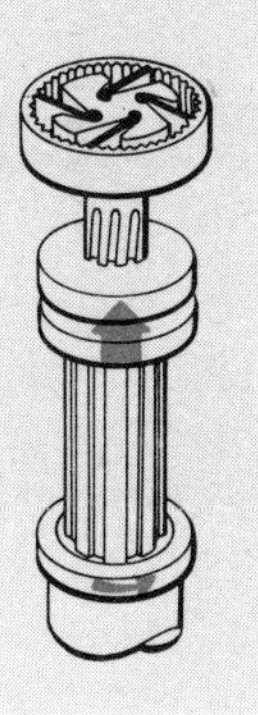

Nearing the end of its downstroke, piston uncovers exhaust port. Air rushes out. Valve flow routes air to bottom inlet port (left side of cylinder), blocked by piston.

Piston rod strikes top of drill bit. Shock and residual air under piston bounce it up enough to uncover bottom inlet port, admitting air to push piston up.

Rising piston can't turn rifle bar, as pawls lock, so must turn itself and bit. Moving up, it uncovers upper port (also on left side) sending air under the valve to lift it.

Air exhausts from big right-side port immediately after. The raised valve now routes high-pressure air to the top of the piston again for the next downstroke.

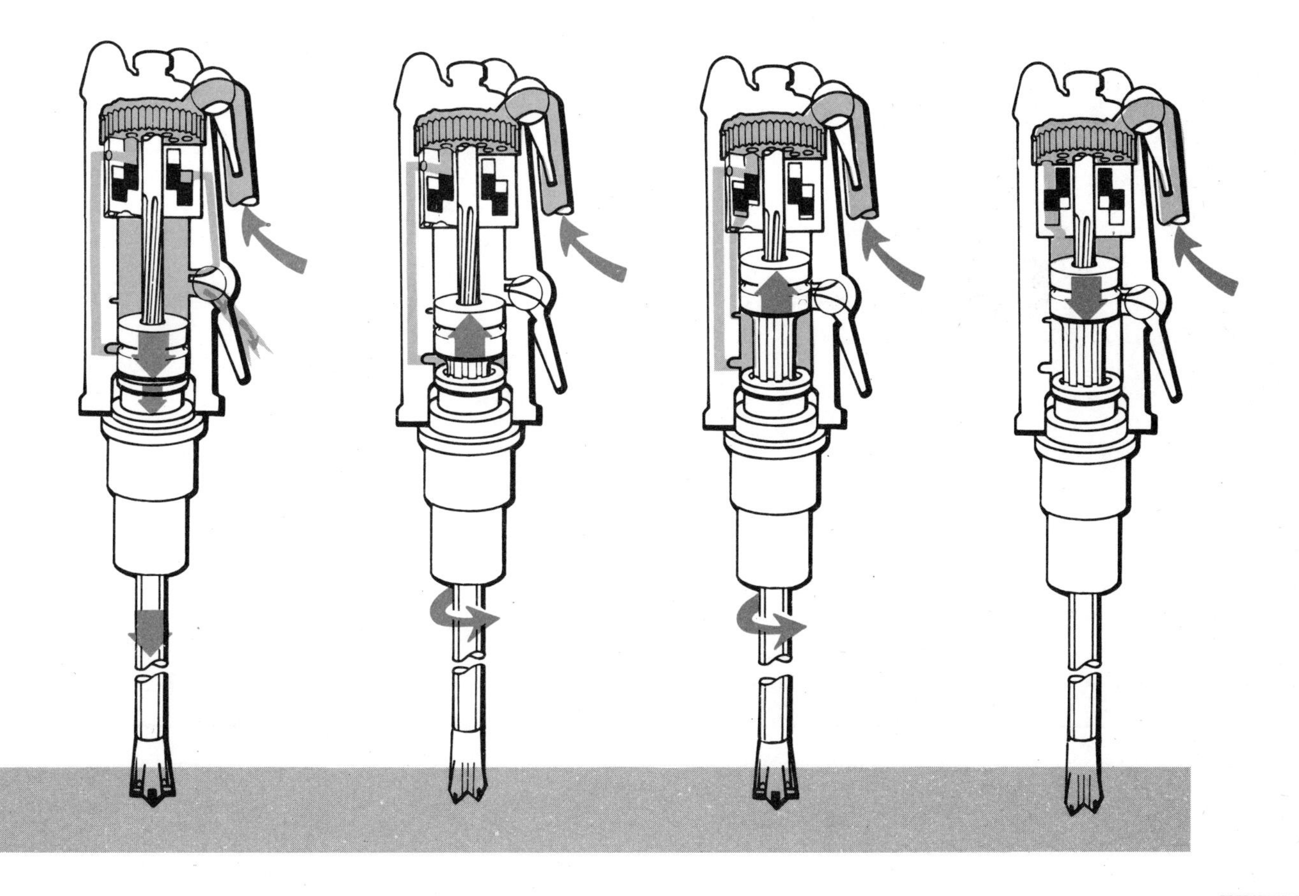

Retractable Ball-point Pen

The principle behind today's ball-point pen dates back to a crude implement patented in 1888 by John J. Loud, an American. But credit for development of the first practical ball-point pen goes to Ladislow and Georg Biro, Hungarian brothers, whose pen was first mass-produced in Argentina in the mid-1940s. Unlike the simple ball-point, today's retractable model employs a replaceable ink tube and a retraction mechanism. Bodies of both the simple and retractable "ball-points" house a plastic or metal ink tube with an 1/8-inch bore. Open at the top, the tube is filled with ink of a gooey consistency. The ink feeds down the tube, primarily pulled by gravity with the help from atmospheric pressure, and opens into a ball and socket assembly that forms the "point" of the pen.

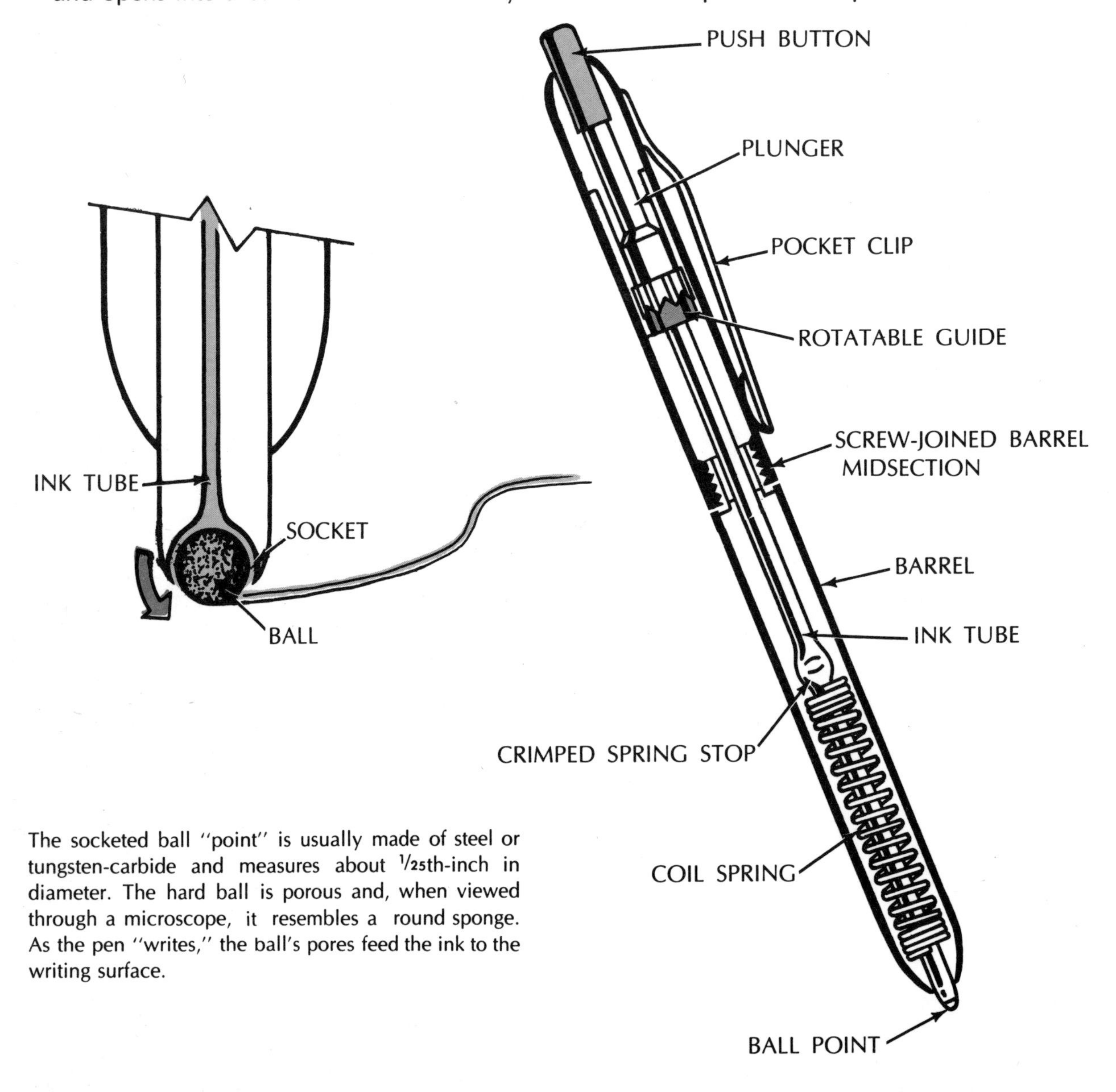

The socketed ball "point" is usually made of steel or tungsten-carbide and measures about 1/25th-inch in diameter. The hard ball is porous and, when viewed through a microscope, it resembles a round sponge. As the pen "writes," the ball's pores feed the ink to the writing surface.

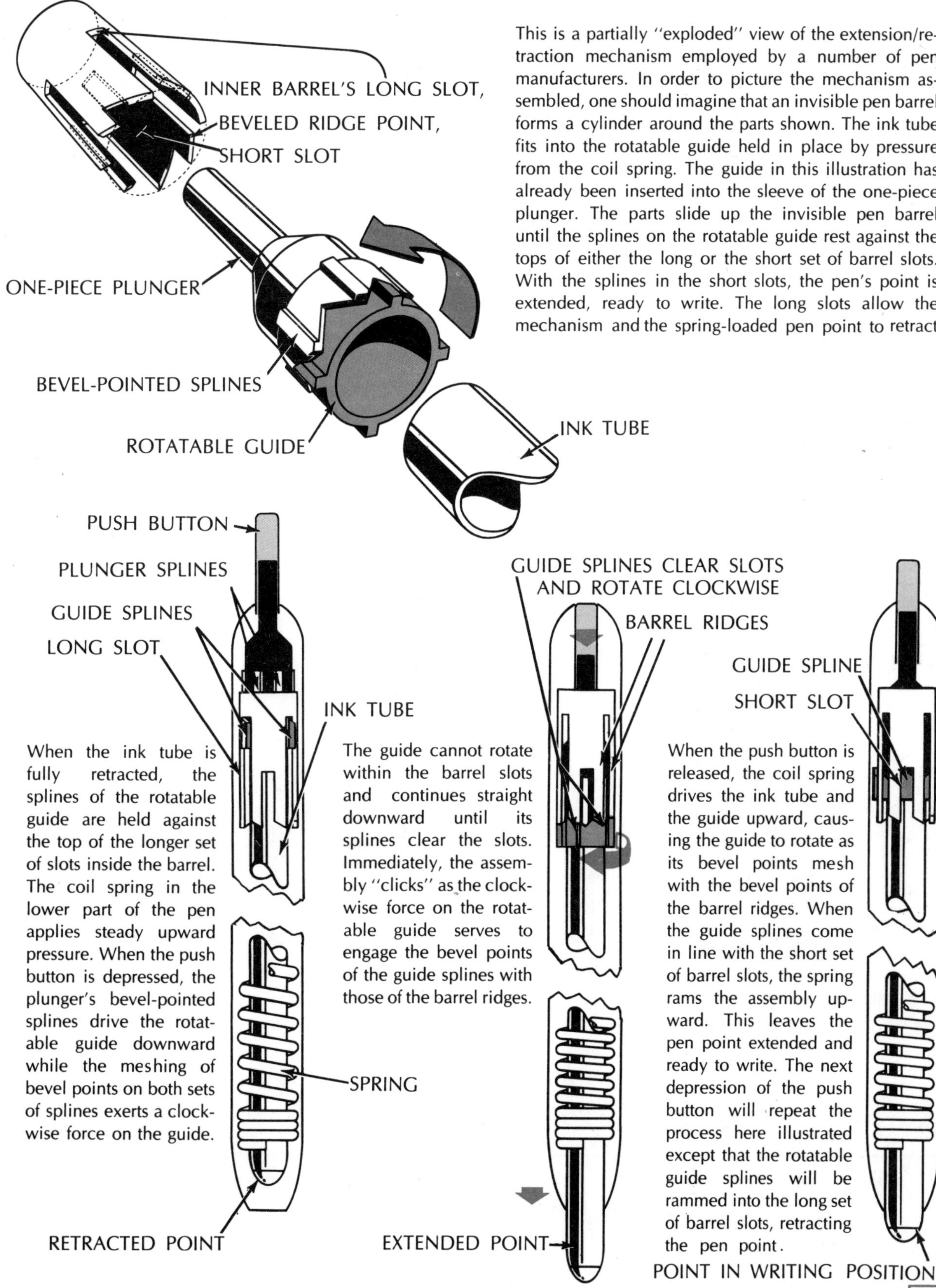

This is a partially "exploded" view of the extension/retraction mechanism employed by a number of pen manufacturers. In order to picture the mechanism assembled, one should imagine that an invisible pen barrel forms a cylinder around the parts shown. The ink tube fits into the rotatable guide held in place by pressure from the coil spring. The guide in this illustration has already been inserted into the sleeve of the one-piece plunger. The parts slide up the invisible pen barrel until the splines on the rotatable guide rest against the tops of either the long or the short set of barrel slots. With the splines in the short slots, the pen's point is extended, ready to write. The long slots allow the mechanism and the spring-loaded pen point to retract.

When the ink tube is fully retracted, the splines of the rotatable guide are held against the top of the longer set of slots inside the barrel. The coil spring in the lower part of the pen applies steady upward pressure. When the push button is depressed, the plunger's bevel-pointed splines drive the rotatable guide downward while the meshing of bevel points on both sets of splines exerts a clockwise force on the guide.

The guide cannot rotate within the barrel slots and continues straight downward until its splines clear the slots. Immediately, the assembly "clicks" as the clockwise force on the rotatable guide serves to engage the bevel points of the guide splines with those of the barrel ridges.

When the push button is released, the coil spring drives the ink tube and the guide upward, causing the guide to rotate as its bevel points mesh with the bevel points of the barrel ridges. When the guide splines come in line with the short set of barrel slots, the spring rams the assembly upward. This leaves the pen point extended and ready to write. The next depression of the push button will repeat the process here illustrated except that the rotatable guide splines will be rammed into the long set of barrel slots, retracting the pen point.

Home Movie Editor

Ideally, once the home movie maker has shown his first experimental films, he recognizes the need to eliminate photographic blunders and bring together his successful film segments. And if successes are few, cutting and splicing at least allow the reorganization of film segments so that sequences follow logically or concentrate on only a few subjects. The answer is a mechanical movie editor that allows viewing, cutting, and splicing with ease. Home movies, themselves, started modestly in 1923 when the first portable, 16-mm. movie camera for amateurs appeared. In 1936 the availability of eight-mm. cameras converted motion picture production from a diversion of the wealthy to a hobby within the dollar reach of most families.

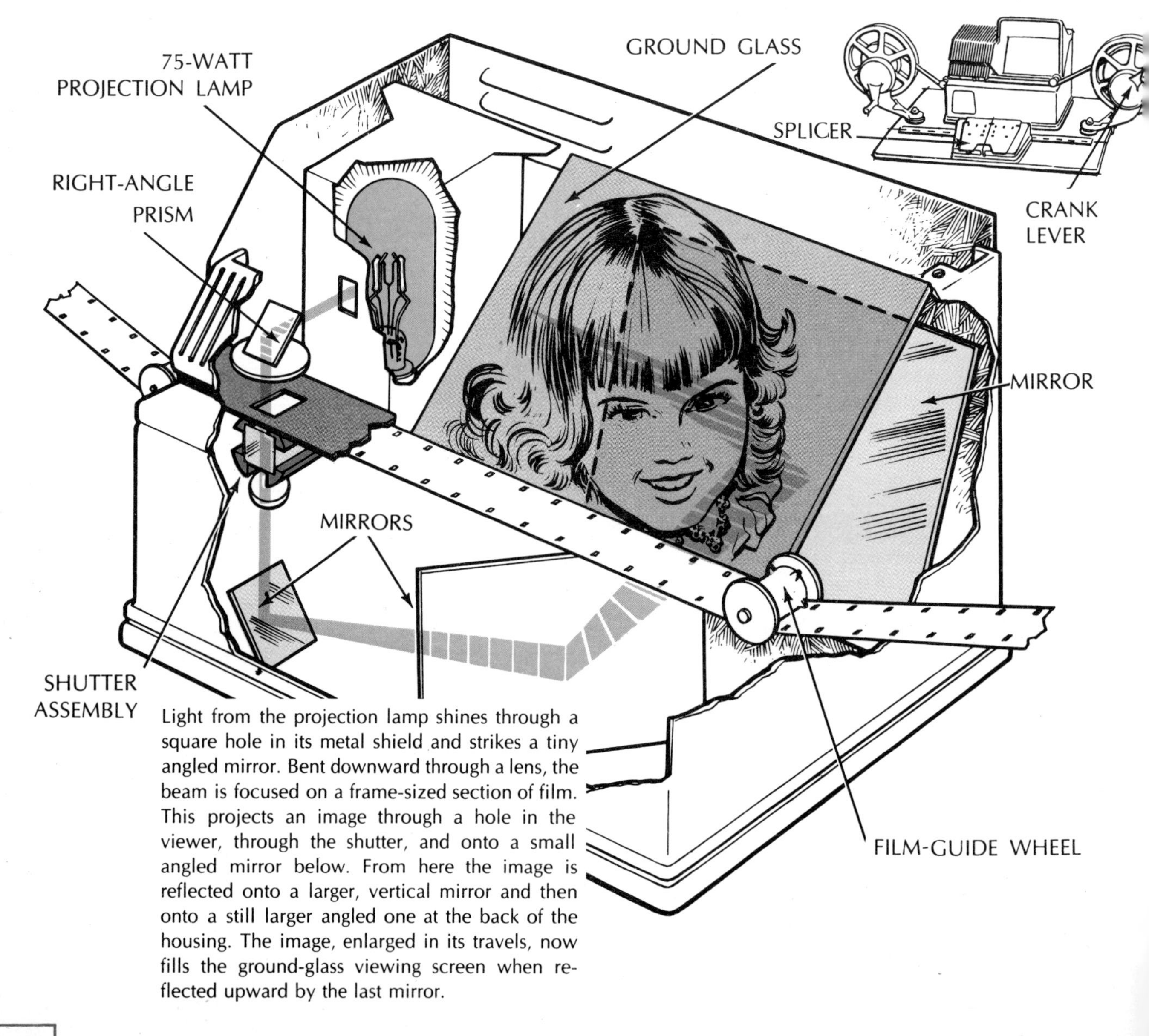

Light from the projection lamp shines through a square hole in its metal shield and strikes a tiny angled mirror. Bent downward through a lens, the beam is focused on a frame-sized section of film. This projects an image through a hole in the viewer, through the shutter, and onto a small angled mirror below. From here the image is reflected onto a larger, vertical mirror and then onto a still larger angled one at the back of the housing. The image, enlarged in its travels, now fills the ground-glass viewing screen when reflected upward by the last mirror.

Shutter Mechanism

1st Frame

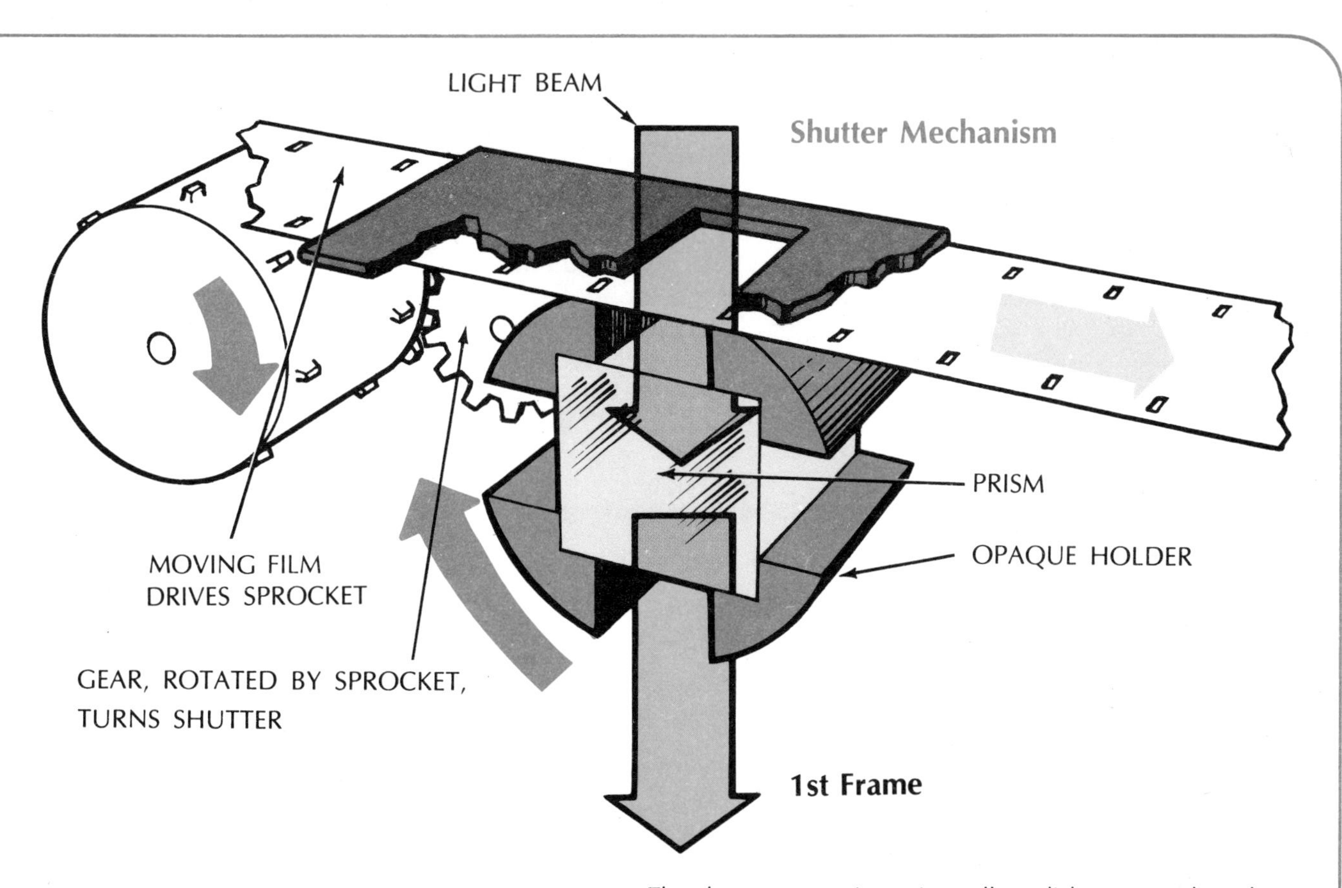

The shutter, a rotating prism, allows light to pass through when the middle of each frame is centered in the light beam. It closes off light as the lines between frames pass through the beam. Thus, only the picture part of each frame appears on the ground glass. The lines are blocked out, causing an interval of darkness between pictures. If the film moves fast enough, the eye blends the sequence of still pictures together. This gives roughly the same impression of continuous motion as a movie projector does. Shutter and film speed are synchronized automatically. As the film moves through the viewer, it turns a sprocket to rotate the shutter. The shutter makes a quarter-turn as the film advances one frame. This leaves the mechanism in the position shown at the top of the page.

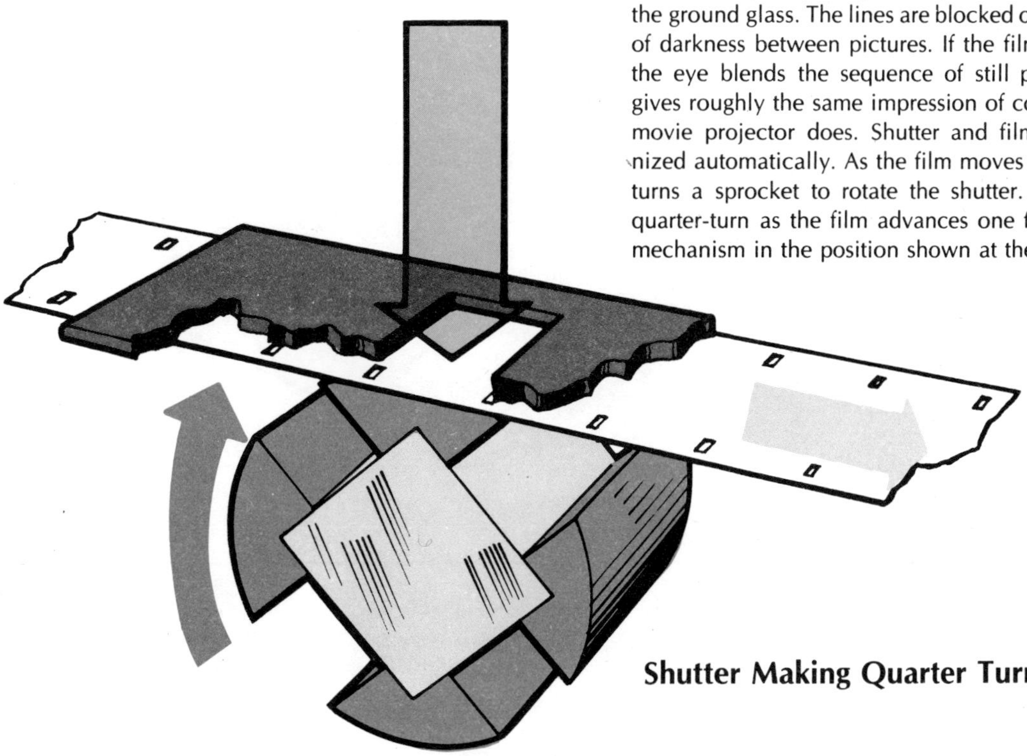

Shutter Making Quarter Turn

Traffic Signal

Mechanical traffic-light controllers like the one shown here can be run manually or automatically and can also follow orders generated remotely by a digital computer. If a traffic officer mixes up the signals manually, controllers can straighten themselves out when switched back to automatic control. Inside the control box is a timer that works the light switches electrically, through a motor-driven camshaft. One timer drum revolution changes lights for a complete signal cycle. Keys on four of the drum's 100 slots determine what percentage of a cycle each light will be "on." A choice of gears can vary cycle time from 30 to 120 seconds in 5-second steps. When operated remotely, the timing may be varied continually to match traffic flow.

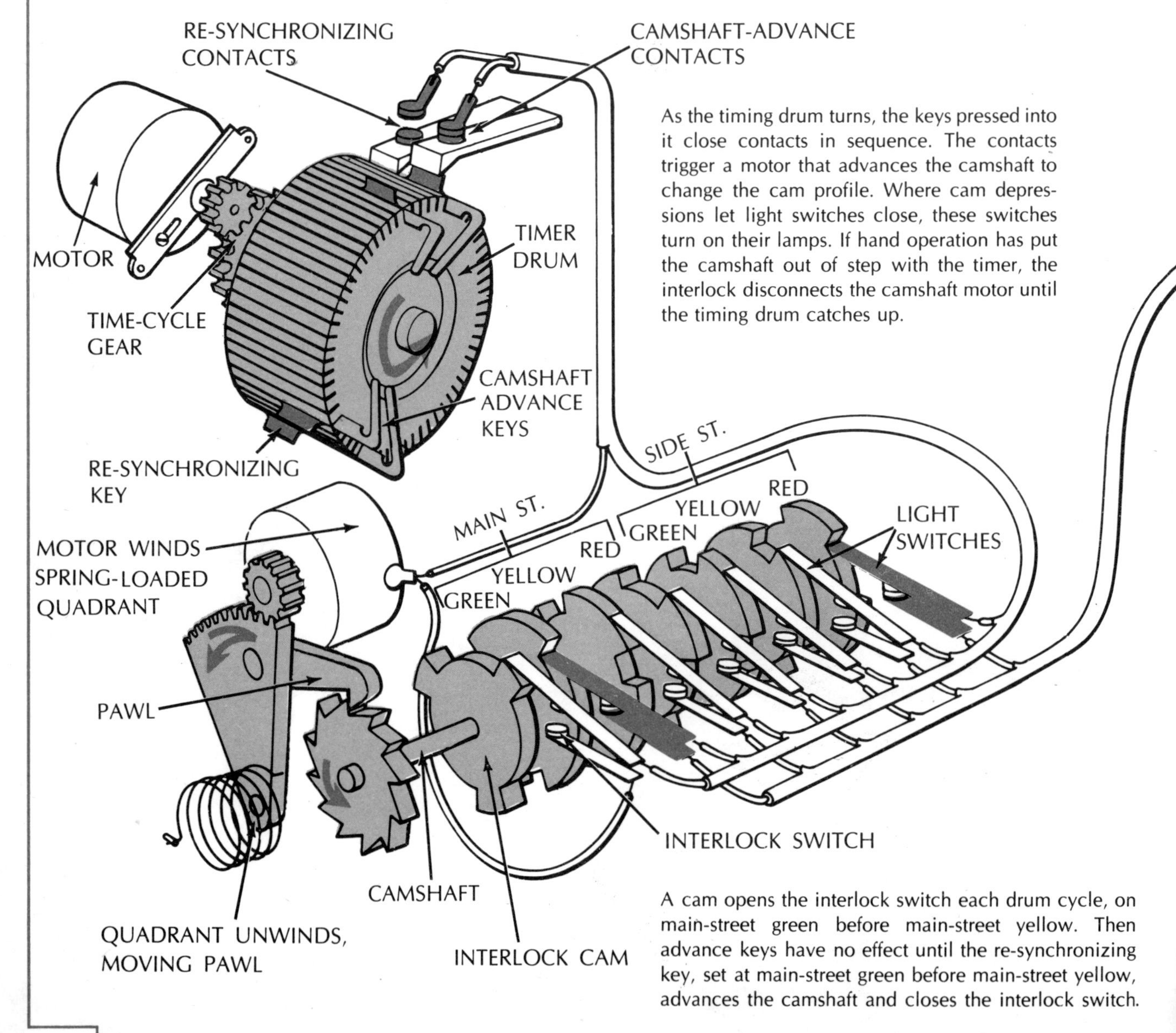

As the timing drum turns, the keys pressed into it close contacts in sequence. The contacts trigger a motor that advances the camshaft to change the cam profile. Where cam depressions let light switches close, these switches turn on their lamps. If hand operation has put the camshaft out of step with the timer, the interlock disconnects the camshaft motor until the timing drum catches up.

A cam opens the interlock switch each drum cycle, on main-street green before main-street yellow. Then advance keys have no effect until the re-synchronizing key, set at main-street green before main-street yellow, advances the camshaft and closes the interlock switch.

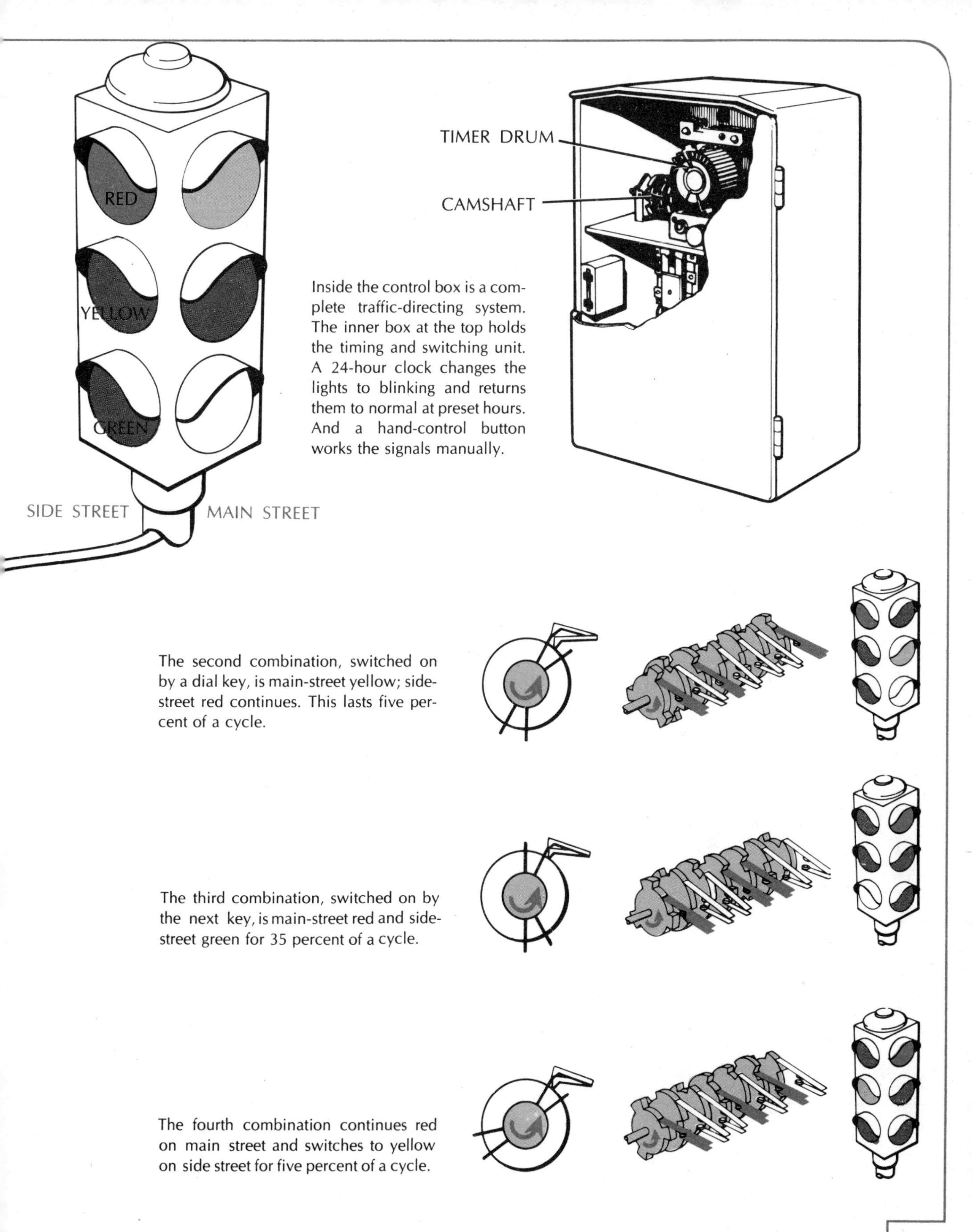

Inside the control box is a complete traffic-directing system. The inner box at the top holds the timing and switching unit. A 24-hour clock changes the lights to blinking and returns them to normal at preset hours. And a hand-control button works the signals manually.

The second combination, switched on by a dial key, is main-street yellow; side-street red continues. This lasts five percent of a cycle.

The third combination, switched on by the next key, is main-street red and side-street green for 35 percent of a cycle.

The fourth combination continues red on main street and switches to yellow on side street for five percent of a cycle.

Automatic Orange Juicer

Some onlookers value the automatic juicer as much for its spellbinding showmanship as for its orange juice. In operation, the juicer is reminiscent of some "Rube Goldberg" contraptions—oranges rolling down ramps, arms pushing and flipping, rubber pressure cups squashing orange halves, and juice flowing into an awaiting container. Yet the juicer is a precision mechanism that can slice and squeeze over 1200 oranges an hour—one every three seconds—while straining the juice and discarding the pulp. For maximum efficiency the automatic juicer must be able to fully squeeze each orange half completely without bruising the skin and causing secretion of bitter oils.

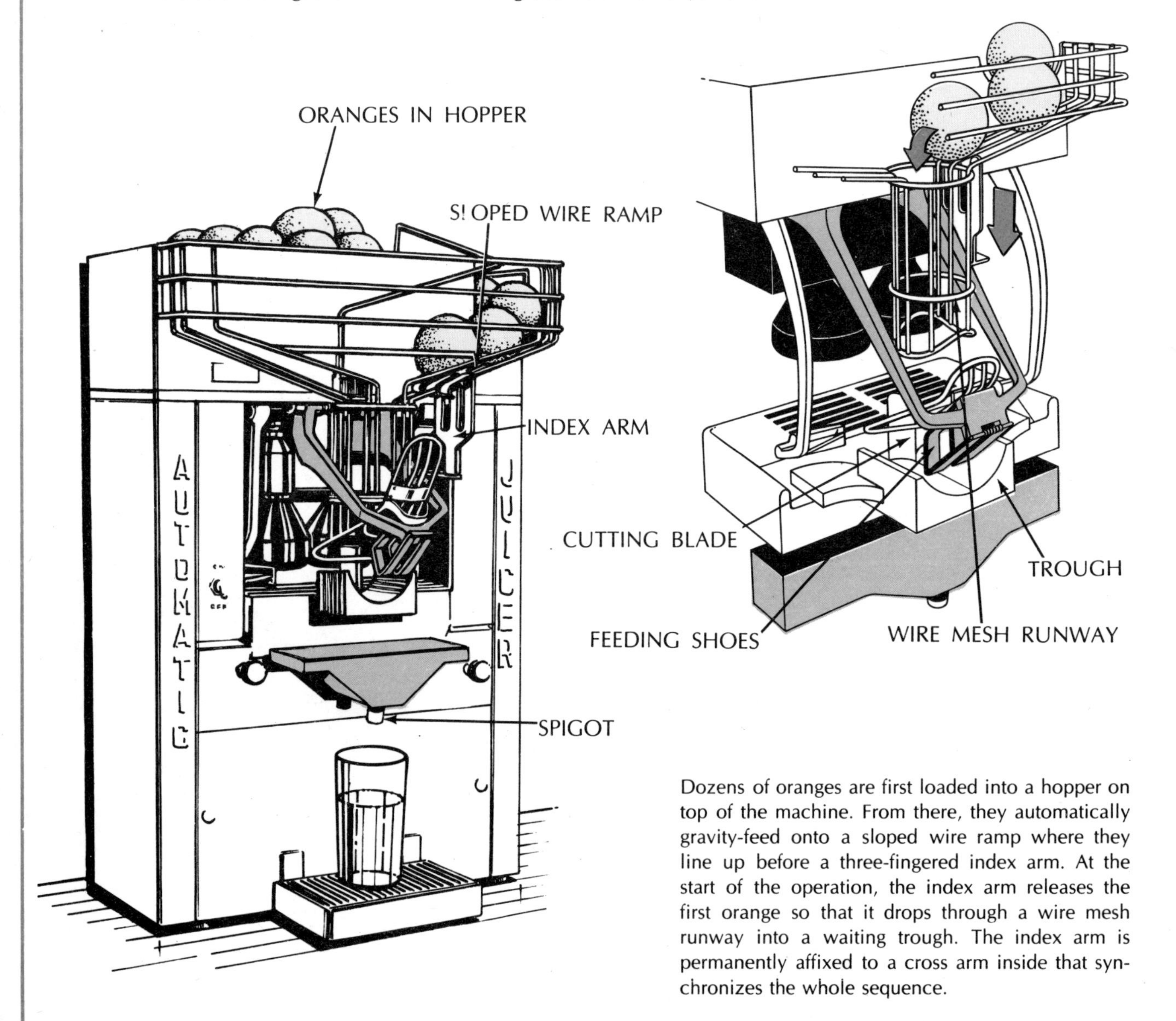

Dozens of oranges are first loaded into a hopper on top of the machine. From there, they automatically gravity-feed onto a sloped wire ramp where they line up before a three-fingered index arm. At the start of the operation, the index arm releases the first orange so that it drops through a wire mesh runway into a waiting trough. The index arm is permanently affixed to a cross arm inside that synchronizes the whole sequence.

After the orange drops through the wire mesh runway, it lands neatly centered before a vertical cutting blade. Two feeding shoes, attached to a U-shaped feeding arm, now push the orange against a sharp blade which divides the orange neatly into halves. The feeding arm pivots on the top plate.

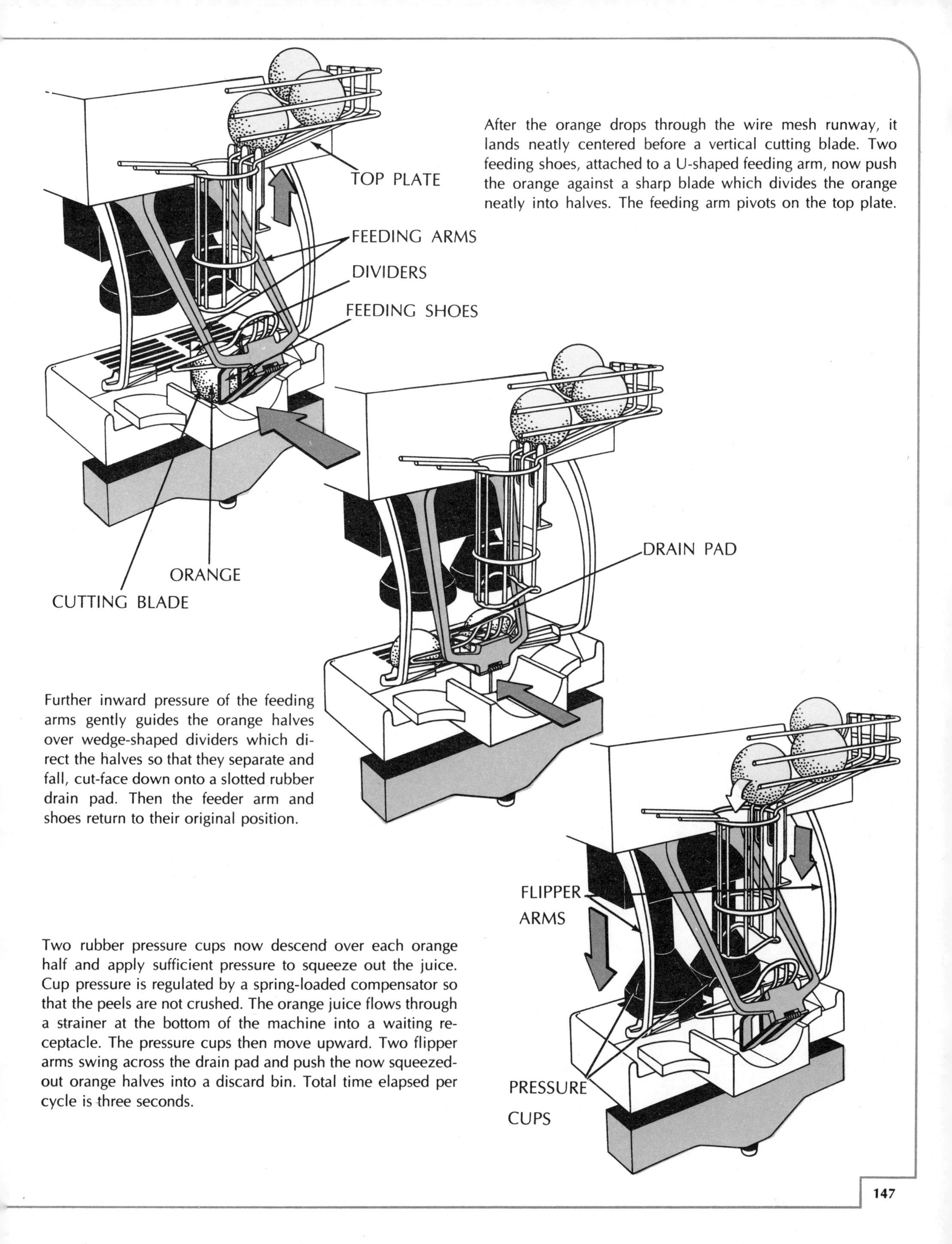

Further inward pressure of the feeding arms gently guides the orange halves over wedge-shaped dividers which direct the halves so that they separate and fall, cut-face down onto a slotted rubber drain pad. Then the feeder arm and shoes return to their original position.

Two rubber pressure cups now descend over each orange half and apply sufficient pressure to squeeze out the juice. Cup pressure is regulated by a spring-loaded compensator so that the peels are not crushed. The orange juice flows through a strainer at the bottom of the machine into a waiting receptacle. The pressure cups then move upward. Two flipper arms swing across the drain pad and push the now squeezed-out orange halves into a discard bin. Total time elapsed per cycle is three seconds.

Spinning Reel

When spinning reels were introduced from Europe in 1932, fishermen employing revolving-spool type reels and educated thumbs were slow to change. But, using spinning gear, any novice fisherman with a few minutes practice can learn to cast a lure repeatedly over 100 feet. Such feats with the early revolving-spool type fishing reel were rare and were often thwarted by the spool's revolving faster than the line rolled off. This caused a frustrating tangle of line known as the backlash. With practice, most spin casters can accurately drop a lure as light as 1/16 ounce as far as 150 feet away. Spinning reels are easy to use because the spool stands still as the line spins off it. Thus there is very little friction and never a backlash.

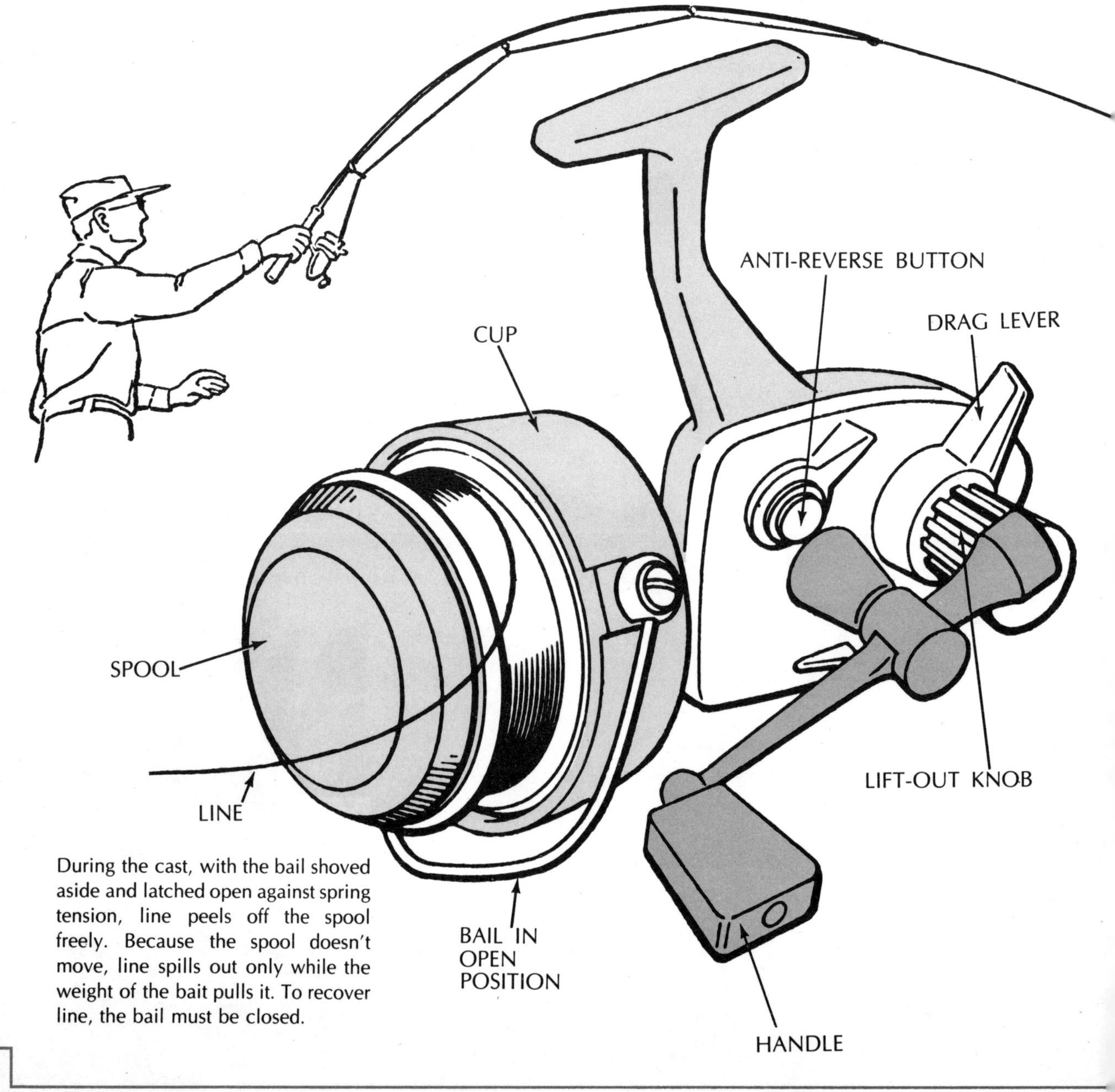

During the cast, with the bail shoved aside and latched open against spring tension, line peels off the spool freely. Because the spool doesn't move, line spills out only while the weight of the bait pulls it. To recover line, the bail must be closed.

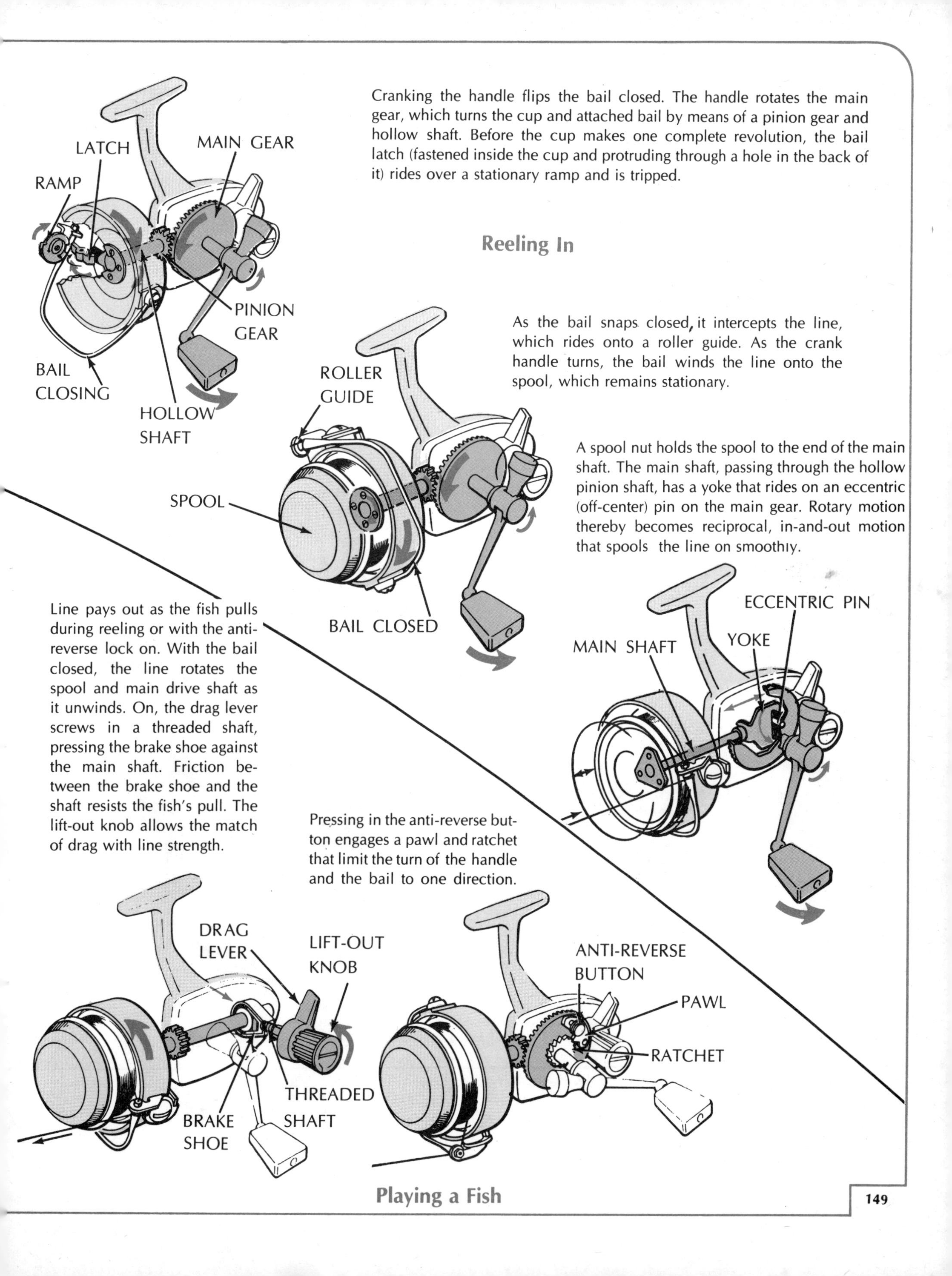

Cranking the handle flips the bail closed. The handle rotates the main gear, which turns the cup and attached bail by means of a pinion gear and hollow shaft. Before the cup makes one complete revolution, the bail latch (fastened inside the cup and protruding through a hole in the back of it) rides over a stationary ramp and is tripped.

Reeling In

As the bail snaps closed, it intercepts the line, which rides onto a roller guide. As the crank handle turns, the bail winds the line onto the spool, which remains stationary.

A spool nut holds the spool to the end of the main shaft. The main shaft, passing through the hollow pinion shaft, has a yoke that rides on an eccentric (off-center) pin on the main gear. Rotary motion thereby becomes reciprocal, in-and-out motion that spools the line on smoothly.

Line pays out as the fish pulls during reeling or with the anti-reverse lock on. With the bail closed, the line rotates the spool and main drive shaft as it unwinds. On, the drag lever screws in a threaded shaft, pressing the brake shoe against the main shaft. Friction between the brake shoe and the shaft resists the fish's pull. The lift-out knob allows the match of drag with line strength.

Pressing in the anti-reverse button engages a pawl and ratchet that limit the turn of the handle and the bail to one direction.

Playing a Fish

Photographic Exposure Meter

Some exposure meters employ a "self-generating" photovoltaic cell, often called a solar cell, that produces electrical energy in proportion to the amount of light striking its surface. Thereby, it provides current to move the needle of a sensitive microammeter. Other exposure meters use a photoconductive cell that is a "light-sensitive" resistor in series with a battery and the microammeter. All exposure meters have external dials that calculate f stops and shutter speeds based on the meter reading. Some meters measure reflected light and are held near the camera and aimed at the subject. Other meters measure incident light—the light falling upon the meter when the meter is placed next to the subject and aimed at the camera. Combination meters can measure either reflected or incident light.

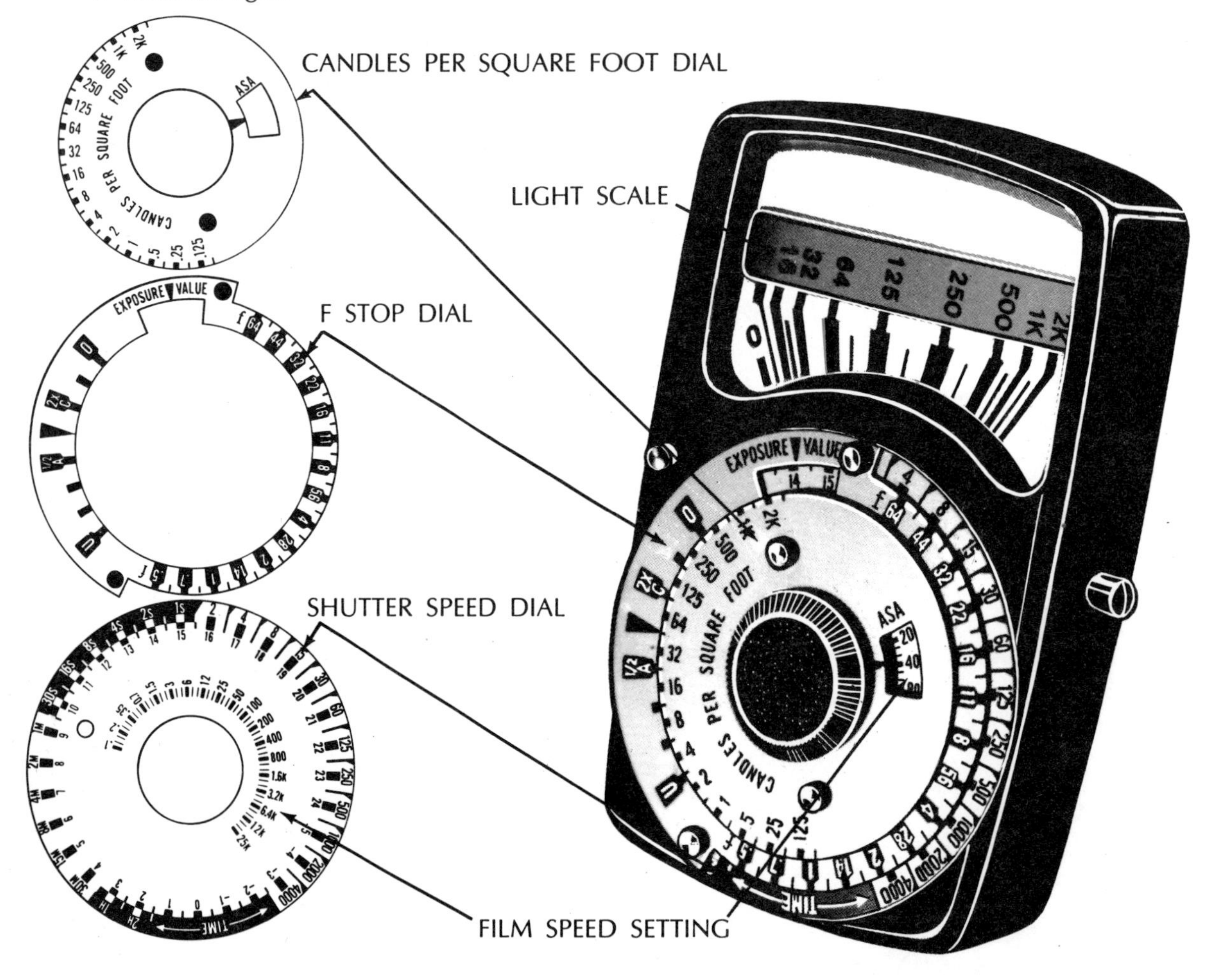

Calculator dials of a meter operate like those of a circular slide rule. First the photographer sets the film's exposure index and then "dials-in" the meter light reading on a second dial. Now he can select combinations of lens opening and shutter speed from the many that are indicated. The exposure indexes marked on the meter shown here range from .1 to 25,000 ASA. Lens openings vary from f/.5 to f/64. Shutter speeds vary from 1/4000 second to two hours. There is also a window which gives the proper light value for Polaroid-Land cameras.

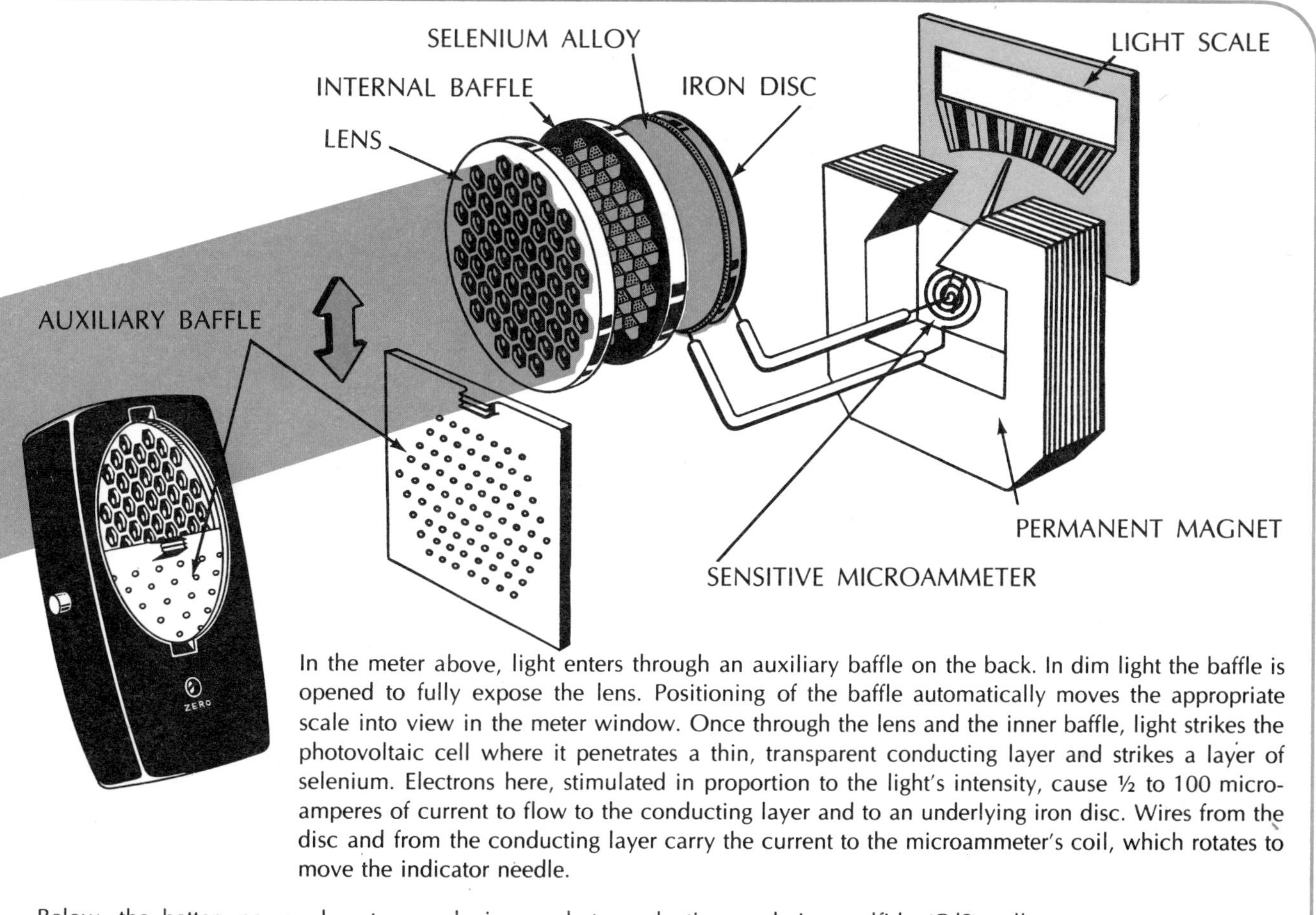

In the meter above, light enters through an auxiliary baffle on the back. In dim light the baffle is opened to fully expose the lens. Positioning of the baffle automatically moves the appropriate scale into view in the meter window. Once through the lens and the inner baffle, light strikes the photovoltaic cell where it penetrates a thin, transparent conducting layer and strikes a layer of selenium. Electrons here, stimulated in proportion to the light's intensity, cause ½ to 100 microamperes of current to flow to the conducting layer and to an underlying iron disc. Wires from the disc and from the conducting layer carry the current to the microammeter's coil, which rotates to move the indicator needle.

Below, the battery-powered meter employing a photoconductive, cadmium sulfide (CdS) cell provides accurate readings even in low-light situations. The CdS cell acts as a light-sensitive, variable resistor and is connected in a series circuit with the battery, a push-button switch, and the coil of the microammeter. Current through the coil moves the needle across the indicator dial. Light energy striking the cell's CdS molecules stimulates them, freeing electrons in large quantities. The free electrons lower the cell's resistance, causing an increase in electric current. A reduction in light falling upon the cell increases the cell's resistance, and this reduces current flow. The more light, the lower the cell's resistance, and the high the microammeter reading.

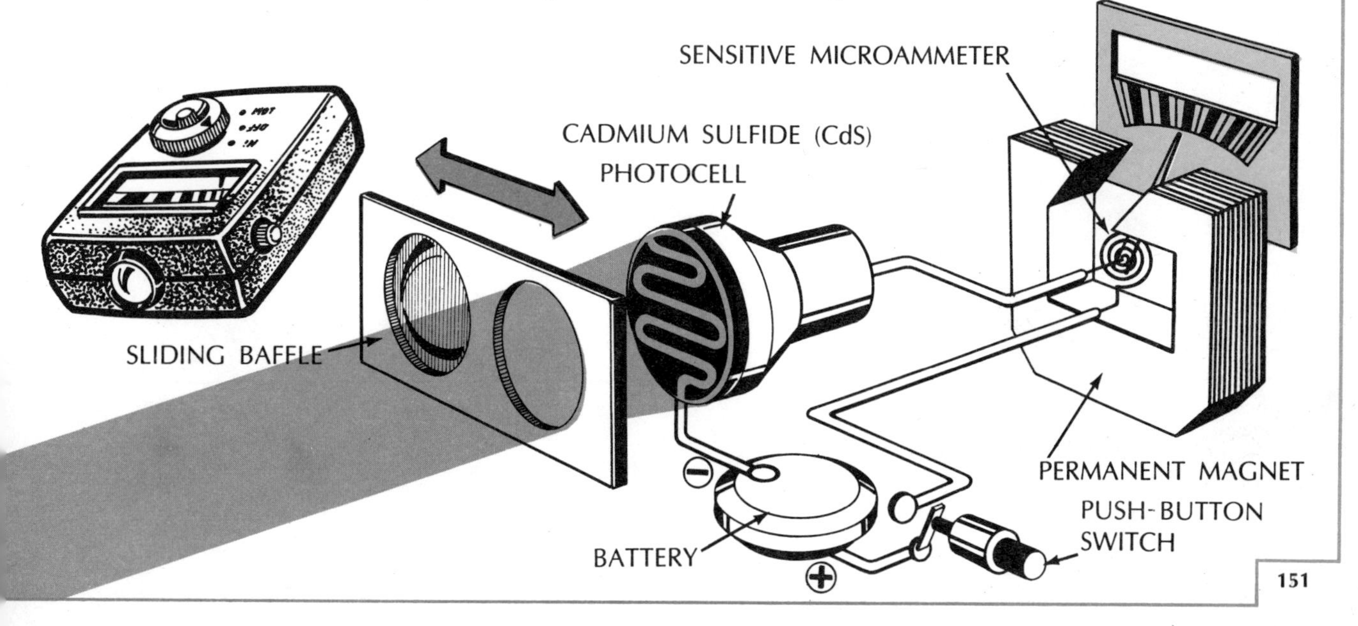

Propane Torch

The propane torch is a simple but precisely made tool handy for such diverse household chores as soldering copper tubing, thawing pipes, removing paint, and even igniting charcoal for a backyard barbecue. The torch consists of a replaceable fuel tank and a burner unit. The cylindrical tank holds liquid propane. Propane is a gaseous petroleum by-product readily converted to a liquid by moderate pressure. When it leaves the pressurized tank, the liquid propane expands more than 250 times and again becomes a gas. A typical tank provides some eight hours of burning time. The burner unit regulates the emerging gas to deliver a flame as hot as 3500° F.

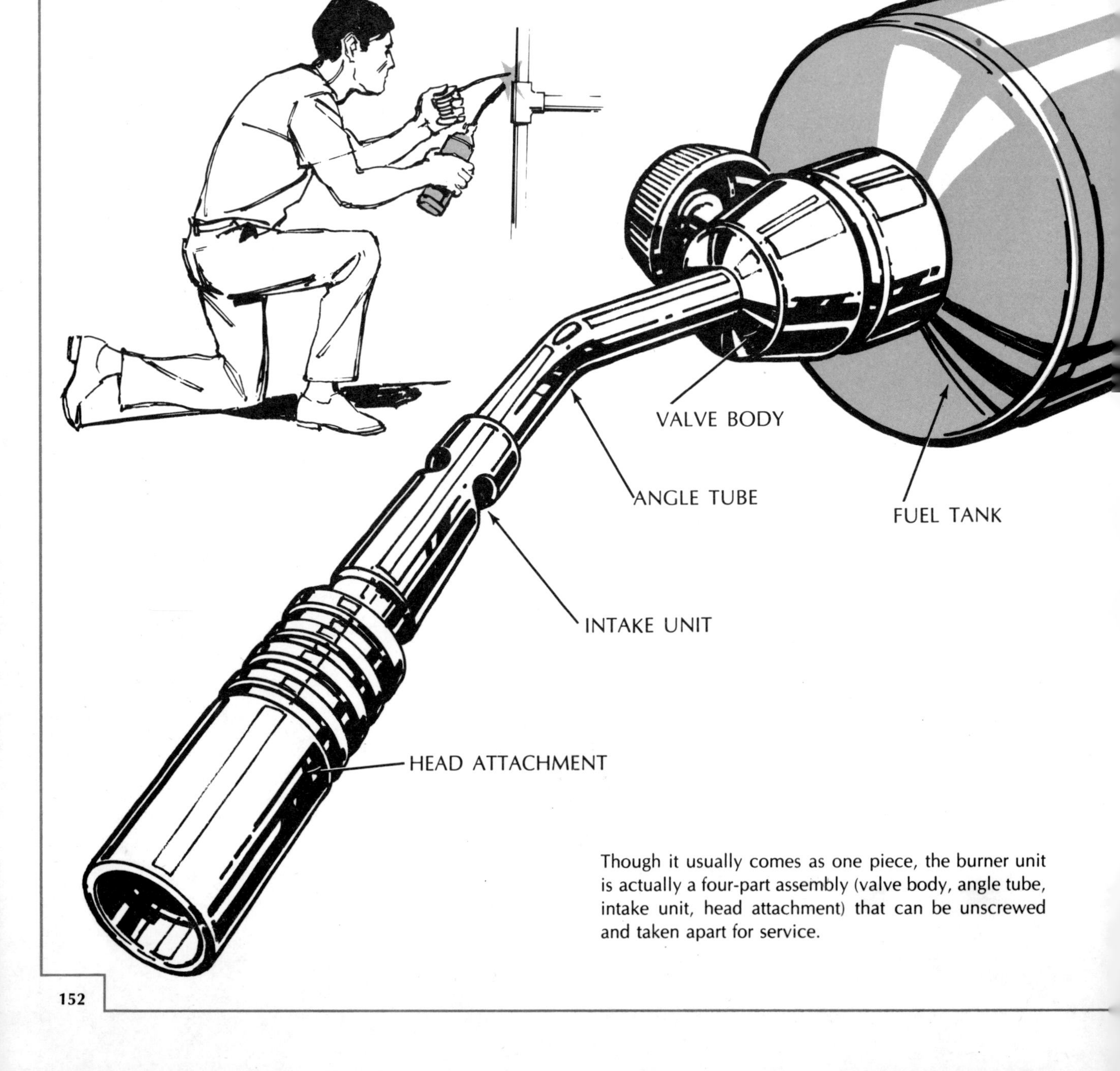

Though it usually comes as one piece, the burner unit is actually a four-part assembly (valve body, angle tube, intake unit, head attachment) that can be unscrewed and taken apart for service.

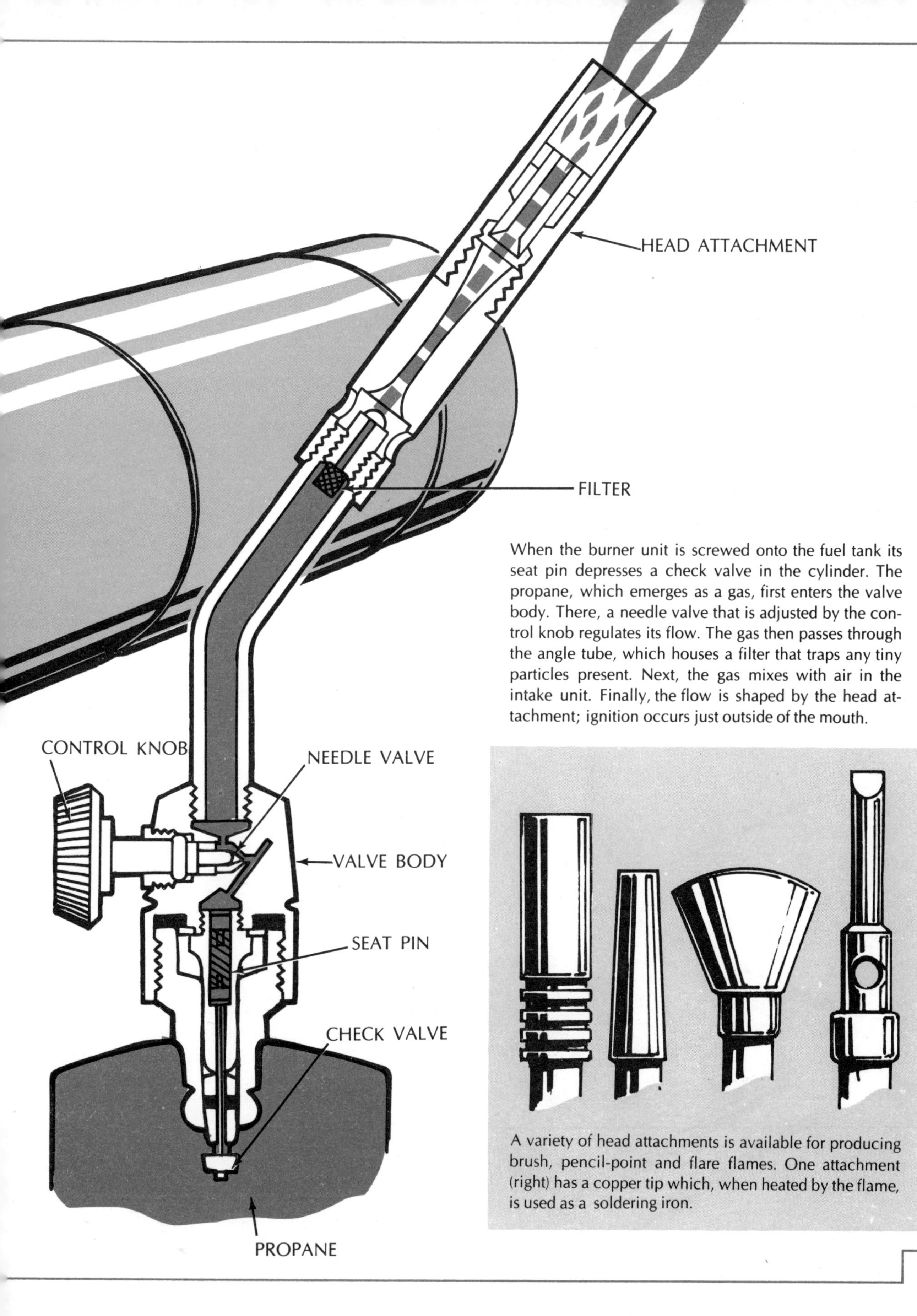

When the burner unit is screwed onto the fuel tank its seat pin depresses a check valve in the cylinder. The propane, which emerges as a gas, first enters the valve body. There, a needle valve that is adjusted by the control knob regulates its flow. The gas then passes through the angle tube, which houses a filter that traps any tiny particles present. Next, the gas mixes with air in the intake unit. Finally, the flow is shaped by the head attachment; ignition occurs just outside of the mouth.

A variety of head attachments is available for producing brush, pencil-point and flare flames. One attachment (right) has a copper tip which, when heated by the flame, is used as a soldering iron.

Ultrasonic Cleaner

Sound is produced by the vibrations of solids, liquids, and gases. *Ultrasound* and thus the descriptive term *ultrasonic* mean sound that is beyond the range of human hearing, with vibrational frequencies of over 20,000 cycles per second. Ultrasonic cleaning is widely favored by manufacturers and users of optical devices and other sensitive instruments. Machines that clean ultrasonically range in size from industrial types as large as a car to handy desk-top units popular among jewelers. Ultrasound literally shakes foreign particles loose, yet it is so gentle that a piece of jewelry dropped into an ultrasonic tank is cleaned without any visible agitation or movement.

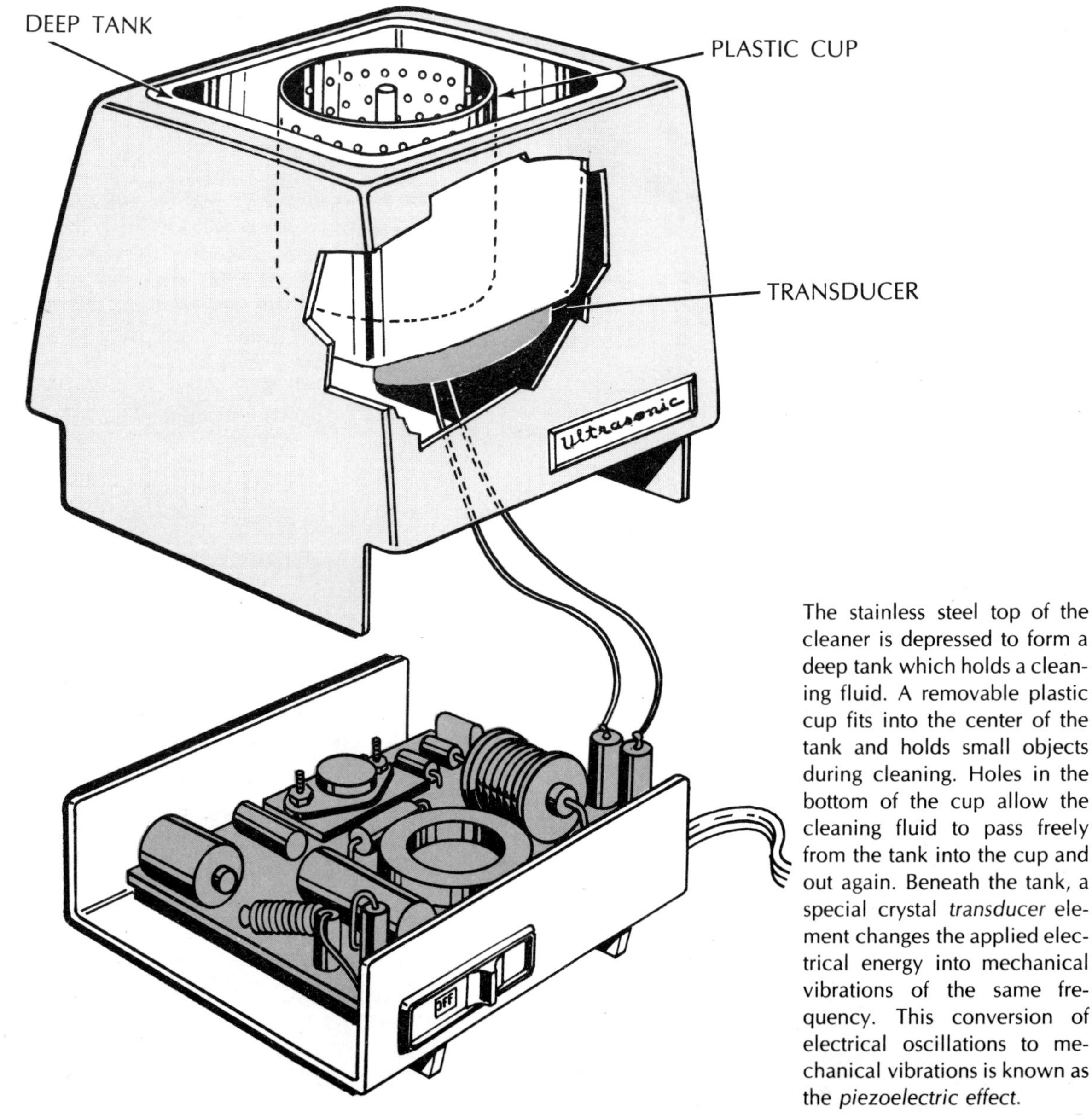

The stainless steel top of the cleaner is depressed to form a deep tank which holds a cleaning fluid. A removable plastic cup fits into the center of the tank and holds small objects during cleaning. Holes in the bottom of the cup allow the cleaning fluid to pass freely from the tank into the cup and out again. Beneath the tank, a special crystal *transducer* element changes the applied electrical energy into mechanical vibrations of the same frequency. This conversion of electrical oscillations to mechanical vibrations is known as the *piezoelectric effect*.

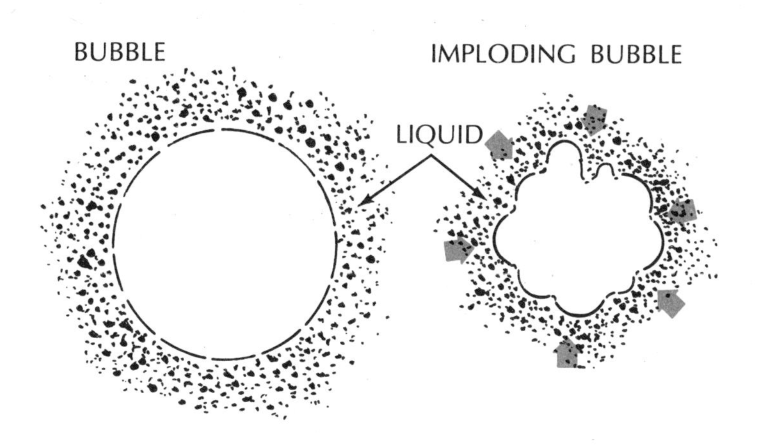

When high-frequency vibrations are transmitted through the tank into the cleaning fluid, the fluid begins a kind of "cold boiling," called *cavitation*. Subjected to rapid vibration, liquid molecules collide violently with one another causing the formation of countless millions of microscopic vacuum bubbles which *implode* (burst inward) almost immediately upon forming. Pressures up to 20,000 pounds per square inch are exerted in infinitesimal space by these implosions. Cavitation accomplishes a microscopic scrubbing action since the imploding bubbles can infiltrate into even the tiniest of pores that resist the entry of conventional liquid cleaners. With the correct cleaning agent and ultrasonic techniques, virtually any material can be freed of foreign particles, dirt, dust, grease, or other contaminants.

The *piezoelectric* effect that changes electrical energy into mechanical vibrations was discovered by Pierre Curie in 1880. He found that if an alternating current is connected to opposite faces of certain crystalline substances or polarized ceramic substances, the current's regularly-varying voltage stresses the substance so that it begins to vibrate at the frequency of the applied voltage. In the ultrasonic cleaner, high frequency alternating current is applied to the transducer element which in turn physically vibrates the tank.

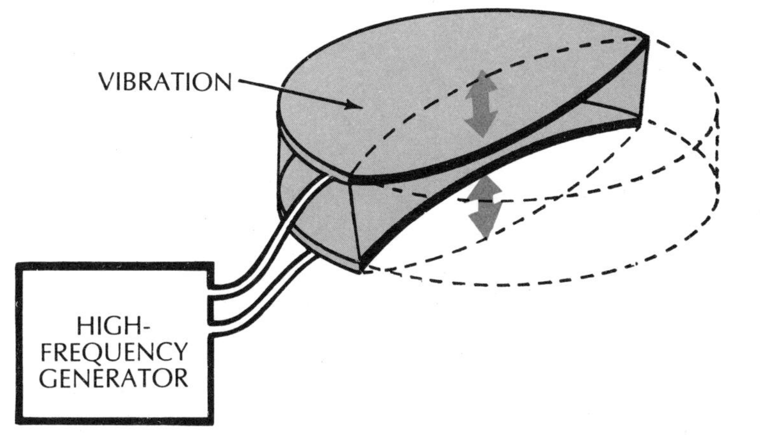

The cleaner's ac generator is a transistorized high-frequency oscillator mounted on a component board set into the base of the cleaner. The oscillator is powered from the standard 110-volt, 60 cycles-per-second ac line and generates a high-level electrical signal that varies at the rate of about 40,000 cycles per second.

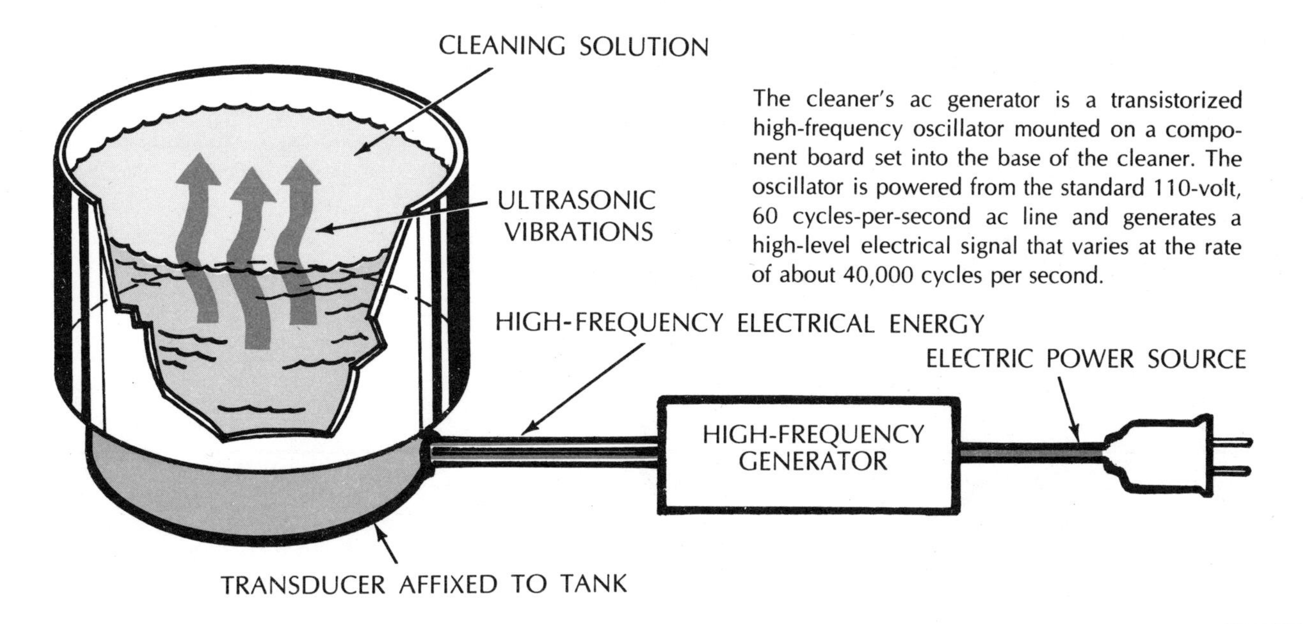

Ice Crusher

The key to the chilling power of crushed ice is its surface area. When an ice cube is broken up into shards, its liquid-to-ice contact increases hundreds of times. Crushing enables the ice to draw heat from a liquid quickly, before the ice absorbs a significant amount of heat from air at the liquid's surface. When the machine's top lid is raised, hidden workings come alive with a whirring noise. As ice cubes are placed onto a spring-hinged drop door, they disappear inside the device and become staccato "clunkety-clunk" sounds. Then the ice emerges out an exit chute and into an awaiting bucket, crushed and ready for the mixologist.

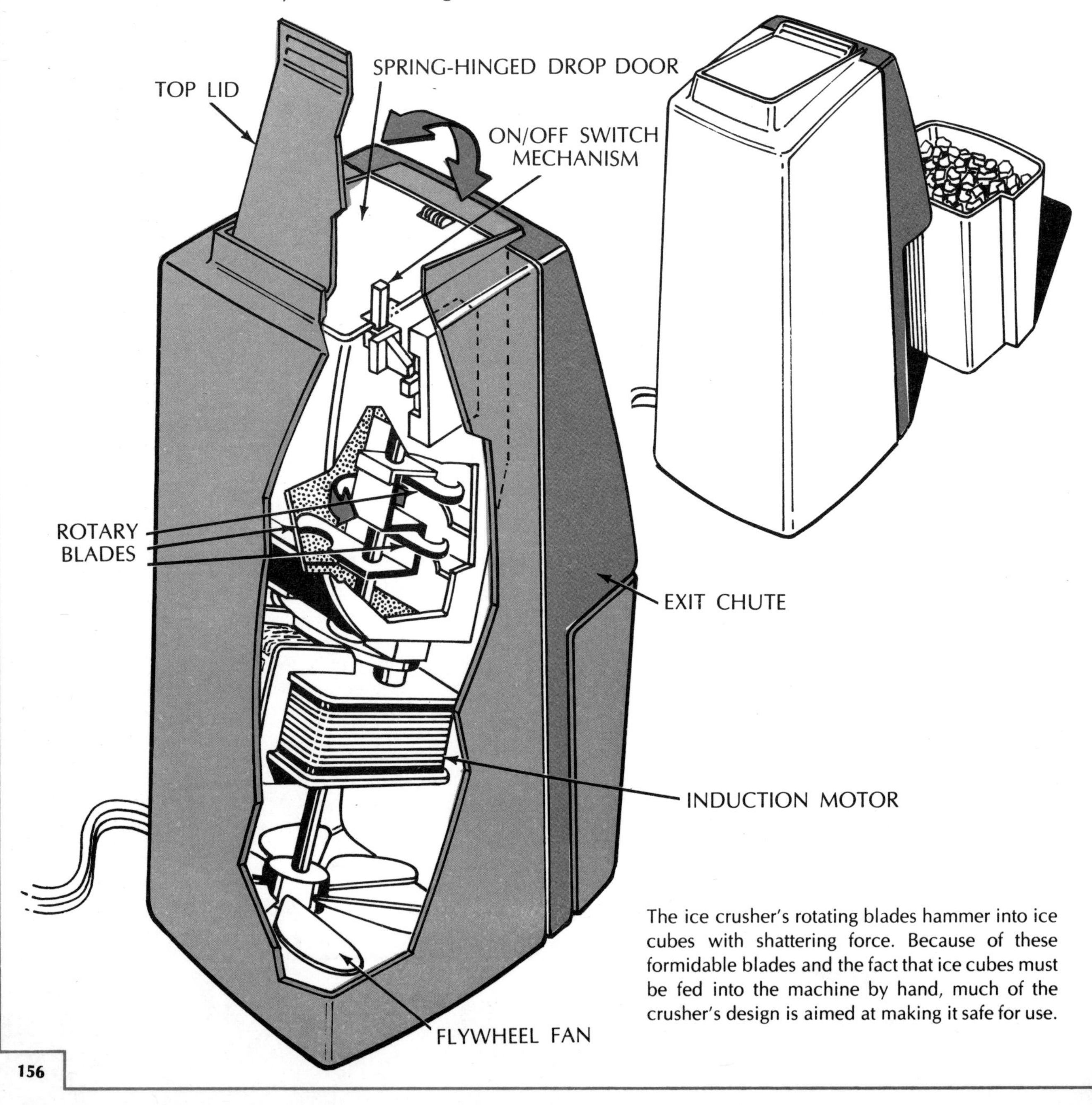

The ice crusher's rotating blades hammer into ice cubes with shattering force. Because of these formidable blades and the fact that ice cubes must be fed into the machine by hand, much of the crusher's design is aimed at making it safe for use.

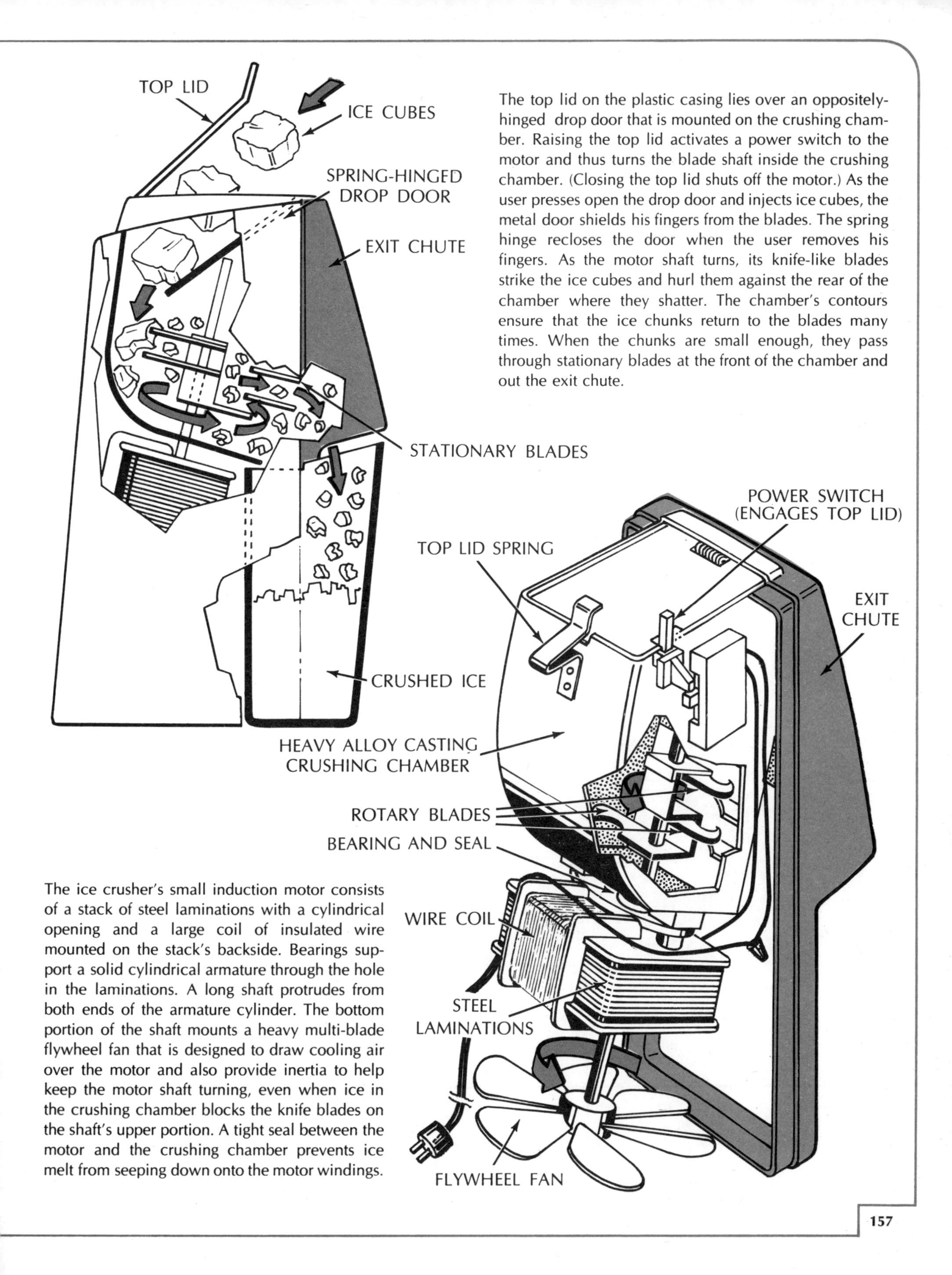

The top lid on the plastic casing lies over an oppositely-hinged drop door that is mounted on the crushing chamber. Raising the top lid activates a power switch to the motor and thus turns the blade shaft inside the crushing chamber. (Closing the top lid shuts off the motor.) As the user presses open the drop door and injects ice cubes, the metal door shields his fingers from the blades. The spring hinge recloses the door when the user removes his fingers. As the motor shaft turns, its knife-like blades strike the ice cubes and hurl them against the rear of the chamber where they shatter. The chamber's contours ensure that the ice chunks return to the blades many times. When the chunks are small enough, they pass through stationary blades at the front of the chamber and out the exit chute.

The ice crusher's small induction motor consists of a stack of steel laminations with a cylindrical opening and a large coil of insulated wire mounted on the stack's backside. Bearings support a solid cylindrical armature through the hole in the laminations. A long shaft protrudes from both ends of the armature cylinder. The bottom portion of the shaft mounts a heavy multi-blade flywheel fan that is designed to draw cooling air over the motor and also provide inertia to help keep the motor shaft turning, even when ice in the crushing chamber blocks the knife blades on the shaft's upper portion. A tight seal between the motor and the crushing chamber prevents ice melt from seeping down onto the motor windings.

Stapling Gun

The stapling gun works like a pile driver, in miniature, and whams in two-legged nails. Powered by a hand lever, the driver attains a speed of 36 m.p.h. before impact and wallops the staple with a 200-pound force. This beefed-up version of the desk-top stapler has long been an industrial tool and ultimately became available to home handymen for chores such as mating screens to window frames and fastening acoustic ceilings and wallboards. One gun can last long enough to drive a quarter-million staples, provided it receives proper use—always butted firmly against the work surface before impact so that the driving force transfers into the staple, itself. Playful firing into the air impacts the driver against the stapler's ram stop and causes excessive wear.

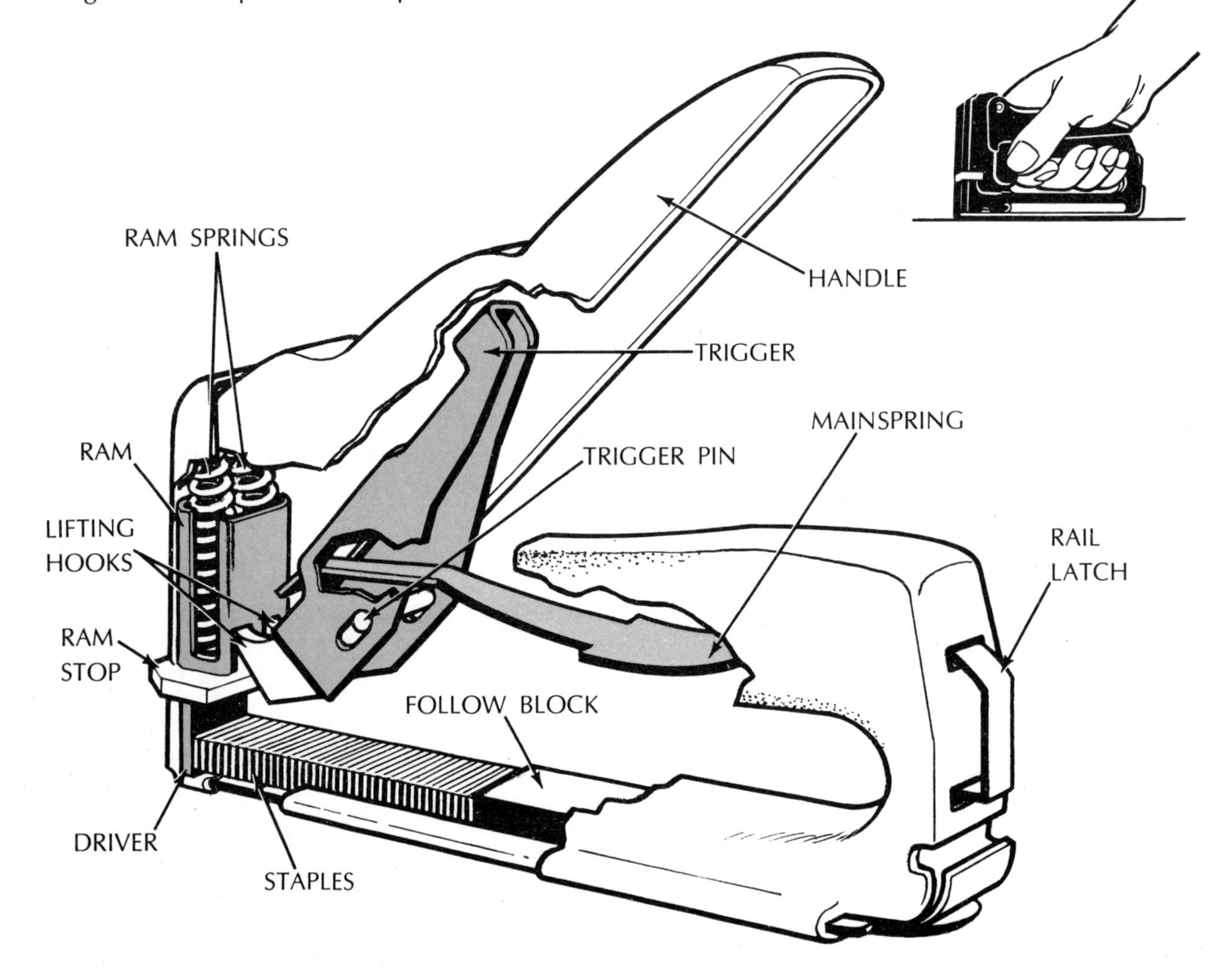

A staple must penetrate about 1/4 inch into wood before it will hold. Long legs allow the fastening of thicker materials. Staples come in sizes from 3/16 inch to 9/16 inch. Smaller sizes are made of low-carbon steel, while the larger staples come in both low- and high-carbon steel—high carbon for toughness. The entire row of staples can be inserted as one unit, since the staples are bonded together with an acetone cement that releases each staple as it is struck by the ram.

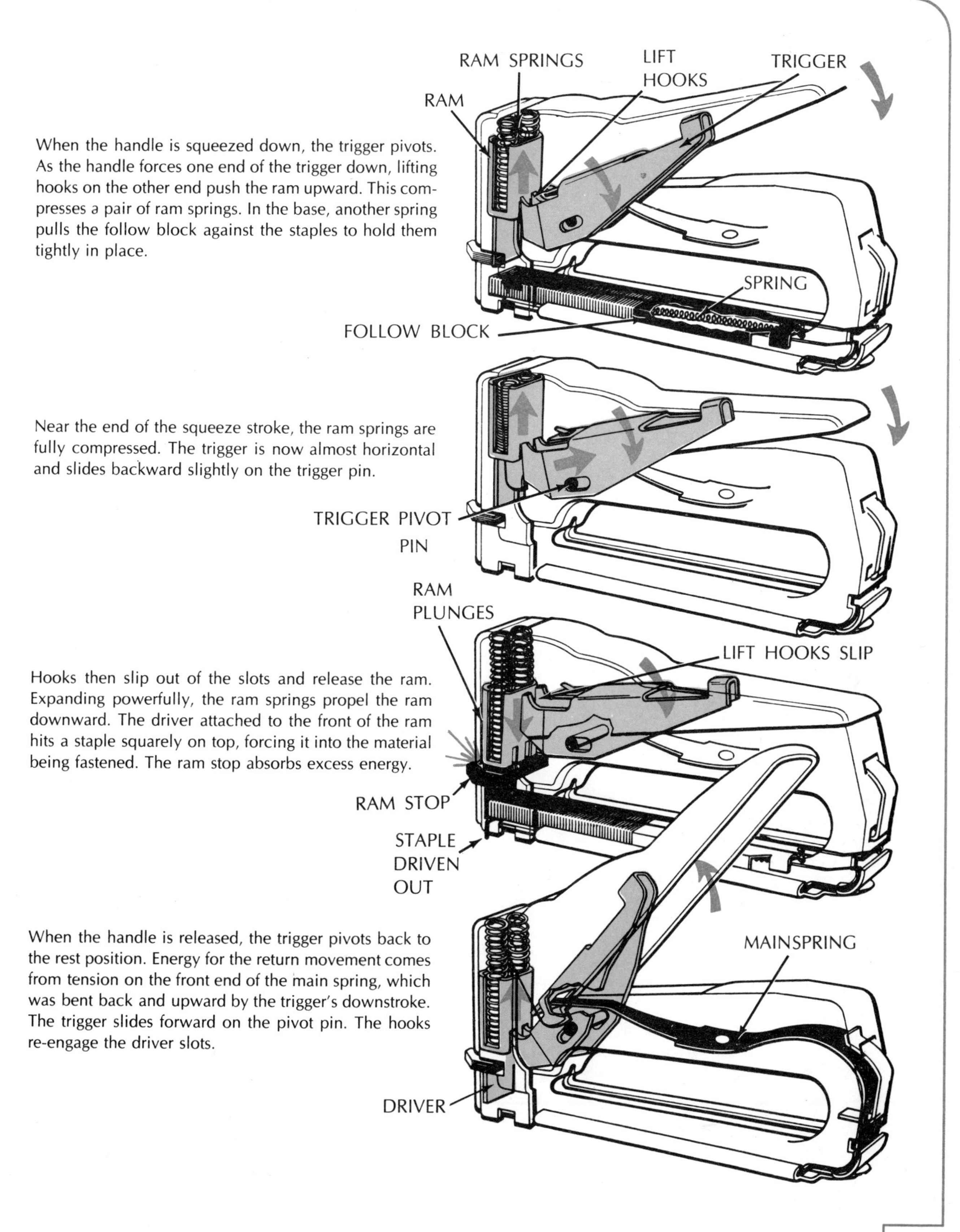

When the handle is squeezed down, the trigger pivots. As the handle forces one end of the trigger down, lifting hooks on the other end push the ram upward. This compresses a pair of ram springs. In the base, another spring pulls the follow block against the staples to hold them tightly in place.

Near the end of the squeeze stroke, the ram springs are fully compressed. The trigger is now almost horizontal and slides backward slightly on the trigger pin.

Hooks then slip out of the slots and release the ram. Expanding powerfully, the ram springs propel the ram downward. The driver attached to the front of the ram hits a staple squarely on top, forcing it into the material being fastened. The ram stop absorbs excess energy.

When the handle is released, the trigger pivots back to the rest position. Energy for the return movement comes from tension on the front end of the main spring, which was bent back and upward by the trigger's downstroke. The trigger slides forward on the pivot pin. The hooks re-engage the driver slots.

Electric Shaver

When Julius Caesar was dividing all Gaul into three parts, he introduced a novel defensive ploy into the science of hand-to-hand combat—*the clean shaven face*. Caesar had observed that a beard made a convenient hand grip that a combatant might grab to throw his foe off balance. To deny the enemy this advantage, Caesar ordered his soldiers to scrape off their chin whiskers before battle using a keen-edged short sword. Today's electric razors offer less hazard and more convenience than the short sword—and a cleaner shave too. In modern electrics, beard hairs pass through the perforations in a thin screen. Blades behind the screen shear off the whiskers below the skin surface.

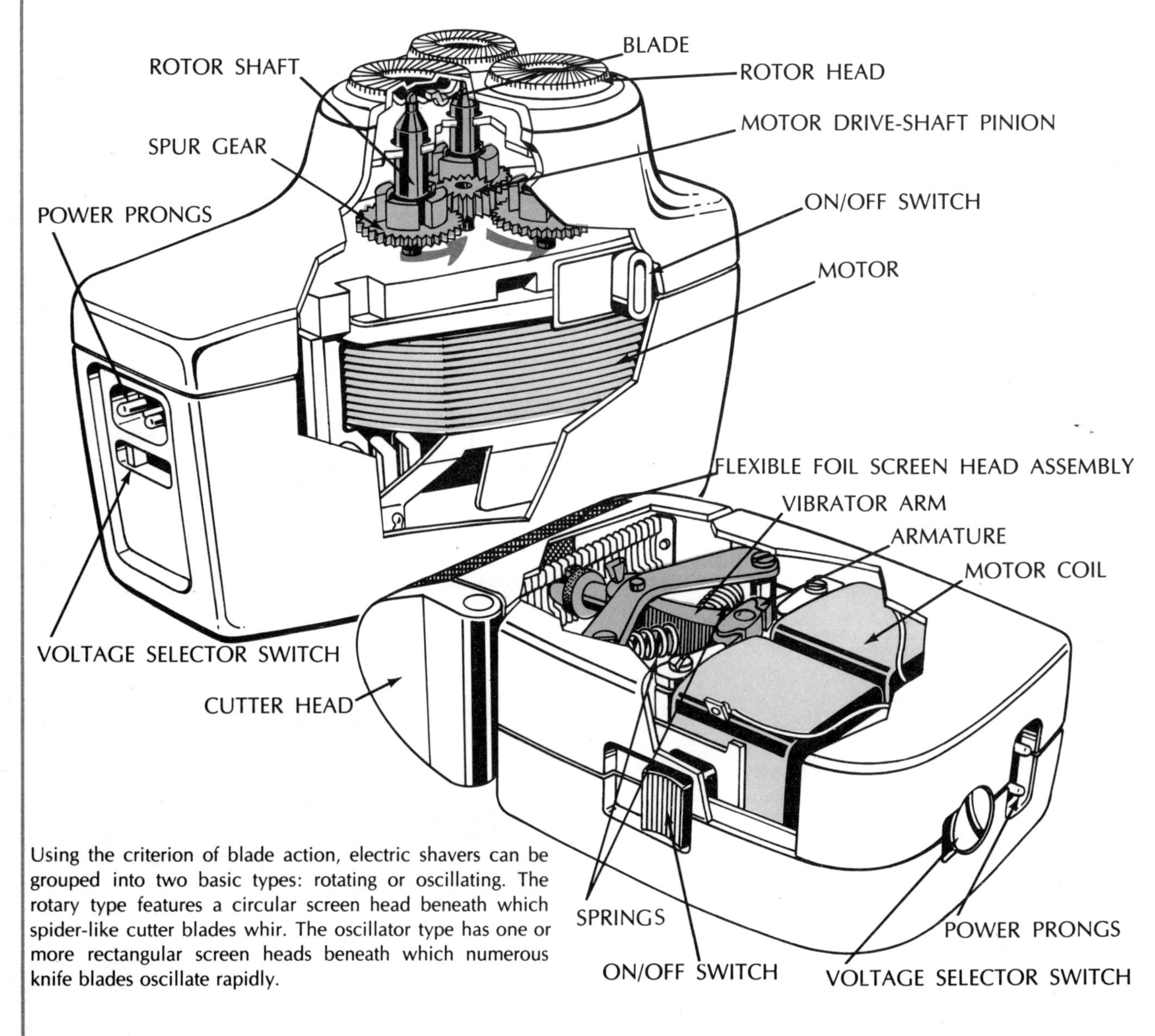

Using the criterion of blade action, electric shavers can be grouped into two basic types: rotating or oscillating. The rotary type features a circular screen head beneath which spider-like cutter blades whir. The oscillator type has one or more rectangular screen heads beneath which numerous knife blades oscillate rapidly.

The rotary cutter head consists of a surgical steel "spider" with six angled arms that end in flat, honed cutting edges. The spider fits inside a head with comb-like slots. The center of the spider mounts a rotor shaft leading back to a pinion on the motor drive shaft (see opposite page). The motor's pinion gear engages separate larger gears on each of the three drive shafts. This gearing produces a rotational step-down ratio so that the shafts and the cutter heads actually spin slower than the motor does. The spider whirs around inside the head, with all six blades continuously shearing off beard hairs protruding through the comb slots. The loosely fitted heads "float" on the spring-tensioned drive shafts and follow facial contours, even around the chin.

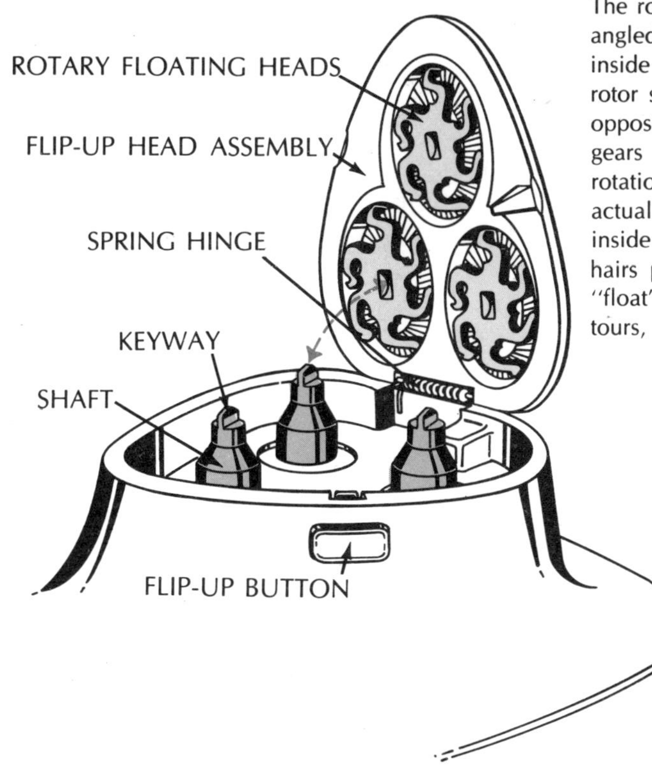

HEAD

"SPIDER" BLADES

SHAFT TOP

This oscillator shaver's cutter head has 34 vertical blades curved to fit the inner contour of the foil screen. The head sits on a swivel mount that is connected to a pivoted metal arm. At the bottom of the arm, a U-shaped armature swings close to the pole pieces of side-by-side coils. Heavy springs on each side of the arm bear against the arm and opposite sides of the motor housing. With the motor on, the alternating electromagnetic push and pull of the pole pieces on the armature cause the armature to swing back and forth, compressing the springs alternately. The springs return the arm as the current passes through zero. Thus, spring and electromagnetic actions vibrate the arm and in turn activate the cutter head.

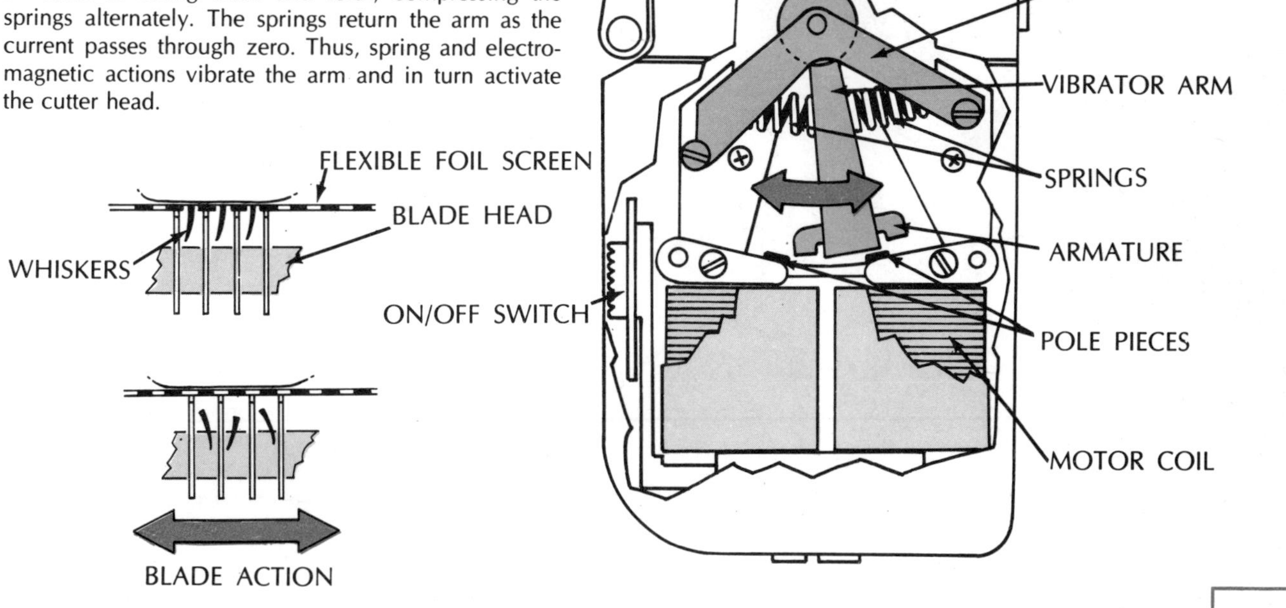

Gas Station "Pump"

The gas dispensing unit at service stations has been called a "pump" ever since the days when gas was pumped manually. Though many dispensing units still contain a pump, an increasingly great number of dispensers or "pumps" contain no pump at all. Rather, two or more dispensing units may rely on a remote pump located near or within the gas storage tank. These remote pumps mainly employ positive pressure, produced by impellers, to *force* gas from the storage tank to the dispenser. However, pumps located within the dispenser, itself, rely wholly on negative pressure, or suction, to *draw* gas through the line.

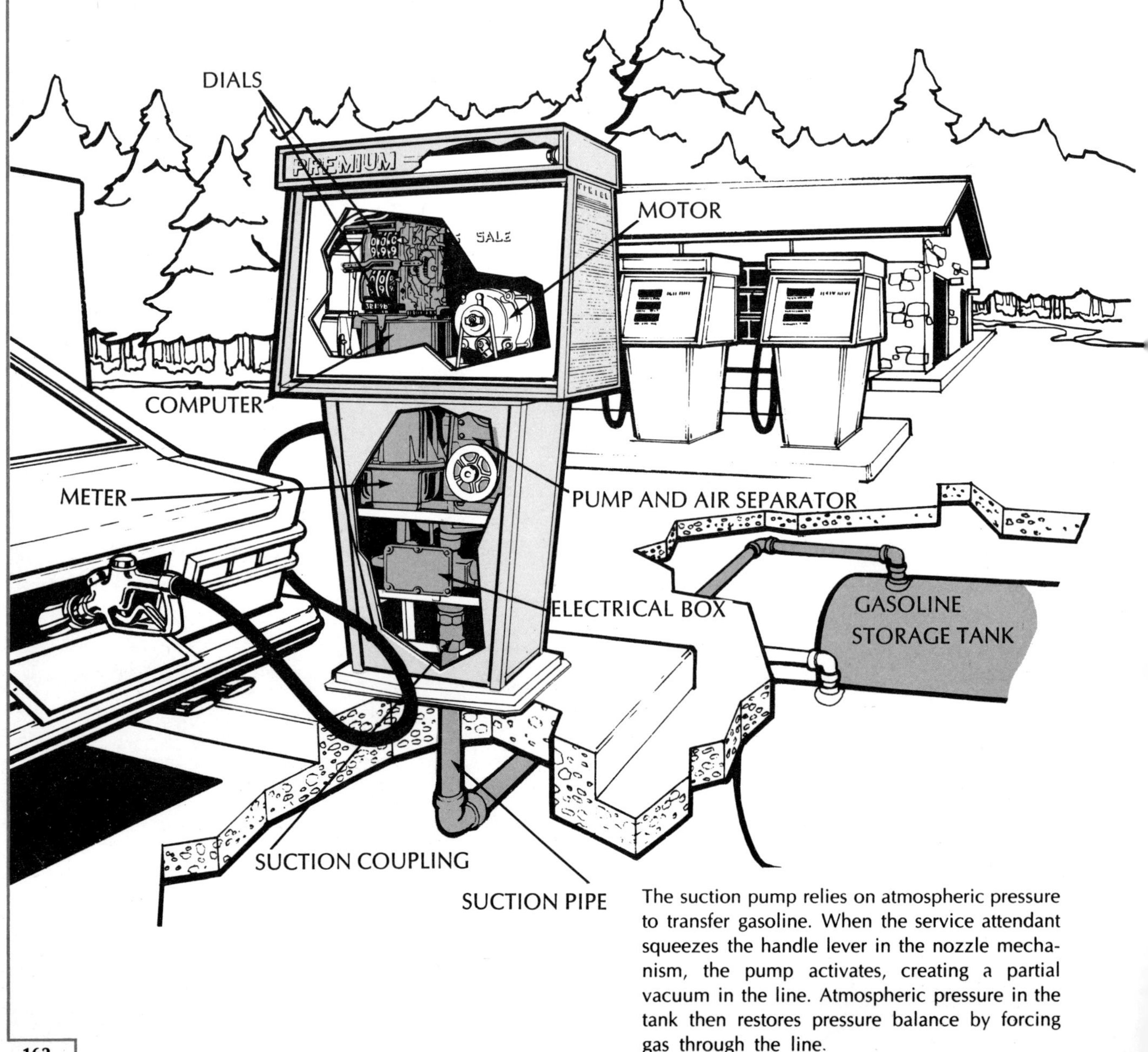

The suction pump relies on atmospheric pressure to transfer gasoline. When the service attendant squeezes the handle lever in the nozzle mechanism, the pump activates, creating a partial vacuum in the line. Atmospheric pressure in the tank then restores pressure balance by forcing gas through the line.

1. The station attendant first removes the hose nozzle from the side of the dispenser and then turns an interlock lever that prevents delivery of gas until all indicator drums return to "zero." This action starts a small motor belted to a rotary pump. The pump builds up pressure until a pump bypass valve allows gas to return to the suction side of the pump, thus preventing excess pressure buildup. When the attendant opens the nozzle, gasoline flows, reducing pressure in the pump and allowing the bypass valve to close.

2. Gas then enters the air separator—a kind of settling tank. Here a port in a suction tube draws vapor and air up to a sump. Back in the separator, deaerated gasoline is forced out.

3. Pressure in the line holds open a two-way pressure-regulating valve. Gasoline proceeds to the meter, turning vanes and a crankshaft. The crankshaft turns gear trains in the computer that revolve number drums, showing gallonage and cost.

4. At the top, the sump is vented to the atmosphere. At the bottom, a float-controlled valve opens to permit return of gasoline into the pump. The sump also serves as a pressure release outlet when gasoline in the system expands, owing to an increases in temperature.

5. From the meter, the gasoline enters the safety-fill nozzle, which can be set by means of a hand-lever clip to deliver gas until a vent tube automatically shuts off flow when gas in the car's tank covers an air inlet port.

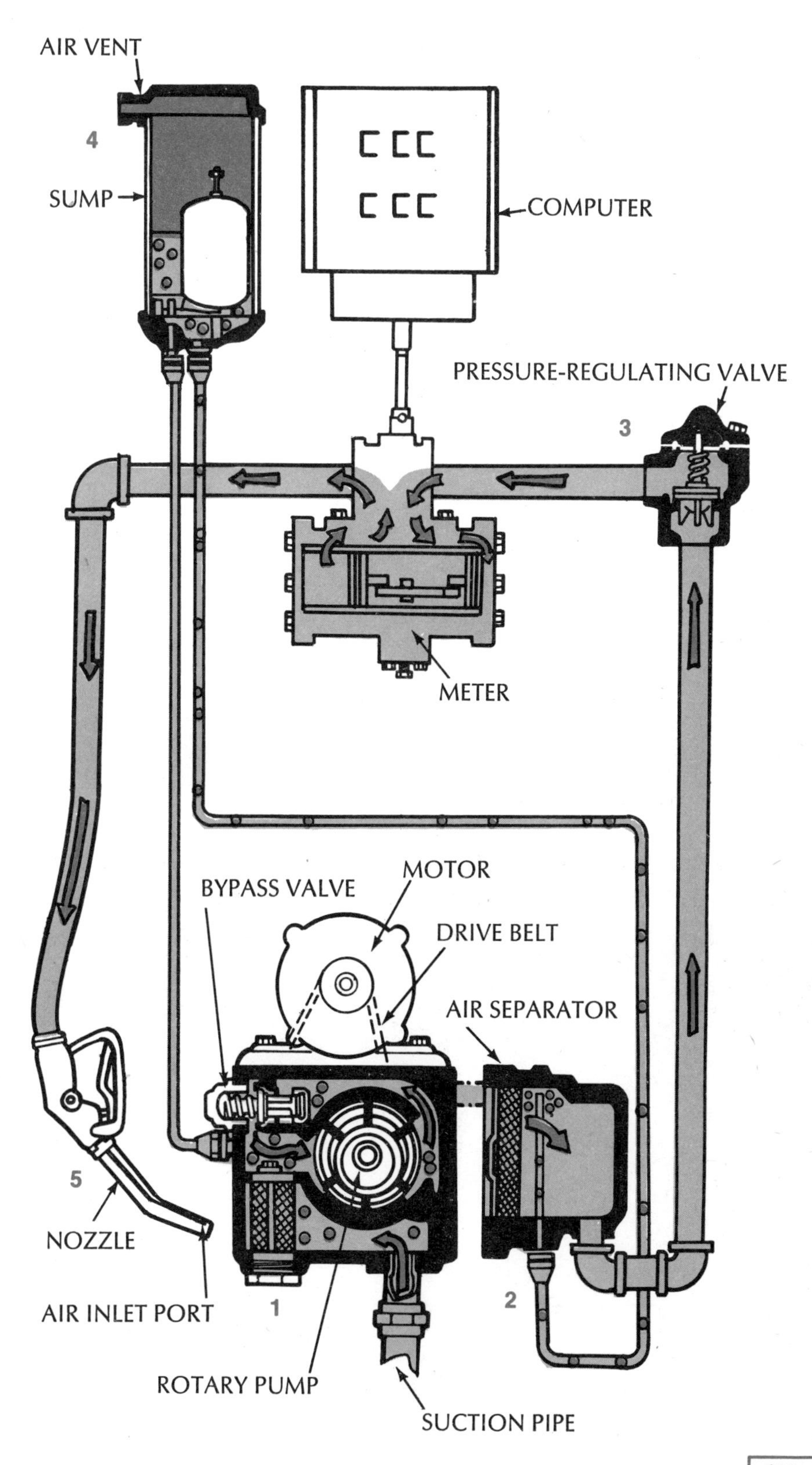

Clock Radio

Since its introduction in 1947, the clock radio has sold by the millions annually. The clock and timer mechanism shown here is used by a number of manufacturers. Two clock knobs in front and one in back can perform a variety of services. Among other things, the right front knob can be set to wake the sleeper with music and later buzz an alarm if the music doesn't do the job. The left front knob can be set to play the radio for a desired amount of time before shutting off the radio automatically, often after the owner falls asleep. The knob in back, as with most clocks, sets time and alarm hands.

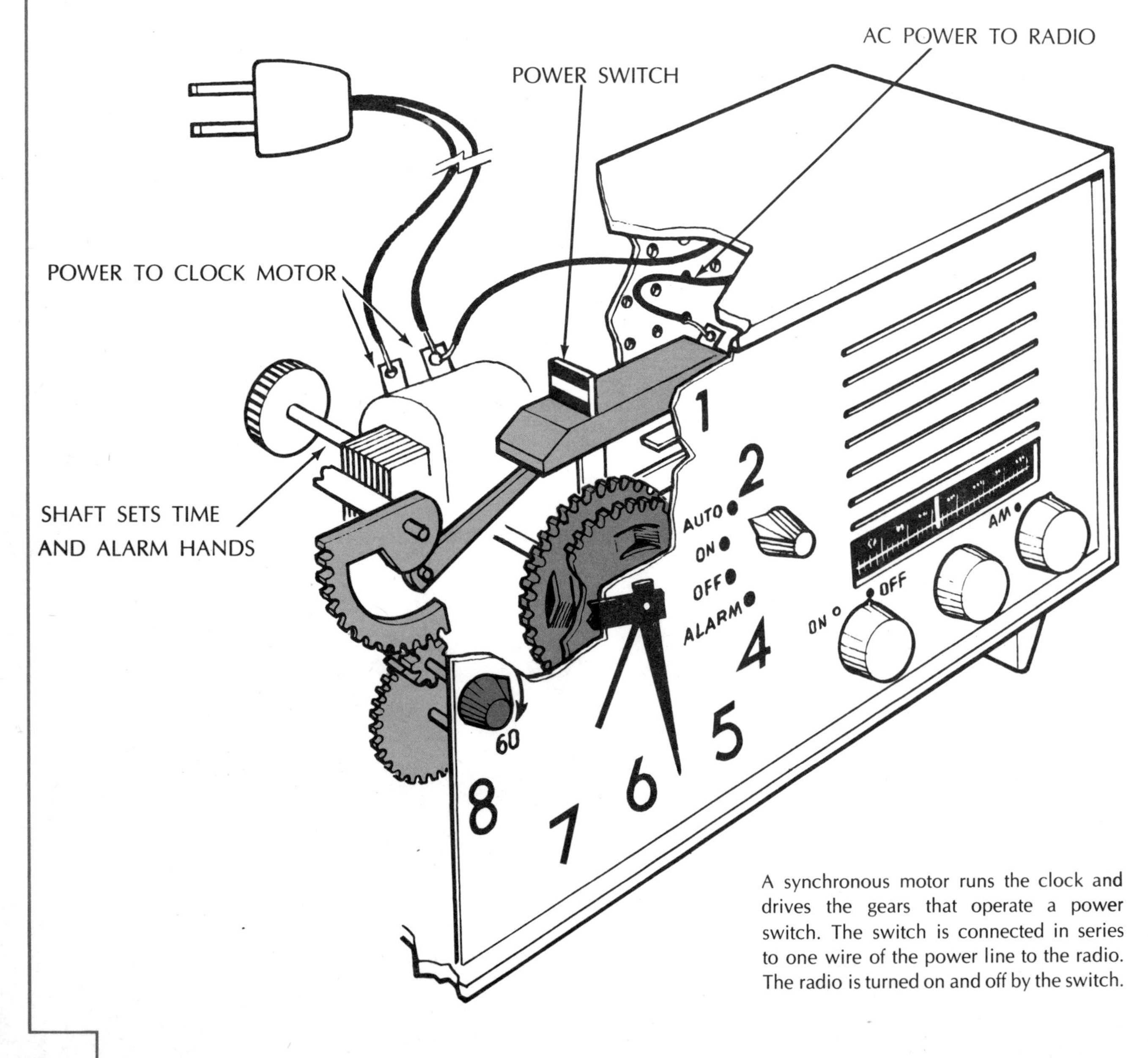

A synchronous motor runs the clock and drives the gears that operate a power switch. The switch is connected in series to one wire of the power line to the radio. The radio is turned on and off by the switch.

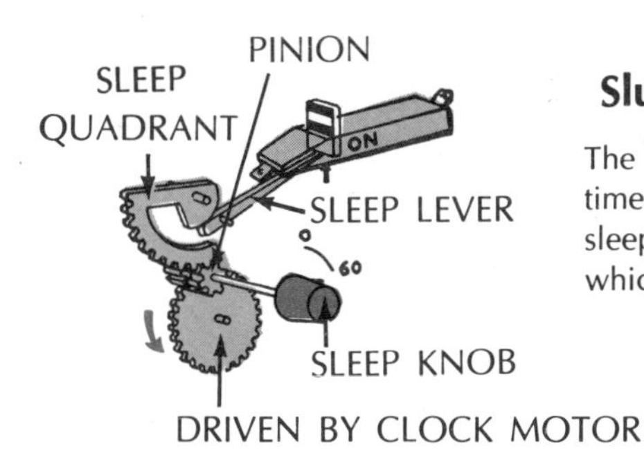

Slumber, Shut-off Mechanism

The automatic radio shut-off can be set for times up to 60 minutes. A pinion on the sleep-knob shaft rotates the sleep quadrant, which pivots a lever against the switch.

The radio shuts off when the sleep quadrant rides off the clock-driven pinion. The quadrant flips, tilting the sleep lever down. This releases the switch, which drops back into the OFF position.

Wake-up Mechanism

Turning the selector knob to "ALARM" and the alarm hand to the wake-up time desired sets up the system. This moves the switch-control lever onto its bottom notch and the alarm-gear ramp a proper distance from the hour-gear ramp.

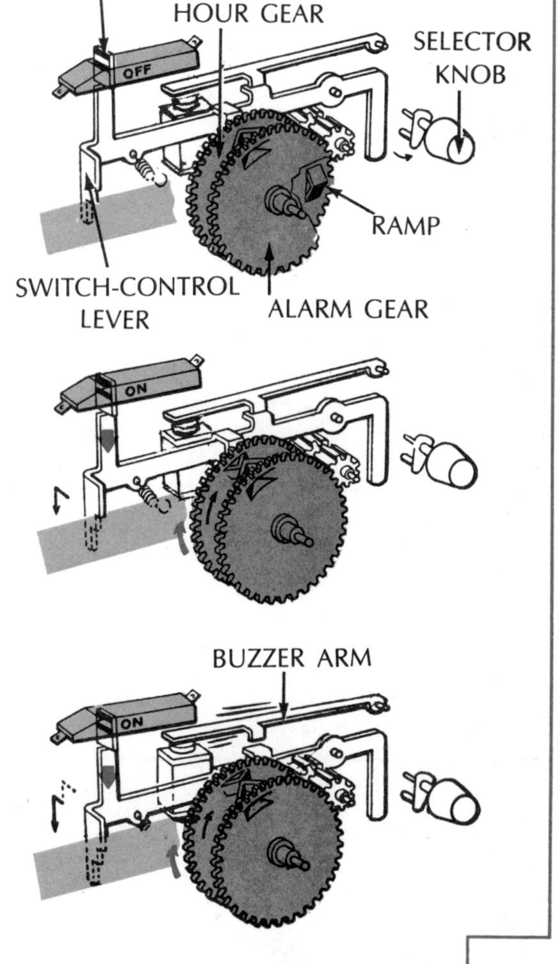

At wake-up time the alarm- and hour-gear ramps meet. Forced back, the hour gear nudges the switch-control lever resting against it. The lever drops onto its top notch. This allows the switch to drop into the lower ON position.

Ten minutes later, as the ramps touch at their widest points, the hour gear nudges the switch-control lever farther back and off its top notch. When the lever drops, the buzzer arm vibrates freely.

Lawn Sprinkler

Most homeowners who like green grass find the sound of a lawn sprinkler deeply satisfying. In some areas of the country, it is possible to maintain healthy lawns and gardens with only infrequent sprinkling. In other areas heavy sprinkling is essential. For small plots, a sprinkling can or the hand-held hose provides an aerated spray that approximates gentle rainfall. But if time and lawn size discourage personal attendance at the sprinkling rites, a mechanical lawn sprinkler is the answer. The beauty of the sprinkler mechanism is that its power source is the very water it disperses. What could be more efficient? And mechanical sprinklers provide a uniformly sufficient moisture content at soil depths of five to six inches, where permanent grass roots reach.

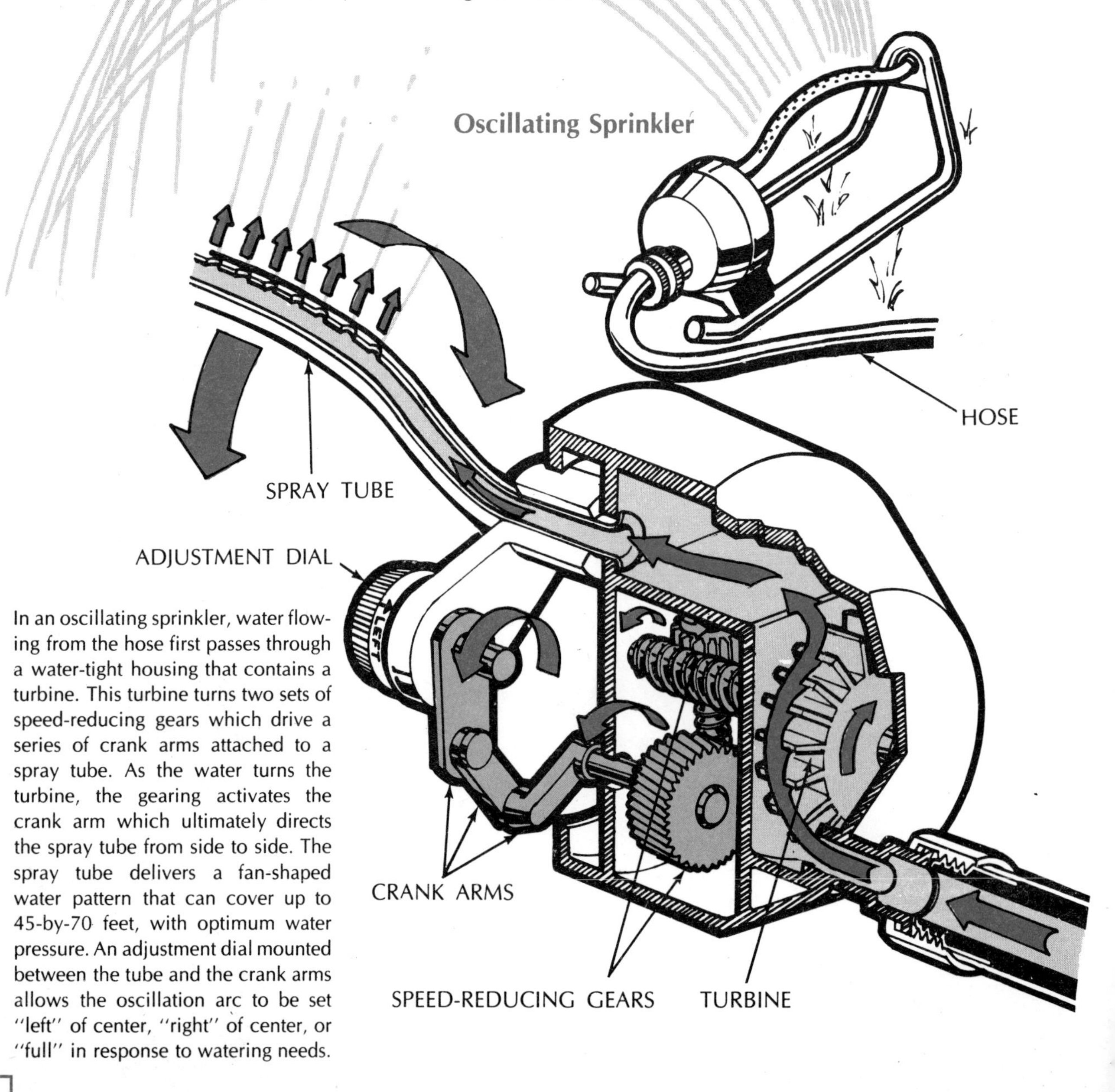

In an oscillating sprinkler, water flowing from the hose first passes through a water-tight housing that contains a turbine. This turbine turns two sets of speed-reducing gears which drive a series of crank arms attached to a spray tube. As the water turns the turbine, the gearing activates the crank arm which ultimately directs the spray tube from side to side. The spray tube delivers a fan-shaped water pattern that can cover up to 45-by-70 feet, with optimum water pressure. An adjustment dial mounted between the tube and the crank arms allows the oscillation arc to be set "left" of center, "right" of center, or "full" in response to watering needs.

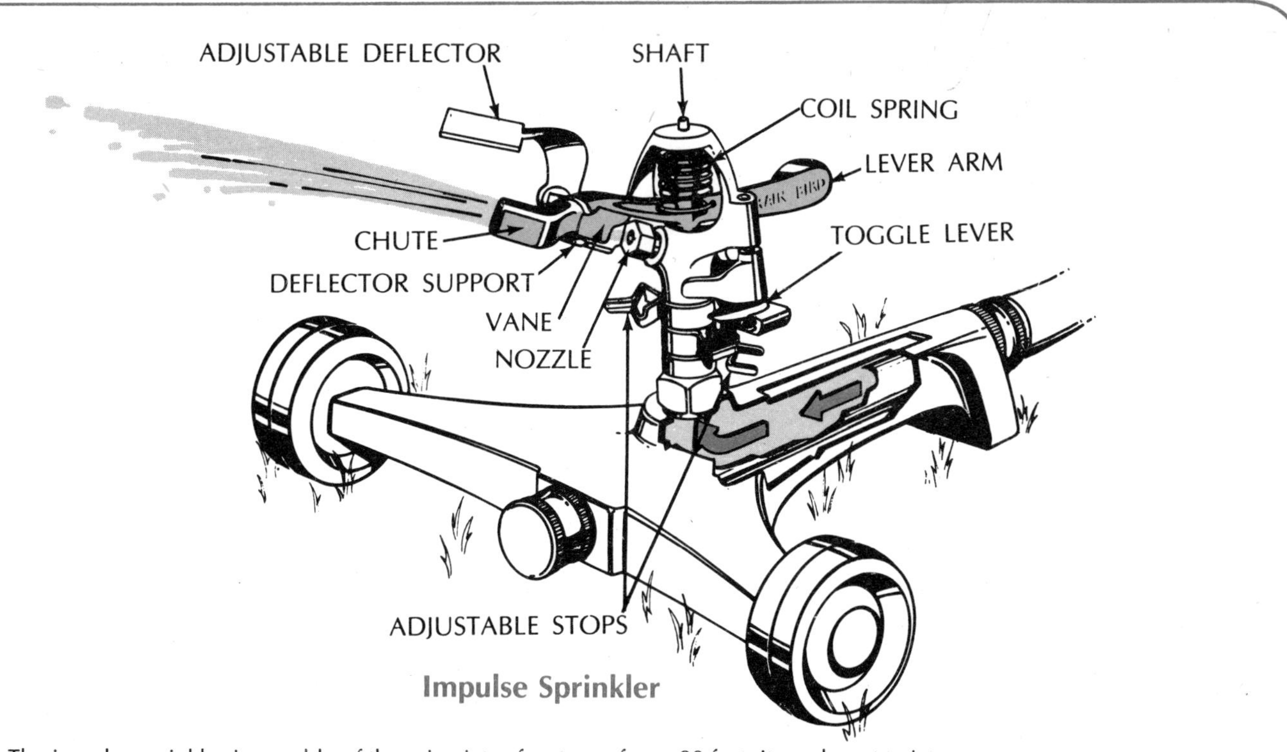

Impulse Sprinkler

The impulse sprinkler is capable of throwing jets of water as far as 80 feet. It can be set to jet circular water patterns or circle segments. Essentially, the steady jet of water from the nozzle is cyclically deflected into a chute on a weighted lever arm that swings in and out of the water's path. The lever pivots on a shaft in the sprinkler head and is attached to the shaft by a coil spring. To illustrate further: **1.** Water from the nozzle strikes a wedge-shaped vane on the lever and escapes through the lever's chute. **2.** Force of the water against the vane pushes the lever counterclockwise, winding the coil spring. **3.** Spring tension returns the lever through the jet, the lever impacting against a deflector support with enough force to step the sprinkler head a few degrees clockwise. **4.** Force of the water against the vane again pushes the lever counterclockwise. **5.** Spring tension again returns the lever, impacting it against the deflector support to step the head a few degrees clockwise. **6 & 7.** This cycling continues until the nozzle has stepped around to the point where stops and levers cause the impulse action of the weighted lever to step the sprinkler head back to its starting position. Then the sprinkling cycle repeats.

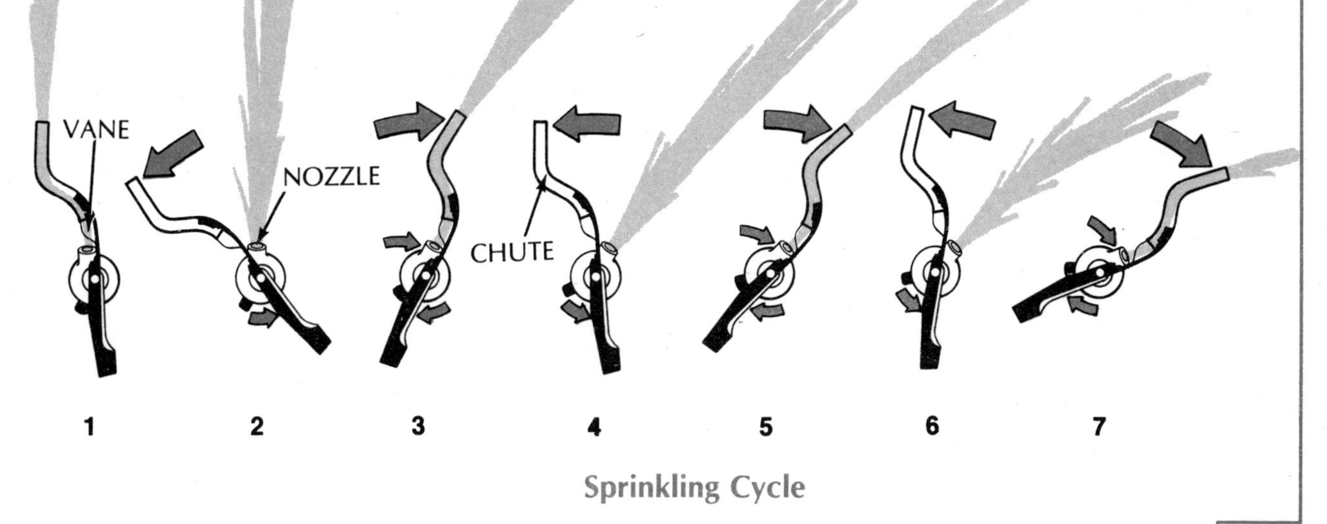

Sprinkling Cycle

Automatic Shower Valve

Using the conventional hand-controlled valves feeding a tee, the shower bather sometimes gets a surprise blast of hot or cold water when someone elsewhere in the household turns on a faucet. But with an automatic shower valve, the water temperature set on the selector dial is maintained by a thermostat (as close as supply permits). When the water from the shower head drops below the desired temperature, the thermostat causes the amount of cold water to decrease. If the cold water supply fails, the thermostat cuts off the hot water.

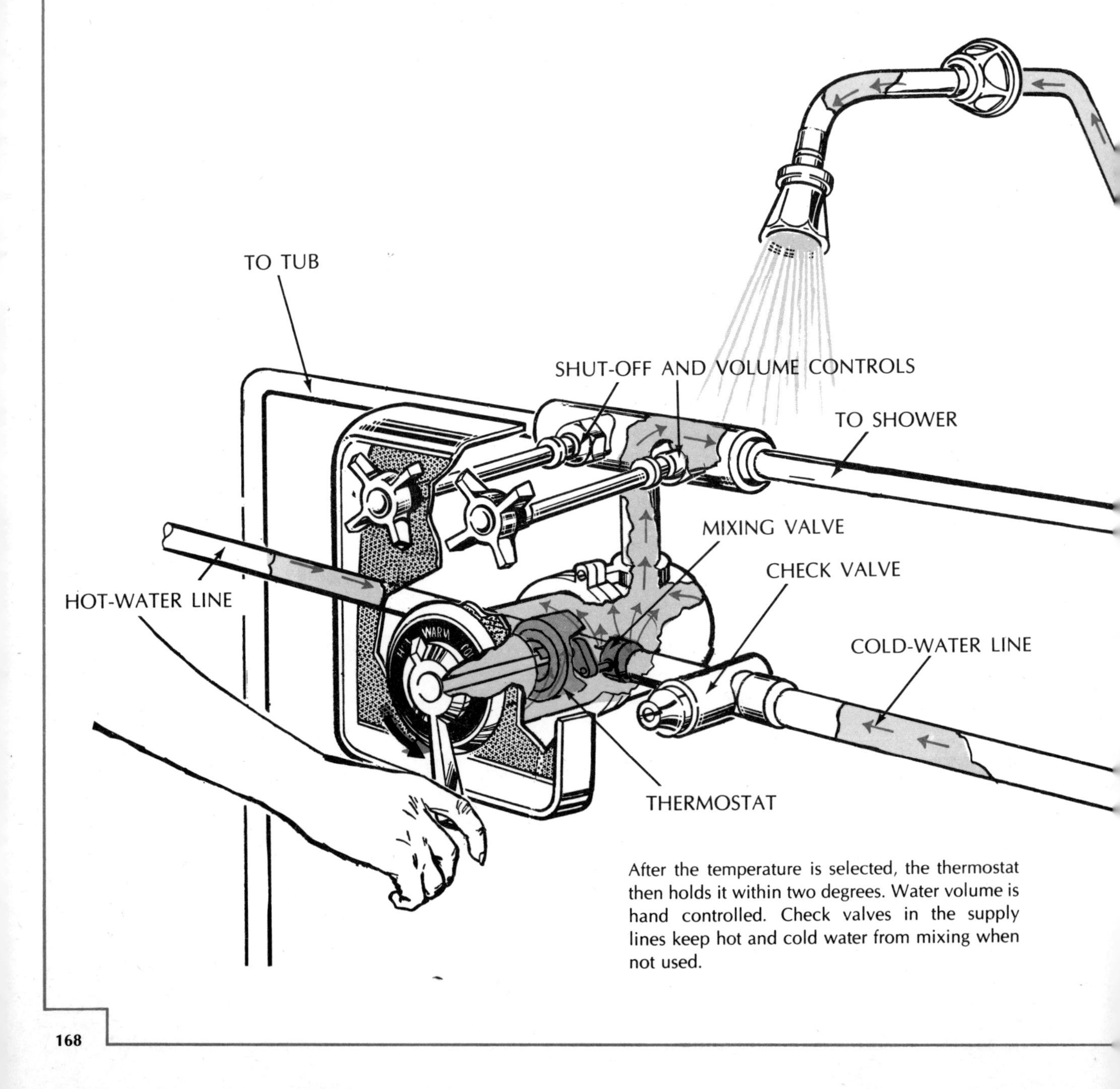

After the temperature is selected, the thermostat then holds it within two degrees. Water volume is hand controlled. Check valves in the supply lines keep hot and cold water from mixing when not used.

Turning the selector to COLD rotates a shaft fixed to the inner turn of a spiral bimetal thermostat. The other end of the coil is fastened to a short crank arm with a hole at its outer end. In this hole is a ball-ended stud set in a small sleeve. The sleeve slides on a short cylinder connected at one end to the hot-water line and at the other to the cold, with a center divider inside and four elongated ports at each end. The sleeve determines how much hot or cold water passes through the ports. With the selector at cold, the sleeve is held over to cover the hot-water ports entirely, opening the cold ones wide. Thus only cold water flows into the shower for those hot-day refreshers.

The thermostat takes over when the selector is moved to a warm setting. The bimetal coil follows and moves the sleeve right, closing the cold and opening the hot ports. The hot water expands the coil, moving the crank arm left. This partly closes the hot ports and partly opens the cold. But the resulting somewhat cooler mixture again opens the hot ports more. Thus, the thermostat hunts that temperature at which the coil (its position determined by the selector setting) stabilizes and holds the sleeve steady. The action is almost instantaneous. If the selector setting is altered, coil tension will again shift the sleeve until the new temperature restabilizes it, providing water at the desired temperature.

A sudden drop in the hot-water supply, such as occurs when hot water is turned on elsewhere, lowers the shower temperature only two degrees before the thermostat reacts by moving the crank arm to the right. This closes the cold-water ports farther or all the way, admitting proportionally more hot water to maintain the temperature. The same thing happens if the temperature, but not the volume, of hot water falls off. The thermostat makes maximum use of what hot water there is. If water temperature falls below the selector setting, the valve delivers the warmest water available. Conversely, if the cold should suddenly fail, the valve prevents a surprise scalding by quickly cutting off the hot-water flow.

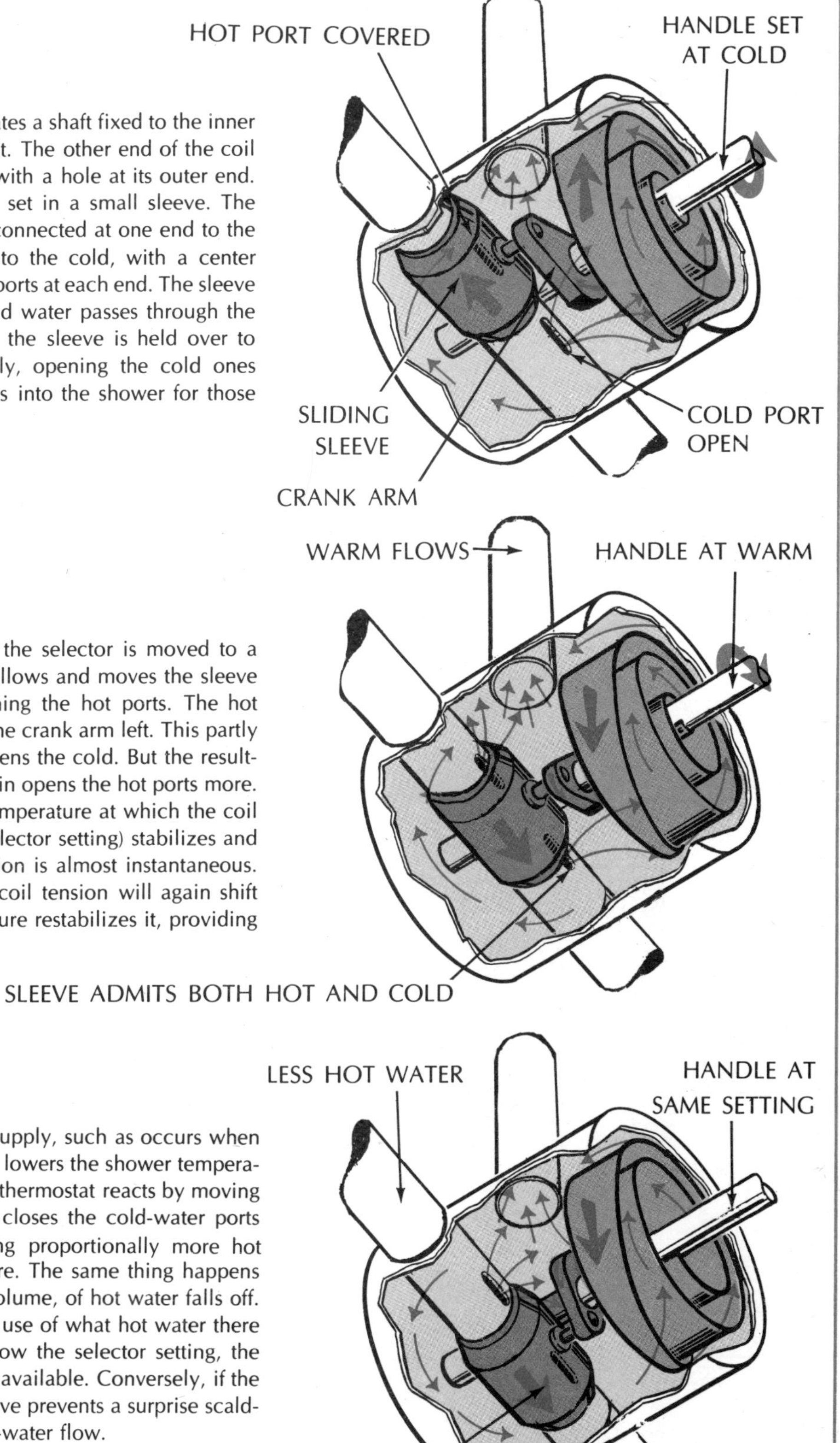

Electric Motor

(*Ed. Note: Electric motors power most of the products presented in this book, and they drive about half of the electrically-operated products in the home. Recent surveys indicate that the typical homeowner possesses about 35 electric motors. The next four pages are designed as a reference aid, devoted to common electric motors and their operating principles.*) Basically, an electric motor is a machine that changes electrical energy into rotary motion that can perform useful work. Most motors consist of two sets of coil windings that carry electric current that generates electromagnetic fields of force. The stationary outer winding is called a *stator* or *field*. The inner rotatable winding is called a *rotor* or *armature*. The two major types of electric motors are *universal motors* and *induction motors*.

Universal Motor

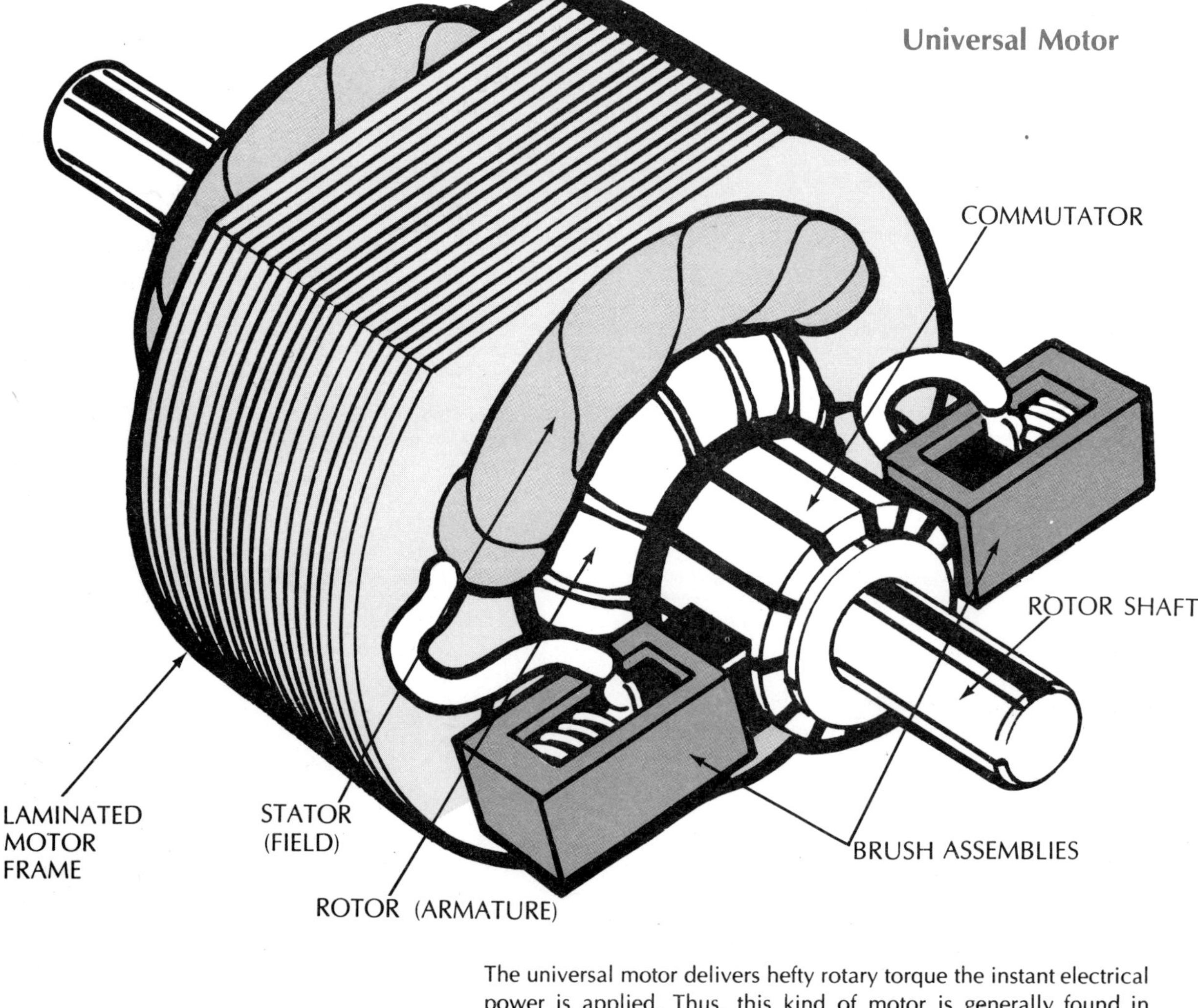

The universal motor delivers hefty rotary torque the instant electrical power is applied. Thus, this kind of motor is generally found in electric drills, blenders, mixers, hedge trimmers, sanders, saber saws and other devices that require the motor to come up to full speed almost instantly. The universal motor operates equally well on ac (alternating current) or dc (direct current) and is used in all battery powered equipment.

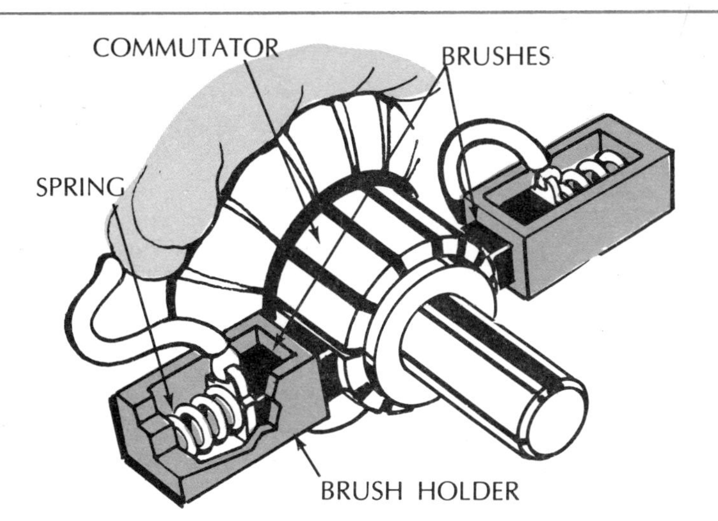

The field winding is made up of many layers of fine wire. The rotor consists of numerous independent wire loops, the ends of which are attached to a *commutator,* which is comprised of many copper segments insulated from one another. Each segment has a counterpart on the opposite side of the commutator. Two ends of a rotor loop are connected between pairs of these opposite segments.

Spring-loaded carbon pieces called *brushes* make electrical contact with the commutator segments. The brushes are machined to ride smoothly on the surface of the commutator. Wires from the field and rotor windings connect through a switch to the power cord.

STATOR COILS

TO AC OR DC POWER SOURCE

ROTOR

CURRENT FLOW

OPPOSING MAGNETIC FIELDS

BRUSHES

COMMUTATOR

STATOR COIL INSULATION

When the power switch is closed, current flow through the series-connected windings sets up magnetic fields which oppose each other. Since the stator is stationary and the rotor is free to rotate, the opposing magnetic fields turn the rotor to a position where magnetic repulsion is minimized. At this point, the brushes contact two new faces on the commutator and thereby "switch in" the next loop of the rotor winding. This once again sets up opposing magnetic fields. The sequence repeats continuously, causing the rotor to revolve at high speed.

Electric Motor (Continued)

Unlike universal-type motors described on the two preceding pages, induction motors have no brushes and no commutator. Nor do they have electrical connections to the rotor. Induction motors run only on ac and require auxiliary starter means. When ac is applied to the stator (stationary) winding, the current sets up a magnetic field which "induces"—hence the name "induction"—a current in the rotor, magnetizing it. Normal ac reverses the direction of the magnetic fields in the stator 120 times a second. As a result, the rotor's induced magnetic field completes a cycle 60 times per second so that the rotor and the stator alternately attract and repulse. The combined effect of rapidly changing fields in the rotor and stator creates the push that causes rotation. The most widely used induction motors are *split-phase, capacitor-start,* and *shaded-pole*.

Split-phase Motor

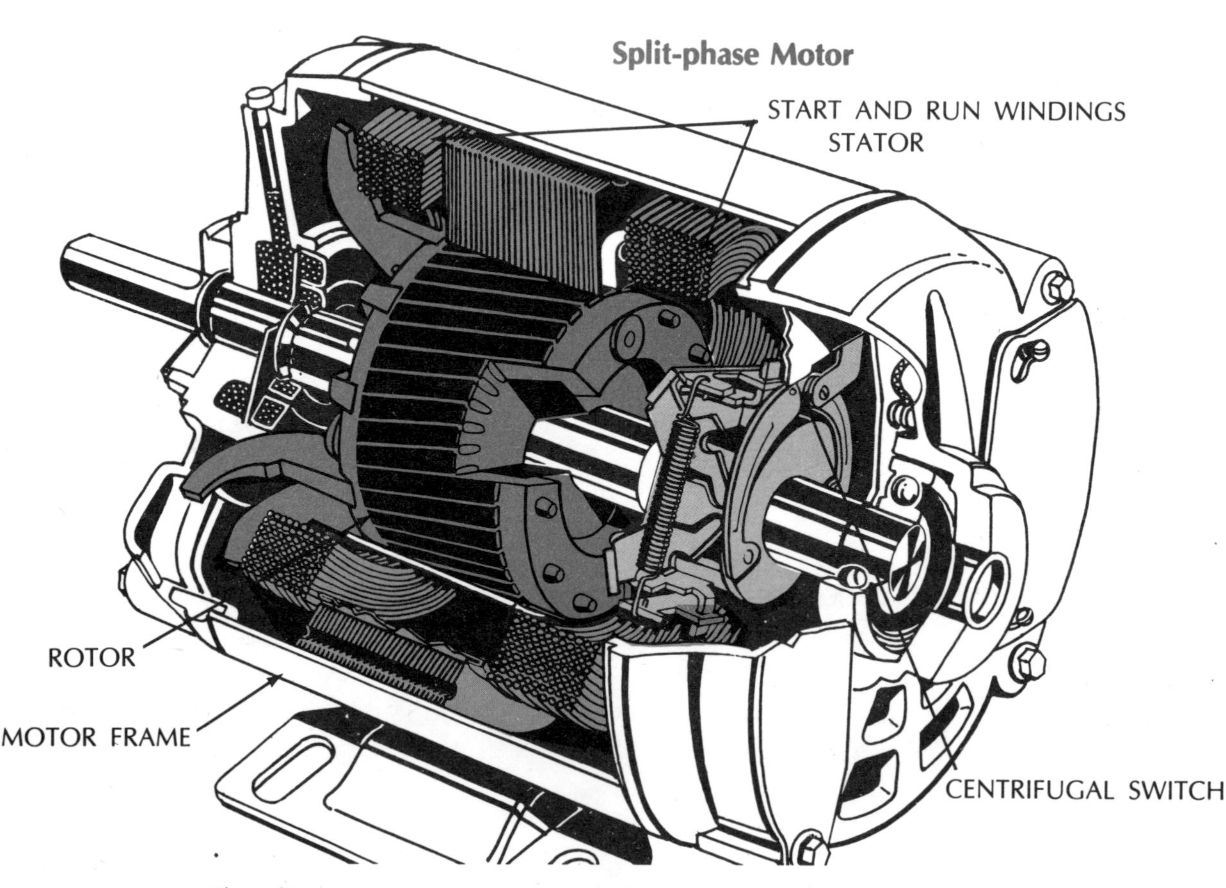

The split-phase motor uses an extra winding called a *start winding* made up of a few turns of heavy wire on top of the fine-wire *run winding*. Both receive power initially. This produces a heavy magnetic field that assures turning of the rotor. Once the rotor attains full speed, a *centrifugal switch* disconnects the start winding from the circuit. Split-phase motors are used for fans, power saws, and when slow buildup of output speed can be tolerated. A capacitor-start motor looks much the like split-phase, shown here, except that it employs an energy storage component called a *capacitor* that produces a brief surge of current through a start winding to produce power. Capacitor-start motors can start under heavy loads like those occurring in pumps and compressors.

In motors powered by ac, the electromagnetic field reverses 120 times per second. One complete cycle is made up of negative and positive halves, measured from "zero" current on the current wave to the matching point in the next cycle. Thus, a field reversing 120 times per second does so in response to 60 cycle ac.

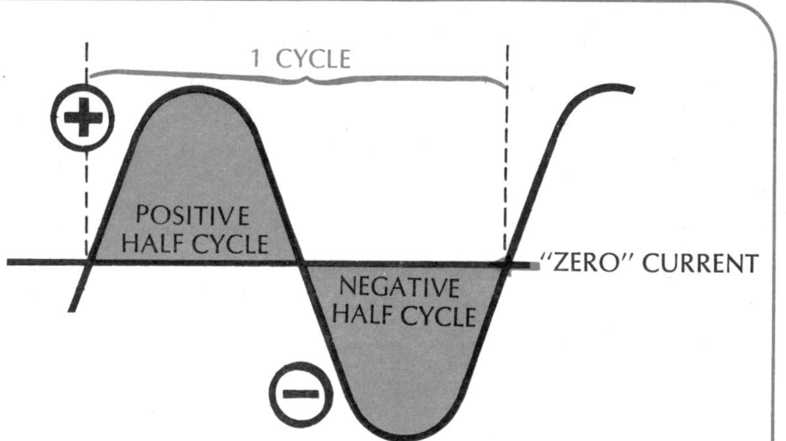

The repulsion forces generated by the electromagnetic fields between the rotor and the stator are here illustrated with simple bar magnets. Like poles repulse and unlike poles attract. Induction motors are designed so that repulsion occurs continuously between the rotor and the stator as ac is applied. This turns the rotor.

Shaded-pole Motor

STATOR WINDING

ROTOR

LAMINATED CORE

POLES "SHADED" WITH COPPER WIRE

The shaded-pole motor has only one stator winding and is generally used in fans, record players, or other light-load products. The solid metal rotor passes through the center of a U-shaped core. The open ends of the core encircling the rotor are called poles. A single turn of heavy copper wire is wound on one edge of each pole. When ac is applied to the stator winding, these "shaded" portions of the poles produce an opposing magnetic field. The rotor responds to the opposing fields by starting to rotate the instant power is applied.

Index